Advanced Fossil Energy Utilization

Topical Conference at the 2009 AIChE Spring National Meeting

Tampa, Florida
26 – 30 April 2009

ISBN: 978-1-61567-263-9

Advanced Fossil Energy Utilization

Topical Conference at the 2009 AIChE Spring National Meeting

Tampa, Florida
26 - 30 April 2009

ISBN: 978-1-61567-263-9

AIChE

Printed from e-media with permission by:

Curran Associates, Inc.
57 Morehouse Lane
Red Hook, NY 12571

Some format issues inherent in the e-media version may also appear in this print version.

Printed by Curran Associates, Inc. (2009)

For permission requests, please contact AIChE
at the address below.

AIChE
3 Park Avenue
New York, NY 10016-5991

Phone: (203) 702-7660
Fax: (203) 775-5177

www.aiche.org

Additional copies of this publication are available from:

Curran Associates, Inc.
57 Morehouse Lane
Red Hook, NY 12571 USA
Phone: 845-758-0400
Fax: 845-758-2634
Email: curran@proceedings.com
Web: www.proceedings.com

TABLE OF CONTENTS

Applications of Non-Equilibrium Plasmas for Plasma-Chemical Fuel Conversion And Hydrogen Production 1
Alexander Fridman

Plasma Catalyzed JP-5 Reformer Scaleup for Navy Shipboard Fuel Cell Systems 2
Joseph J. Hartvigsen, S. Elangovan, Anthony Nickens, Lyman Frost, Piotr Czernichowski, Michele Hollist

Characterization of a New Glid-Arc Plasmatron On-Board Reformer for An Integrated Solid Oxide Fuel Cell Auxiliary Power Unit 3
M.J. Gallagher Jr., A. Polevich, A. Rabinovich, Alexander Fridman

Plasma Catalyzed JP-8 Reformer Development for Military 10kW SOFC APU 4
Joseph J. Hartvigsen, S. Elangovan, Anthony Nickens, Lyman Frost, Piotr Czernichowski, Michele Hollist

Non-Thermal Plasma Assisted Decomposition of Light Hydrocarbons to Hydrogen-Rich Gas And Carbon 5
Nazim Muradov, Franklyn Smith, Gary Bokerman

Hydrogen Produced Via Nonequilibrium Plasma Reactions 6
Mingwei Li, Xiaomei Gao, Yiling Tian

Theoretical and Experimental Analysis of Non-Thermal Arc Discharge Assisted Reforming of Gasoline and E85 7
Guillaume Petitpas, JosÈ Gonzalez-Aguilar, Adeline Darmon, Laurent Fulcheri

Development of ATR Fuel Processor for Reformation of Logistic Fuels 16
Christian Junaedi, Subir Roychoudhury, Dennis Walsh, Rich Mastanduno

Multi-Fuel Direct Steam Reforming Process for SOFC Applications 17
Aly Shaaban, Christian D. Rasmussen, Ragab M. Moheisen, Reza Salavani

Combined Pre-Reforming-Desulfurization of High-Sulfur Diesel for Hydrogen Fueling Station Applications 27
Nazim Muradov, Karthikeyan K. Ramasamy, Cunping Huang, Ibrahim Adebiyi, Franklyn Smith, Clovis Linkous, Ali T-Raissi, James Stevens

Catalytic Partial Oxidation of a Surrogate Diesel Fuel Mixture Using Pyrochlores: Effect of Reforming Metal 28
Daniel J. Haynes, David A. Berry, Dushyant Shekhawat, Mark Smith, James J. Spivey

Dry Catalytic Partial Oxidation of Diesel-Fuel Distillates Into Syngas 29
M.V. Mundschau, David A. Gribble Jr., Paul N. Plassmeyer, Lee M. Henton, Nicholas R. Jentzsch

Effect of Catalyst Layer Formation and Character On Partial Oxidation of Liquid Hydrocarbons In the Presence of Oxygen-Conducting Supports 38
Mark Smith, David A. Berry, Dushyant Shekhawat, Daniel J. Haynes, James J. Spivey

Kinetic Studies of the Autothermal Reforming of Tetradecane Over Pt/Al_2O_3 Catalyst In a Fixed-Bed Reactor 39
Rajesh D. Parmar, Arunabha Kundu, Christopher P. Thurgood, Brant. A. Peppley, Kunal Karan

Natural Gas Direct Carbon Fuel Cell for Hydrogen and Electricity Production (NGDCFC) 48
Meyer Steinberg

Dry Reforming of Biogas to Syngas And Hydrogen for Fuel Cell Applications 53
Nazim Muradov, Franklyn Smith, Nathaniel Garceau

Non-Catalytic, Partial Oxidation for Syngas Production 54
Janet L. Ellzey, Ingmar Schoegl, Colin H. Smith

Novel High Stability Hydrogen Production from Ethanol Steam Reforming Over Catalyst of Nano-NiO Particles Based on Flowerlike CeO_2 Microspheres for EH Hybrid Vehicle Application 55
Jie Sun, Yuguang Wang, Xinping Qiu, Liquan Chen, Jigang Li

Ni-Fe Binary Metal Catalysts for Hydrogen Production Via Auto-Thermal Reforming of Ethanol 63
Lihong Huang, Jian Xie, Rongrong Chen, Deryn Chu, Andrew T. Hsu

Design and Pilot-Scale Tests of Monolithic Catalysts for Autothermal/Steam Reforming of Natural Gas and Biofuels 64
Julian Ross, Sergei Beloshapkin, Vladislav A. Sadykov, Natalia V. Mezentseva, Galina M. Alikina, Zakhar Vostrikov, Yulia Fedorova, Vladimir. Pelipenko, Vladimir Usoltsev, Sergei Tikhov, Oleg Smorygo, Vladimir Yu. Ulyanitskii

Thermodynamic and Kinetic Analysis of the Thermal Partial Oxidation Process of Ethanol 71
Zayed Al-Hamamre

Pure Hydrogen and Pure Carbon Dioxide From Gaseous Hydrocarbons Via Chemical Looping Reforming 83
Tobias Proell, Johannes Bolhar-Nordenkampf, Philipp Kolbitsch, Klemens Marx, Hermann Hofbauer

Sorbent-Enhanced Water Gas Shift Reaction Studies 91
Robert W. Stevens Jr., Abolghasem Shamsi, Stephen Carpenter, Ranjani V. Siriwardane

High-Pressure Hydrogen Purification Using Electrical Swing Adsorption 97
John P. Hurley, Gregory T. Dvorak, Dustin P. McNally, Nikhil Patel

Optimization of Oxygen Feed Membranes In Autothermal Steam-Reformers 98
J. Hüppmeier, S. Barg, M. Baune, J. Thöming

A Membrane Reactor Concept for Simultaneous Production of Pure Hydrogen and Synthesis Gas Over Ceramic Materials 105
Antigoni P. Evdou, Lori Nalbandian, Vassilis T. Zaspalis

Hydrogen Separation with Palladium-Copper Membranes on Porous Stainless Steel with Sol-Gel Derived Ceria as Diffusion Barrier 113
Ailing Qiao, Ke Zhang, Ye Tian, Zebao Rui, Yongdan Li

Hydrogen Production by Methane Decomposition Using Ordered Mesoporous Carbons as Catalysts with Remarkable Activity And Stability 120
D. P. Serrano, J. A. Botas, J.L.G. Fierro, R. Guil-López, P. Pizarro, G. Gómez

Study On a Compact Reactor for Methane Steam Reforming In Fuel Cell Systems 131
Liwei Pan, Zhongshan Yuan, Changjun Ni, Cheng Zhang, Deyi Li, Shudong Wang

Coal Liquefaction Using High Energy H_2 and O_2 From Electrolysis Assisted by a Nuclear Source 132
Gary Otterstrom

Carbon Dioxide Capture by Pressure Swing Adsorption 133
Amal Mehrotra, Armin D. Ebner, James A. Ritter

Preparation of New Organic-inorganic Hybrid Adsorbents for Post-combustion CO_2 Capture 136
Sunho Choi, Jeffrey H. Drese, Daniel J. Fauth, McMahan L. Gray, Christopher W. Jones

Evaluation of Solid Sorbents As A Retrofit Technology For CO_2 Capture 137
Sharon M. Sjostrom

The Effect of Hydration on the Adsorption of Carbon Dioxide with Tertiary Amidines on Activated Carbon 145
W. Richard Alesi, John R. Kitchin, McMahan L. Gray

A Regenerative Process for CO_2 Removal and Hydrogen Production in Integrated Gasification Combined Cycle (IGCC) Processes 147
Javad Abbasian, Armin Hassanzadeh-Khayyat

A First Principles Evaluation of the Role of Substituent Effects on the Interaction of Carbon Dioxide with Tertiary Amines 148
W. Richard Alesi, John R. Kitchin

Modeling CO_2 Capture Using Ab Initio and Statistical Mechanical Methods 149
J. Karl Johnson, Jin-Chen Liu, Rees B. Rankin, De-Li Chen, Hong-bin Xie, Bo Zhang

Irreversible Capture of Mercury 150
David A. Atwood

Update on Global Mercury Emissions and Controls 151
Leonard Levin

Specification Metrics for Activated Carbons Used for Mercury Control 152
Sid Nelson Jr.

Screening of Sorbents for Mercury Capture Using Ab initio Computations 153
Anubhav Jain, Seyed-A. Seyed-Reihani, C. Fischer, D. Couling, G. Ceder, William H. Green

Feasibility of Removal of Trace Impurities from Coal Derived Raw Syngas to Produce Ultra-Clean Syngas 155
Ty Nguyen

Effect of Flue Gas Oxygen and Water Vapor Content on High Temperature Capture of Mercury by MinPlus Sorbent 156
Sung Jun Lee, Jost O.L. Wendt

CO_2 Capture Using Nanoparticle Ionic Materials and Task-Specific Ionic Liquids 157
Ah-Hyung Alissa Park, Kunyi Andrew Lin

Reversible Ionic Liquids as Double-Action Solvents for Efficient CO_2 Capture 158
Vittoria M. Blasucci, Ryan Hart, Cerag Dilek, Hillary A. Huttenhower, Veronica Llopis Mestre, Pamela Pollet, Eduardo Vyhmeister, Charles Eckert, Charles L. Liotta

Accelerated Carbonation of Steel Slag via Two-Stage Carbon Mineral Capture Technology 159
Ah-Hyung Alissa Park, Laura Simone, Dong Hyun Lee

Ramgen's Low-Cost, High-Efficiency CO_2 Compressor Technology 160
Pete Baldwin

Ceramic and Coal: ITM Oxygen for Power Generation with Reduced CO_2-Emissions 215
John M. Repasky, VanEric E. Stein, Phillip A. Armstrong

New Technology Integration Approach for Energy Planning with Carbon Emission Considerations 232
Sajjad Ahmed, Juan Du, B. Erik Ydstie, Peter L. Douglas, Ali Elkamel

Development of an Integrated Carbon Capture, Separation and Geologic Storage Pilot: A Technical Research Project of the Southeast Regional Carbon Sequestration Partnership 233
Gerald Hill, Richard Esposito, Richard Rhudy, Kimberly Sams

Two-Stage Stochastic Programming and Financial Risk Management for Energy Planning with Uncertainty 236
Sajjad Ahmed, Juan Du, B. Erik Ydstie, Peter L. Douglas, Ali Elkamel

Activated Carbon Injection for Mercury Control: Overview 237
Sharon M. Sjostrom

MetalliFix: A Novel Scrubber Additive for the Capture and Sequestration of Oxidized and Elemental Mercury 241
Mike Cooper

Slipstream Tests of Palladium Sorbents for High Temperature Capture of Mercury, Arsenic, Selenium and Phosphorous from Fuel Gas 242
P. Vimalchand, Frank C. Morton, Subhash Datta, Robert C. Lambrecht, Evan J. Granite, Henry W. Pennline, Dennis Stanko, Hugh Hamilton, Liz Rowsell, Stephen Poulston, Andrew Smith, Wilson Chu

Palladium Sorbents for High Temperature Capture of Mercury, Arsenic, Selenium and Phosphorus from Fuel Gas 243
Evan J. Granite, Henry W. Pennline, Dennis Stanko, John P. Baltrus, Hugh Hamilton, Liz Roswell, Stephen Poulston, Andrew Smith, Wilson Chu

Surface Characterization of Palladium-Alumina Sorbents for High Temperature Capture of Mercury and Arsenic From Fuel Gas 244
John P. Baltrus, Evan J. Granite, Henry W. Pennline, Dennis Stanko, Hugh Hamilton, Liz Roswell, Stephen Poulston, Andrew Smith, Wilson Chu

Author Index

1a Applications of Non-Equilibrium Plasmas for Plasma-Chemical Fuel Conversion And Hydrogen Production

Alexander Fridman, Mechanical Engineering and Mechanics, Drexel University, Philadelphia, PA 19104

This talk will give a general overview of the applications of non-equilibrium plasmas used for the conversion of various hydrocarbon fuels into synthesis gas. Non-equilibrium plasmas have the advantage of operating at lower gas temperatures while stimulating vibrationally and electronically excited species which are used as chain carriers in many reacting flow situations. This is the basis of the so-called plasma-catalytic effect that is observed when combining plasmas with fuel conversion reactions. There are many types of non-equilibrium plasmas used in fuel conversion processes; however, the simplest and most effective discharge is a gliding discharge, commonly known as gliding arc. The Gliding Arc discharge is a high energy density transitional plasma with moderate gas temperatures (2500-4000K) that is capable of providing high productivity while maintaining good chemical selectivity. The Gliding discharge plasma has relatively low current (<0.5A) and is considered a "warm" non-thermal plasma that is much like a glow discharge where cathode processes are controlled by secondary electron emission. Applications of Gliding arc discharges include partial oxidation of heavy hydrocarbons, biomass, and coal to syn-gas, steam reforming of methane to syn-gas, and H2S dissociation.

The Drexel Plasma Institute has been working in collaboration with others to produce synthesis gas using non-equilibrium gliding discharge in a reverse vortex flow ("tornado") reactor. The plasma reformer could be combined with an integrated system that utilizes syn-gas for direct use in a solid oxide fuel cell for onboard conversion liquid fuels into electricity. This integrated system can be thought of as an advanced power generator that can satisfy electrical needs in remote areas where only hydrocarbon fuel sources are available. This system has applications in the transportation industry (long-haul trucking) and could have many potential military applications.

17a Plasma Catalyzed JP-5 Reformer Scaleup for Navy Shipboard Fuel Cell Systems

Joseph J. Hartvigsen, S. Elangovan, Anthony Nickens, Lyman Frost, Piotr Czernichowski, and Michele Hollist. Ceramatec, Inc., 2425 South 900 West, Salt Lake City, UT 84119

The US Navy is studying various options of incorporating distributed power generation using fuel cells onboard future generations of warships. Currently, one of the major barriers to adoption of distributed shipboard power generation using fuel cells is the size and complexity of the fuel processing system. Ceramatec has shown promising results with its GlidArc plasma reformer successfully reforming JP-8 at the 5kW-10kW scale. The Office of Naval Research has supported the scaling of this technology to the 60kW scale. Due to our background as an SOFC developer, Ceramatec has long been a strong proponent of exploiting the high system efficiencies that can be obtained by thermal and process integration of the reformer and SOFC compared with standalone POx reformers. However in this case, the Navy has not committed to the selection of an SOFC as the platform for their shipboard power system. Therefore, the Ceramatec 60kW JP-5 reformer was designed to allow the flexibility of testing the reformer in modes compatible with either a high temperature fuel cell integrated configuration or a low temperature fuel processing integrated configuration. The 60kW test article design, fabrication and operational results will be presented.

17b Characterization of a New Glid-Arc Plasmatron On-Board Reformer for An Integrated Solid Oxide Fuel Cell Auxiliary Power Unit

M.J. Gallagher Jr.[1], A. Polevich[1], A. Rabinovich[1], and Alexander Fridman[2]. (1) A.J. Drexel Plasma Institute, Drexel University, Philadelphia, PA 19104, (2) Mechanical Engineering and Mechanics, Drexel University, Philadelphia, PA 19104

The non-equilibrium Glid-Arc Plasmatron device for hydrocarbon reforming into synthesis gas was developed over a decade ago and its capabilities are well known [1-3]. Some of these capabilities include fast start time, wide dynamic range of operation (with fuel input rates spanning 0.1 to 3 g/s), small and compact physical dimensions, and low electrical power consumption. The GA-Plasmatron is capable of reforming a wide variety of hydrocarbon fuels including methane, propane, and liquid fuels such as, gasoline, diesel, bio-diesel, and JP-5. Reforming results using spray liquid Diesel fuel have shown that the Glid-Arc Plasmatron is capable of producing H2 and CO yields up to 50% and 60%, respectively. Conversion rates as high as 80% are reported in some cases and this includes some additional light hydrocarbons, such as methane, ethylene, and ethane that are produced as byproducts of the reforming process. A lack of sensitivity to sulfur is an additional benefit of using plasma-assisted reforming technologies, since sulfur is known to have adverse effects alternate reforming technologies, such as steam reforming catalysts.

In this investigation, a new GA-Plasmatron device has been developed to operate as part of an integrated system with a solid oxide fuel cell (SOFC) to create a complete plasma-assisted on-board auxiliary power unit (APU). The GA-Plasmatron APU system recycles part of the SOFC anode gas exhaust stream and recuperates heat from the exhaust at temperatures near 850C using an internal heat exchanger. By recycling heat and a portion of the SOFC exhaust, it is possible to limit the amount of external air used as an oxidant, promote endothermic steam reforming reactions, and increase the overall quality and heating value of the reformate stream. The composition of the SOFC exhaust stream was 5.7% H2, 34.5% H2O, 6.8% CO, 36.3% CO2, 16.7% N2 and was created under laboratory conditions using several mass flow controllers and a water evaporation system. Low sulfur Diesel fuel was used in initial trials and was evaporated and premixed with the preheated SOFC recycle stream just before it entered the plasma reaction chamber. The total recycle stream flow rate was varied as was the oxygen-to-carbon (O/C) ratio and the H2O/C ratio. Preliminary results show some improvement over traditional POX results and are comparable to steam reforming results obtained using traditional catalysts.

References

[1] L. Bromberg et al., Compact Plasmatron-boosted hydrogen generation for vehicular applications. Int. J. Hydrogen Energy, 24, (1999).

[2] L. Bromberg et al., Experimental investigation of plasma assisted reforming of methane I: steady state operation. PSFC/JA-05-10.

[3] A. Lutz et al., Thermodynamic analysis of hydrogen production by partial oxidation reforming. Int. J. Hydrogen Energy, 29, (2004).

17c Plasma Catalyzed JP-8 Reformer Development for Military 10kW SOFC APU

Joseph J. Hartvigsen, S. Elangovan, Anthony Nickens, Lyman Frost, Piotr Czernichowski, and Michele Hollist. Ceramatec, Inc., 2425 South 900 West, Salt Lake City, UT 84119

The need for efficient and quiet power generation on military vehicles, from JP-8 fuel, is well know. The difficulty in reforming and processing JP-8 to fuel cell quality hydrogen or reformate is delaying the implementation of fuel cell technology in a package that meets military operational requirements. Ceramatec, with the support of the US Army (TACOM-TARDEC) is leading the development a fuel processing system to address this need. Ceramatec, the Drexel Plasma Institute, Auburn University and Intramicron are working together to address the process and system integration of a plasma catalyzed reformer and desulfurization system within an SOFC APU. The system design strategy, packaging, subsystem advances, operational results and outstanding challenges will be presented.

17d Non-Thermal Plasma Assisted Decomposition of Light Hydrocarbons to Hydrogen-Rich Gas And Carbon

Nazim Muradov, Franklyn Smith, and Gary Bokerman. Florida Solar Energy Center, University of Central Florida, 1679 Clearlake Road, Cocoa, FL 32922

Single-step decomposition (or pyrolysis, cracking) of light hydrocarbon fuels, such as natural gas (NG) and liquefied petroleum gas (LPG), to hydrogen and carbon is an attractive CO/CO2-free alternative to steam methane reforming (SMR) [1]. The process is moderately endothermic, e.g., in case of methane, the energy input requirement per mole of hydrogen produced (37.8 kJ/mole H2) is considerably less than that for the SMR process (63.3 kJ/mole H2). Furthermore, the process produces a valuable byproduct-clean carbon. However, a major problem with the practical realization of the NG pyrolysis process is that methane is one of the most thermally stable organic molecules, thus, thermal methane decomposition requires very high temperatures (>1200oC). Recently, Kværner Co. (Norway) has developed process for thermal plasma-assisted pyrolysis of NG to hydrogen and carbon black [2]; however, the process is very energy intensive, because electrical energy is used to heat the feedstock to very high temperatures (>5000oC).

In this work, non-thermal plasma is utilized to decompose pure methane, pipeline NG and commercial propane to H2 and carbon at near ambient conditions. In contrast to thermal plasma, non-thermal (or non-equilibrium) plasma is characterized by very high electron temperatures, but low bulk gas temperatures (as a result, the feedstock is not heated to high temperature, and the energy consumption is relatively low). Electronically-excited methane molecules CH4* play a major role in non-thermal plasma assisted conversion of methane to hydrogen and carbon:

(e >10 eV) + CH4 => (CH4)* => •CH3, •CH2, •CH, H• => H2, C

Non-thermal plasma reactors with stationary and moving electrodes for continuous decomposition of light hydrocarbons were designed, fabricated and tested in this study. Different electrode materials (e.g., graphite, stainless steel, Ni, Ni-Cu, Fe) were used in the reactor, and their effect on the reactor performance was evaluated. The main products of plasma-assisted NG decomposition were H2 and carbon, along with smaller amounts of C2+ hydrocarbons. It was found that the distribution of the reaction products was mostly affected by the residence time in the plasma reactor. During plasma decomposition of NG, H2 concentration in the effluent gas reached 40-50 vol.% (the balance: mostly CH4 and C2+ hydrocarbons). The carbon product was characterized by SEM, TEM, XRD, Raman, FTIR methods, and it was shown to be amorphous carbon. Advantageously, the process is not affected by sulfur impurities, which would allow utilizing commercial sulfurous hydrocarbon feedstocks without an expensive pre-treatment stage. After a gas purification stage, the hydrogen-rich gas can be used as a fuel in a PEM fuel cell.

References

[1] N.Muradov, Int. J. Hydrogen Energy 18 (1993) 211.

[2] S. Lynum et al. Proc. 12th World Hydrogen Energy Conf., Buenos Aires, Argentina, 1998, p.637.

17e Hydrogen Produced Via Nonequilibrium Plasma Reactions

Mingwei Li, Xiaomei Gao, and Yiling Tian. Department of Chemistry, Tianjin University, Tianjin, 300072, China

At present, three major processes, including partial oxidation, steam reforming, and CO2 reforming of fossil fuels, may be used to produce synthesis gas. Synthesis gas can be transformed into liquid fuels via Fischer-Tropsch synthesis. Moreover, these processes are also the main routes for producing hydrogen, which is a well-known clean energy carrier. Conventional catalytic reforming methods often meet serious problems. The catalysts are inclined to deactivate due to coke deposition. Additionally, higher reaction temperature and higher energy consumption is a disadvantage for the application of the processes in industry.

Nonequilibrium plasmas possess lower temperature, and nonequilibrium plasma reactions possibly possess an inherent quenching mechanism, i.e., the gases were rapidly cooled as they flowed out of the plasma area. They can be used to produce hydrogen-rich synthesis gas from fossil fuels. We have studied CO2 reforming of methane via corona plasma reactions at atmospheric pressure. The results show that the H2/CO molar ratios in the products strong depend on the CH4/CO2 molar ratios in the feeds. The discharge power, flow rate, and corona types have slight influence on the composition of synthesis gas. In our experiments using AC corona plasma CO2-reforming reactions, H2/CO ratios in the products increased from 0.21 at 0.2 CH4/CO2 to 2.15 at 2.0 CH4/CO2. The conversions of CH4 changed from 67.5 to 90.5% and the conversions of CO2 changed from 45.7 to 78.5%. The conversions increased with increasing discharge power.

Two kinds of representative catalysts, Ni/Al2O3 and HZSM-5 zeolite catalysts were used in the CO2-reforming plasma reactions and they show little effect on the conversions. We propose that rapid free-radical processes play a main role during plasma reactions. However, H2/CO ratios in the products were modified, and more hydrocarbons and oxygenates were formed after the catalysts were used.

Nonequilibrium plasma reactions may be applied to store electric energy from clean nuclear energy into chemical energy. It is estimated that more than 10% of electric energy could be converted to chemical energy stored in H2 and CO in our experiments. In the future, more kinds of reactants, including water and biomass, might be tested to produce hydrogen via plasma reactions.

Theoretical and experimental analysis of non-thermal arc discharge assisted reforming of gasoline and E85

G. Petitpas[1], J. Gonzalez-Aguilar[1], A. Darmon[2], L. Fulcheri[1]

[1]*Center for Energy and Processes, MINES ParisTech, 06904 Sophia Antipolis, France*
[2]*Technocentre Renault, DREAM/DTAA - Service 68240, 78288 Guyancourt Cedex, France*

Abstract

On-board hydrogen production by means of hydrocarbon reforming systems is an attractive option for the development of fuel cells in vehicles. Interest on plasma as an alternative to classical catalytic reforming systems has increased in recent years. In this work a review of recent advances on non-thermal plasma reforming of commercial hydrocarbons: gasoline and E85 (85% ethanol 15% gasoline) carried out in our laboratory is presented.

The plasma technology is based on a low current - high voltage tip/cylinder configuration discharge. Two numerical models have been developed and their results compared to experimental results. The first model aims at describing the mixture flow by a Computational Fluid Dynamics (CFD) program including chemistry. The second model consists in an association of well defined chemical reactors sub-models.

Introduction

A major drawback for the large-scale development of fuel cells for automotive applications is the commercial unavailability of hydrogen by the end users at least at short and mid-term. This problem is caused by the absence of real hydrogen distribution network and the lack of low cost, efficient, onboard hydrogen storage technologies. For example, liquefying hydrogen takes 1/3 to 1/2 of its energy and its storage (LH2) costs $6 per kWh (e.g. almost $1.000 for a full tank !) [1].

An alternative way to overcome these limitations consists in the onboard production of hydrogen from conventional liquid hydrocarbon fuels such as gasoline or diesel fuel (available via existing gas stations) thanks to a process named reforming. This transition solution from a fossil based to a hydrogen based economy offers lower greenhouse gases production than conventional vehicles, whereas production, distribution and storage costs are well known and managed (e.g. gas stations). However, downscaling reforming from large plant to cars is a real challenge: very short transient regimes and small reactors are strongly required.

In order to arise a hydrogen based transportation economy, one solution could thus be the development of compact and fast response reformers. Among existing technologies, interest in plasma as an alternative to classical catalytic reforming systems has increased in recent years (see review [2]). Together with the experimental tests, the simulation work conducted in our laboratory has helped a lot in the elaboration of a competitive non-thermal plasma assisted reforming reactor out of liquid commercial fuels (gasoline and E85). This paper presents the main results of this investigation.

Experimental set up

The plasma technology is based on a low current - high voltage tip/cylinder configuration discharge [3]. The experimental arrangement that was used in this work has been described in details previously [4] and is briefly described here for clarity.

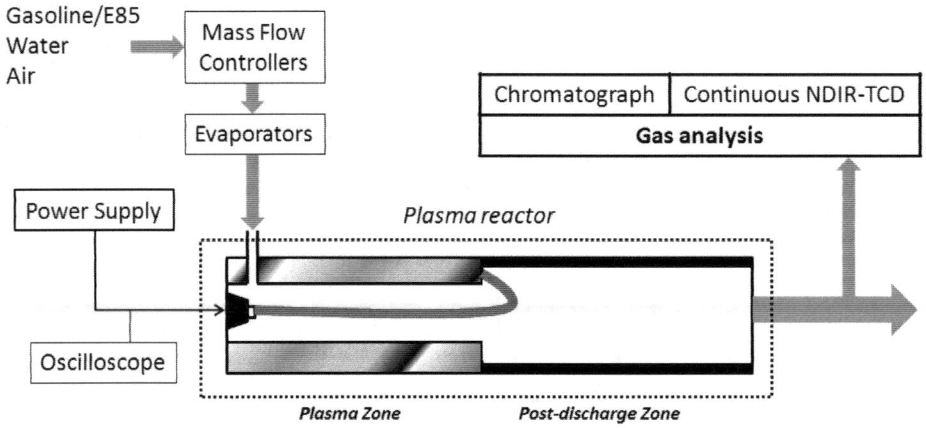

Figure 1. Diagram of the experimental set up.

An overall scheme of the experimental setup is shown in figure 1. The reactor includes a non-thermal arc plasma torch and a post-discharge zone. The plasma torch has a tip-cylinder configuration. The cylinder electrode (also called "nozzle") is at ground potential and its geometry is adjustable over a wide range of dimensions and shapes (length, inner diameter, inlet and tip shape). The plasma torch is supplied by a system based on a resonant converter technology [30], that enables a control of the arc current with a high accuracy in the range 200 – 660 mA and a maximum voltage of 15 kV.

Reformate gas composition is analyzed using a NDIR – TCD analysis bench (Rosemount, NGA 2000), which enables a continuous monitoring of H_2, CO, CO_2 and CH_4 molar fractions (dry sample). The principle of the measurement is based on the assumption that outlet dry gas does not contain other gases than nitrogen and the above-mentioned. This assumption was verified by gas analysis with gas phase chromatograph (Perichrom).

A gas feeding system (mass flow controllers + evaporators) supplies the gas mixture. This device produced mixtures composed of fuel vapor, air and steam in a controlled way. The air flow rate is controlled in the range 15-50 slpm using a mass flow controller while fuel and water mass flow rates are controlled in their liquid phase via rotameters in the range 0.05 – 0.2 g/s and 0 – 0.1 g/s, respectively. The reactants are injected tangentially in a mixing chamber located right in front of the anode tip. Inlet temperature can be freely adjusted between ambient temperature and 500 K.

Modeling section

The modeling approach used is derived from combustion. No charged species is considered. The discharge is modeled by a cylinder of given radius and length in which power density is constant (typically, 5 W/mm^3). Note that the arc discharge is taken into account as a high-density heat source domain which is under thermodynamics equilibrium and thus the non-equilibrium phenomena are not taken into account in the discharge. The kinetic mechanisms employed here are usually used in the study of methane or gasoline oxidation ([5], [6]). Two numerical models were developed in order to understand the major phenomena.

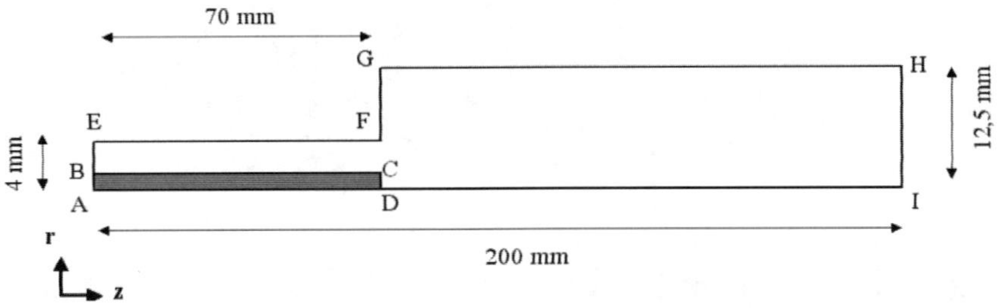

Figure 2. 2D axisymmetric mesh of the domain located downstream inlet/mixture chamber

The first model solves the Navier Stokes and energy equations over the whole geometry. The model has been implemented using the open source Computational Fluid Dynamics (CFD) toolkit OpenFOAM [7]. The ReNormalizationGroup (RNG) κ-ε model has been used for accounting turbulence effects, while the overall rate of reaction is controlled by turbulent mixing. The turbulence-chemistry interaction model used in this study includes detailed chemical mechanisms in turbulent flows. It assumes that reaction occurs in small turbulent structures, called the fine scales. The reactor domain's investigation was performed in two steps : a full 3D grid of the mixture chamber (where no chemistry is considered), and a 2D axisymmetric grid of the domain located downstream (see figure 2). The ABCD domain describes the arc discharge zone, AEFD the plasma zone and DGHI the post-discharge zone.

The CFD model is computationally expensive and can not model reforming of hydrocarbon molecules heavier than methane. As a consequence, a 1D phenomenological model has been developed (using ChemKin II [8]) based upon results obtained with the CFD model. The flow diagram of the system associated with this model is plotted on figure 2. The approach assumes that: only a fraction of the reactants inlet flow really passes through the arc discharge, the remaining fraction being deviated from the arc; there is no mass transfer between the arc zone and the remaining gas fraction in the plasma torch; the two cold and hot streams are instantaneous mixed at the plasma torch exit before entering the post discharge zone. A perfectly stirred reactor (PSR) receiving an input heating power equal to the electric power describes the plasma region (1). Mixture temperature at the plasma torch exit is evaluated from global enthalpy balance from both streams (2). Finally, a plug flow reactor (PFR) model is used to describe the post discharge zone (3).

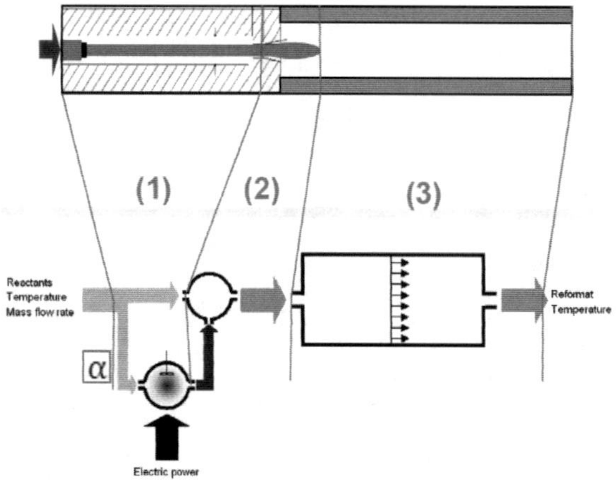

Figure 3. Schematic diagram of the phenomenological model.

The model is based on a parameter that represents the fraction of the reactants passing through the arc discharge: α. Based on geometrical considerations, this parameter can be expressed as a function of an effective arc discharge radius R_{arc} by the formula:

$$\alpha = \frac{Q_{arc}}{Q_{total}} = \left(\frac{R_{arc}}{R_{nozzle}}\right)^2 \qquad (1)$$

with R_{nozzle}, the nozzle radius; Q_{arc}, the mass flow rate passing through the arc discharge, and Q_{total}, the total reactants mass flow rate.

Results and discussion

The simulation work was conducted with different tools in order to improve our understanding of the major phenomena and elaborate an accurate model. The experimental tests, carried out in parallel, helped the investigation and enabled to evaluate the accuracy of the models.

First results concern the study of the flow in the mixture chamber, where no chemistry occurs (temperature is lower than 500 K). The reactants (fuel, air, steam) enter the plasma reactor tangentially in front of the tip electrode, at different velocity magnitudes and through 2 periodically opposed inlets. The contours of the air molar fraction in the whole domain are plotted on figure 4, right. One can observe that the complete mixture establishes quickly after the entrance. Contours of the z-velocity component at the mixture chamber's exit are reported on figure 4 (right): the flow is axially symmetric at this point.

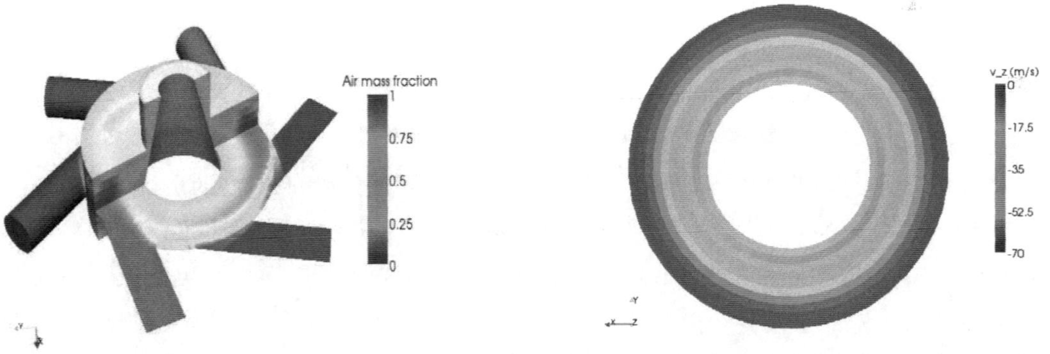

Figure 4. Molar air fraction in the mixture chamber (left), and contours of z-velocity component at the domain's exit (right). Gasoline/air/steam: 1/17/1,5. Total flow rate: 0.02 mol/s.

As a consequence, a 2D axisymetric grid was then chosen to model the interaction between the flow and the chemistry in the rest of the domain (downstream the mixture chamber). Figure 5 shows the molar fraction variations over the 2D domain of the major species (except N_2) for the partial oxidation of methane. The dotted line represents the gap between the arc discharge and the post-discharge zone. Through this region, we can observe lower CH_4 consumption and H_2 production to the benefit of an important production of steam. This transition zone decreases the reformer efficiency: the H atoms combine with O atoms (to produce H_2O) instead of producing H_2.

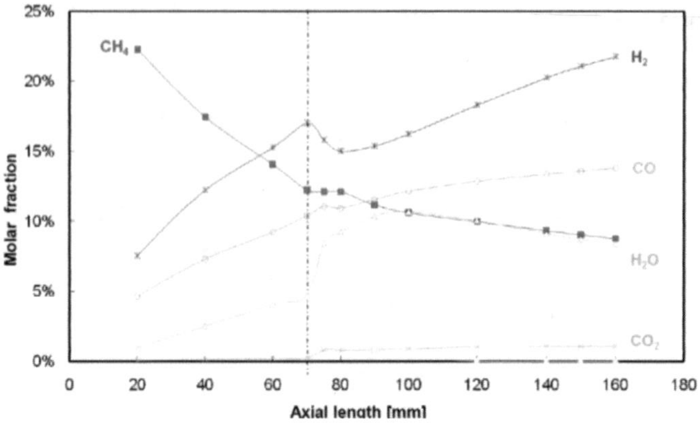

Figure 5. Partial oxidation of methane (CH_4/air=1/2,4). Plasma power: 1000 W. 2D axisymetric model. Molar fraction of majoritary species along the axis (the balance nitrogen N_2 is not represented for clarity).

The 1D phenomenological model was developed considering the viscosity and temperature gradients observed in the CFD approach. A good correlation was observed between both models for the case of methane (not showed). The results from the 1D model were then compared with results from the test bench for the case of gasoline reforming (see figure 6). The reformer's performances were evaluated as function of the conversion rate and efficiency, defined in [4]. As defined in the same reference, the O/C value stands for the mixture ratio between the oxygen from the air and the carbon from gasoline. Here, the air flow rate was increased. Three different values of α were computed (α=0.0625, 0.14 and 0.25, corresponding to 1/16, 1/9 and 1/4). A good fitting between the 1D model and experimental results can be observed at α=1/16, that corresponds to the case where the inner nozzle radius is 4 mm and the arc radius 1 mm. Small discrepancies occur because the calculation does not account for non-equilibrium behavior of the plasma discharge and the thermal losses, that are inevitably present in experiment.

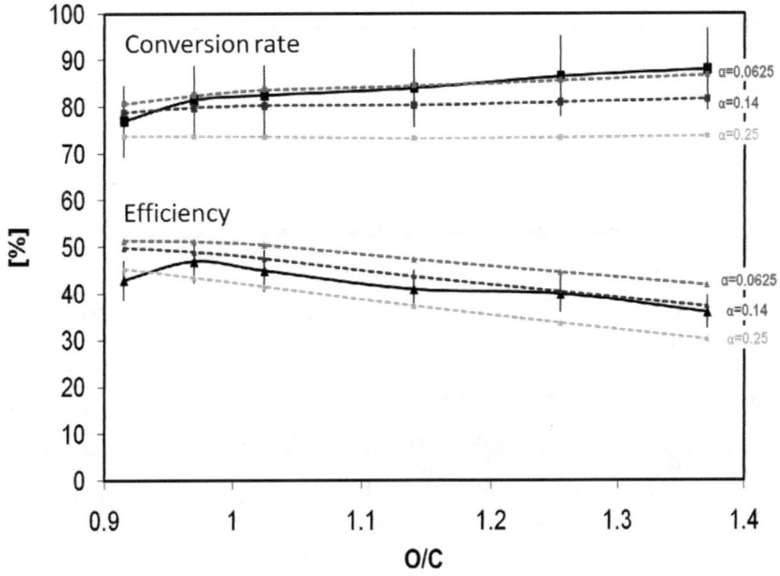

Figure 6. Gasoline reforming performances vs O/C ratio:
experimental (solid lines) and theoretical (dotted lines). Plasma power: 1000 W.

Last result concerns partial oxidation of E85 (no reactive steam). Experimental tests were run with the same configuration as with gasoline, then compared with results from the 1D model with α=1/16. Figure 7 shows the very good correlation of the model. Furthermore, a 50% energy efficiency and about 85% conversion rate were observed experimentally.

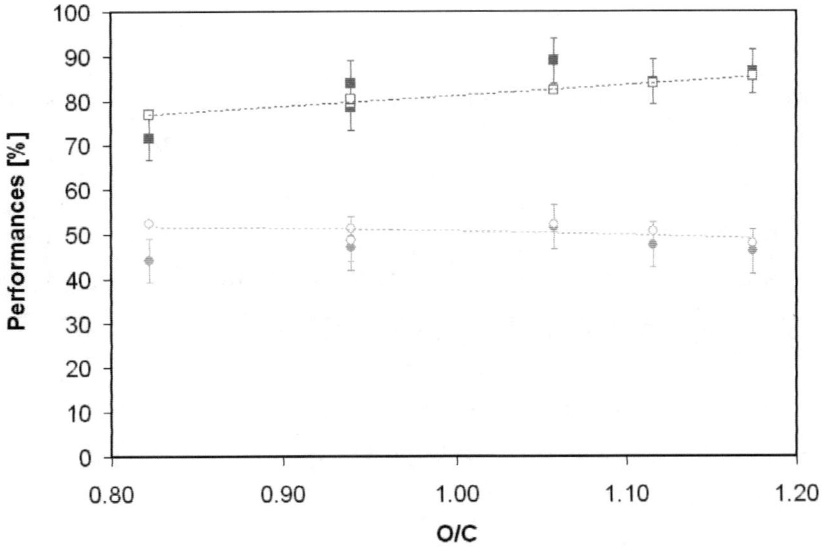

Figure 7. Partial oxidation of E85. Conversion rate (model: □, experimental:■); energy efficiency (model:○; experimental:●) as a function of O/C ratio. α=1/16. Plasma power: 1000 W.

Conclusion

E85 is considered as a renewable fuel of growing importance, entering distribution network in different countries. The use of a non-thermal plasma arc discharge reformer for hydrogen production out of two commercially available fuels (gasoline and E85) is thus relevant, and promising considering our results, for on-board application

.

Different modeling approaches (CFD only, CFD coupled to chemistry, phenomenological) were investigated in order to elaborate an accurate model. Quite a good fitting to experimental was observed, in spite of the stronger simplifications in 1D approach (perfect mixing, well defined plasma zone). The models can simulate the reforming of hydrocarbons like methane, gasoline and E85 over a wide range of conditions. At last, our reformer shows a 50% energy efficiency in partial oxidation conditions.

Next work should now address the thermal insulation of the reformer in order to improve the overall performances, as well as the study of the dynamic behavior.

Acknowledgements

The authors gratefully acknowledge Renault SAS and ADEME (French agency for environment and energy management) for their financial support.

References

[1] Milliken, JoAnn (2002), "Energy Efficiency and Renewable Energy, hydrogen storage", *DOE Committee Meeting, Washington D.C., December 2-3, 2002*

[2] Petitpas, Guillaume, Rollier, Jean-Damien, Gonzalez-Aguilar, José, Darmon, Adeline, Fulcheri, Laurent and Rudolf Metkemeijer (2007), "A comparative study of non-thermal plasma assisted reforming technologies", *Int. J. Hydrogen Energy*, 32, pp. 2848-2867.

[3] Fulcheri, Laurent, Rollier, Jean-Damien, and José Gonzalez-Aguilar (2006), "Design and electrical characterisation of a low current-high voltage compact are plasma torch", *Plasma Sources Sci. Technol.*, 16, pp.183-192.

[4] Rollier, Jean-Damien, Petitpas, Guillaume, Gonzalez-Aguilar, José, Darmon, Adeline, Fulcheri, Laurent and Rudolf Metkemeijer (2008), "Experimental study on gasoline reforming assisted by non thermal arc discharge", *Energy & Fuels,* pp.556-560.

[5] Smooke, M., Puri, I and K. Seshadri (1986), “A comparison between numerical calculations and experimental measurements of the structure of a counterflow diffusion flame burning diluted methane in diluted air”, *Proc. of Twenty-First Symposium (International) on Combustion, The Combustion Institute, Pittsburgh,* pp. 1783-1792.

[6] Glaude, P. A., Warth, V., Fourner, R., Battin-Leclerc, F., Scacchi, G., and G. M. Côme (1998), “Modeling of the oxidation of n-octane and n-decane using an automatic generation of mechanisms”, *International of Chemical Kinetics*, 30, pp. 949-959.

[7] OpenFOAM project web site, http://www.opencfd.co.uk/openfoam/, ©2004-2009 OpenCFD Limited.

[8] Kee, R. J., Rupley, F. M. and J. A. Miller (1989), *Sandia National Laboratories Report SAND 89-8009.*

35a Development of ATR Fuel Processor for Reformation of Logistic Fuels

Christian Junaedi, **Subir Roychoudhury**, Dennis Walsh, and Rich Mastanduno. Precision Combustion, Inc. (PCI), 410 Sackett Point Rd., North Haven, CT 06473

Fuel cell power systems are an attractive solution for future portable power generation in many applications such as battlefield, shipboard, remote construction sites, etc. In developing viable fuel cell-based power generation units, durability of reforming systems to continuously supply hydrogen and to permit stable, coke-free fuel cell stack operation via catalytic reforming of commonly available fuels is a key challenge. Precision Combustion, Inc. (PCI) has developed an Auto Thermal Reformer (ATR) unit and has demonstrated stable, coke-free operation using a low-sulfur, Tier II diesel for ~1100 hours with complete fuel conversion and reforming efficiency of >80% (ratio of LHV of reformate to fuel). In this unit, a distillate fuel (e.g., JP-8 or diesel) was processed with low water usage; therefore, the test demonstrated the feasibility of long-term operation at a low steam-to-carbon ratio while producing SOFC-quality reformate (i.e., <10 ppmv of higher hydrocarbons).

In this paper, we will give an overview of the ATR development at PCI and will show some performance results when operating the reactor using a low-sulfur fuel. Additionally, we will present experimental results obtained from ATR sulfur tests using JP-8 fuels consisting of various sulfur levels. Here, the fuels were doped with an organic sulfur surrogate or contained naturally occurring sulfur compounds. The effect of increasing fuel-sulfur level on the ATR performance, including fuel conversion, reforming efficiency, H2 selectivity, and reformate composition will be discussed. In our study, the sulfur level in the liquid fuel was increased stepwise from 4 ppmw to 3000 ppmw by doping a low-sulfur fuel with dibenzothiophene. At each sulfur level, the ATR reactor was operated for at least 50 hours and the ATR reformate stream was analyzed for H2, CO, CO2, higher hydrocarbons as well as H2S and carbonyl sulfide. This study helped to clarify issues surrounding reformation of sulfur containing fuels and to illustrate one possible approach for reformation of higher sulfur logistic fuels for fuel cell applications.

Finally, long-term durability data on the ATR reforming of JP-8 with 400 ppmw naturally occurring sulfur will also be presented. The data will be compared with the results obtained from operating the ATR reactor using doped sulfur fuels. The evaluation of the reactor performance results from this set of 50-hour tests using different sulfur fuels provides a measure of the ATR performance that can be expected under realistic conditions with readily available fuels. The results also give valuable insights to the system design and operation strategy of the integrated fuel reformer and APU system.

Multi-Fuel Direct Steam Reforming Process for SOFC Applications – ID# 145628

Aly H. Shaaban [1], *Christian D. Rasmussen* [1], *Ragab M. Moheisen* [1], *Reza Salavani* [2]
[1] Applied Research Associates, Inc, 430 5th Street, Panama City, Florida
[2] Air Force Research Laboratory, Airbase Technologies Division, Tyndall AFB, Florida

ABSTRACT

With the high probability of introducing renewable fuels such as synthetic fuels, biodiesel, and syngas into commercial and military markets, the stakes are high to find an electric generator technology that can adequately and efficiently utilize these fuels and meet the EO 13423 and the Energy Policy Act of 2005 to improve energy efficiency and reduce greenhouse gas emissions.

Direct steam reforming process converts hydrocarbon fuels to hydrogen rich gas to power fuel cell electric generators. There are many applications for fuel cell electric generators. For example, an Auxiliary Power Unit (APU) can provide secondary power for start up of larger aircraft main engines or provide power to large trucks' cabins during stops to run air conditioning and refrigerator equipment to minimize CO_2 emissions. Distributed power generation where quiet electric generators are distributed to strategic locations to provide power to residential subdivisions and industrial compounds to cut down on transmission costs, losses, and emissions are applications that well suit fuel cell technology. For practical reasons, these power generators would use the fuel available in the applications. APUs will use the same fuel of the main engine: jet fuel (kerosene) in the case of aircrafts, diesel or biodiesel in the case of large trucks, while distributed power fuel cells will use available fuel such as natural gas (methane) or propane. The ability to reform a multitude of fuels can accelerate the introduction of the more efficient and quiet fuel cell electric generators into power applications.

The direct steam reforming components are divided among three major groups: sulfur removal, steam reforming, and balance of plant. This paper will discuss the results of the direct steam reforming technology in reforming of JP-8, synthetic jet fuel (S-8), diesel, biodiesel, and propane. While in some applications a compact unit is a must, the impact of thin-film catalysis on the performance of fuel processing unit will be introduced.

RESEARCH PROGRAM

The fuel processor, Figures 1-2, consists of a sulfur removal group, direct steam reforming group, balance of plant (BOP) to maintain process temperatures and utilize waste heat, and a process control. The design point is a 4.3g/min JP8 flow to the steam reformer to produce reformate stream to power a 1.5kW SOFC stack. The steam reforming group and the fractionator are housed in a thermally insulated high temperature component housing to maximize the use of waste heat. The unit measures 30"W x 24"D x 24"H and was designed with modification needs in mind; therefore, the unit has excess space to allow for maintenance and easy change of components during laboratory testing. The sulfur removal group includes the fractionator, condenser, heat exchanger, gas-liquid separator, and two sulfur absorption beds. The sulfur removal beds are directly behind the high temperature component housing to be in close proximity to the heat exchanger recovering burners' exhaust heat.

The steam reforming group consists of the two high temperature components the steam generator and the steam reformer. Both are heated by radiant burners, Figure 3. The steam generator is a 4-inch diameter stainless steel coiled tube made of a 20ft 0.25in OD tube with a 0.035in wall thickness. The steam generator was designed to produce 33.5g/min steam in excess of 600°C. The reforming reactor is a 5-inch diameter coiled tube made of a 20ft piece of 0.5in OD Haynes 230 tubing to handle operation above 800C. This coil is packed with a proprietary sulfur tolerant steam reforming catalyst.

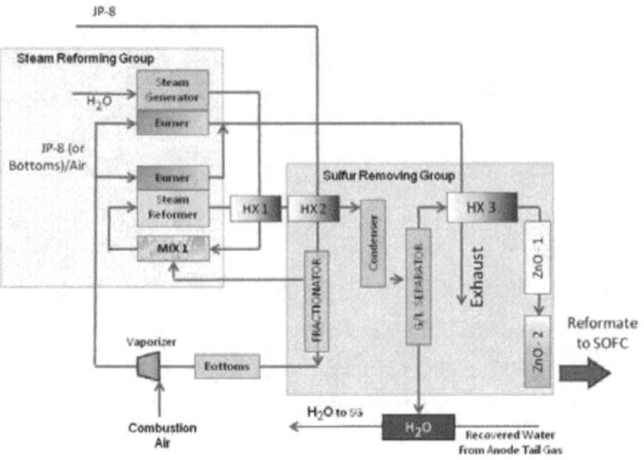

Figure 1: General Fuel Processor Schematic

Figure 2: The Multi-Fuel Processor for a 1.5kW SOFC Stack

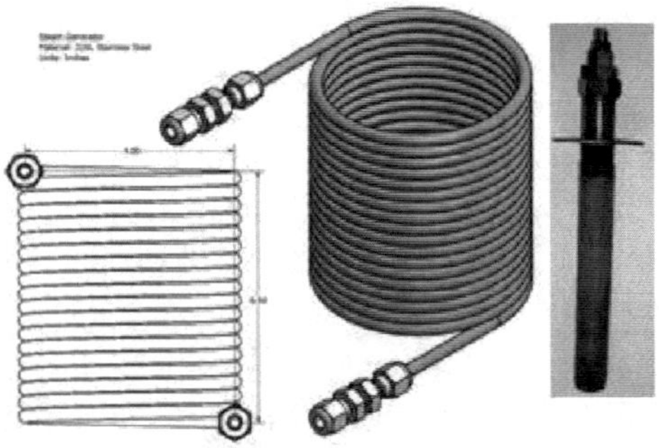

Figure 3: Steam Generator / Steam Reformer Design

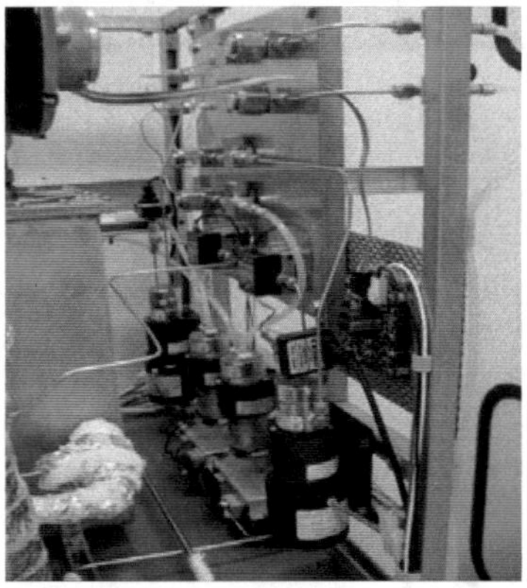

Figure 4: Balance of Plant – Reformer Fuel and water Pumps

The reforming reactor and steam generator are enclosed in a central housing along with their respective radiant burners. The radiant energy transferred to the reformer and steam generator is adjusted by changing the fuel and air delivered to the burners. The fractionator and sulfur removal bed preheat heat exchangers are also located inside this housing. Fuel is vaporized before entering both burners and steam reformer. A Fuel vaporizer, consists of a FeCrAlM alloy mesh-pad, Micron Fiber-Tech Corp – D-mat, housed in a cavity of a stainless steel frame heated with two electric heaters, Chromalox CTR-2061 – 39W/in^2. The vaporizer temperature is controlled and maintained to provide steady evaporation rate for the fuel in use and fuel processor demands. The process JP-8 and water pumps are Tuthill D-Series DGS.19PPPV1NN15000 – 0-38 LPH. The burner fuel pumps are Tuthill U-Series magnetically coupled external gear pumps, Tuthill UBS.02CCNT4SN77000 – 0-5 LPH. The Tuthill burner fuel pumps are driven by a 12 VDC brushless motor, Pittman GM9413-1.

Plant control is accomplished through the use of National Instruments Field Point hardware and LabView software. The data acquisition and control hardware consist of a FieldPoint Ethernet communication module, National Instruments NI FP-1601, an 8 Channel analog voltage output module, National Instruments NI FP-AO-210, a 16 Bit analog voltage input module, National Instruments NI FP-AI-112, and an 8 Channel/16 Bit thermocouple input module, National Instruments FP-TC-210.

Measuring the flow from the process pumps McMillan Flow 100 sensors; with a 0-100mL/min range and maximum full scale error of ±3.0%; were used. Inline 15-micron size sintered metal T-Type filters, Swagelok SS-2TF-15, were installed in all fluid lines to protect the pumps and flow meters. The pressure of the process JP-8 and water line is monitored by 0-200psi Omega Model PX181-200GSV pressure transducers with ±0.3% full scale accuracy. The air needed for the JP-8 combustion in the burners is provided by Ametek Rotron Mini Jammer BLDC low-voltage blowers, Ametek 150016-02 – 24VDC, 2.5A, 34CFM, with an absolute error of <0.1% . The temperature of the different streams were measured using type K Quick Disconnect Thermocouple Probes, Omega Model #: KMQXL-062G-12, with an accuracy of ± 2.2°C for temperatures up to 293°C and ±0.75% of the measured temperature for temperatures exceeding 293°C.

APPROACH AND DISCUSSION

The Multi-Fuel Processor was tested for six different fuels: JP8; diesel; propane; synthetic jet fuel (S8); 50/50 mixture of S8 and JP8 (50/50 mix); and Biodiesel. Synthetic jet fuel and its 50/50 mixture with JP8 was provided by the Air Force Fuel Certification Office, 77AESW/LF. The biodiesel was produced in-house from frying oil using a transestrification procedure. The test matrix is given in TABLE 1. Twelve (12) tests for each fuel were conducted covering three (3) steam-to-carbon ratios (S/C), 4, 5, and 6. Fuel flow to the steam reformer was kept at 2g/min, which represents half of the full load design. Each S/C had four (4) test runs covering a reforming reactor exit temperature range of 400-800°C. For the JP8, twelve (12) tests were conducted at the full load design of 4.3g/min and eleven (11) tests were conducted for the 50/50 mix at the same fuel flow rate covering the same temperature range. TABLE 1 shows the test matrix for the 2g/min fuel flow rate for all six fuels tested and 4.3g/min for JP8 and 50/50 mix fuels.

TABLE 1: Multi-Fuel Test Matrix – 2 and 4.3g/min Fuel Flow Rates

JP8			Diesel			Propane			50/50 S8/JP8			Synthetic: S8			Biodiesel B-100		
S/C	Process Fuel Flow Rate [g/min]	Water Flow Rate [g/min]	S/C	Process Fuel Flow Rate [g/min]	Water Flow Rate [g/min]	S/C	Process Fuel Flow Rate [g/min]	Water Flow Rate [g/min]	S/C	Process Fuel Flow Rate [g/min]	Water Flow Rate [g/min]	S/C	Process Fuel Flow Rate [g/min]	Water Flow Rate [g/min]	S/C	Process Fuel Flow Rate [g/min]	Water Flow Rate [g/min]
4	2.00	10.37	4	2.00	10.51	4	2.00	9.90	4	2.00	10.37	4	2.00	10.37	4	2.00	9.31
4	2.00	10.37	4	2.00	10.51	4	2.00	9.90	4	2.00	10.37	4	2.00	10.37	4	2.00	9.31
4	2.00	10.37	4	2.00	10.51	4	2.00	9.90	4	2.00	10.37	4	2.00	10.37	4	2.00	9.31
4	2.00	10.37	4	2.00	10.51	4	2.00	9.90	4	2.00	10.37	4	2.00	10.37	4	2.00	9.31
5	2.00	12.98	5	2.00	13.16	5	2.00	12.40	5	2.00	12.98	5	2.00	12.98	5	2.00	11.63
5	2.00	12.98	5	2.00	13.16	5	2.00	12.40	5	2.00	12.98	5	2.00	12.98	5	2.00	11.63
5	2.00	12.98	5	2.00	13.16	5	2.00	12.40	5	2.00	12.98	5	2.00	12.98	5	2.00	11.63
5	2.00	12.98	5	2.00	13.16	5	2.00	12.40	5	2.00	12.98	5	2.00	12.98	5	2.00	11.63
6	2.00	15.58	6	2.00	15.81	6	2.00	14.90	6	2.00	15.58	6	2.00	15.58	6	2.00	13.92
6	2.00	15.58	6	2.00	15.81	6	2.00	14.90	6	2.00	15.58	6	2.00	15.58	6	2.00	13.92
6	2.00	15.58	6	2.00	15.81	6	2.00	14.90	6	2.00	15.58	6	2.00	15.58	6	2.00	13.92
6	2.00	15.58	6	2.00	15.81	6	2.00	14.90	6	2.00	15.58	6	2.00	15.58	6	2.00	13.92

JP8			50/50 S8/JP8		
S/C	Process Fuel Flow Rate [g/min]	Water Flow Rate [g/min]	S/C	Process Fuel Flow Rate [g/min]	Water Flow Rate [g/min]
4	4.30	22.30	4	4.30	22.30
4	4.30	22.30	4	4.30	22.30
4	4.30	22.30	4	4.30	22.30
4	4.30	22.30			
5	4.30	27.90	5	4.30	27.90
5	4.30	27.90	5	4.30	27.90
5	4.30	27.90	5	4.30	27.90
5	4.30	27.90	5	4.30	27.90
6	4.30	33.50	6	4.30	33.50
6	4.30	33.50	6	4.30	33.50
6	4.30	33.50	6	4.30	33.50
6	4.30	33.50	6	4.30	33.50

Test Procedure

Fractionation was not used during test runs. Reforming of the entire fuel fractions was more essential for comparison purposes than reducing sulfur content, since S8 and biodiesel have no sulfur. To measure the performance of the fuel processor, fuel and water flow rates, pressures and temperatures of different streams, and test duration were acquired. Gas samples of the reformate stream were analyzed using a gas chromatograph (GC), Agilent Technologies 6890N. The Gas Chromatograph was calibrated to measure molar percent of H_2, CO, CO_2, C_2H_2, C_2H_4, C_2H_6, C_3H_6, and C_3H_8 in the reformate stream.

Test Results and Discussion

Reaction Selectivity: Hydrogen, carbon monoxide, and methane selectivities are shown in Figures 5–22. Figures 5–7 for JP8, Figures 8–10 for diesel, Figures 11–13 for S8, Figures 14–16 for 50/50 mix of JP8 and S8, Figures 17-19 for propane, and Figures 20-22 for biodiesel. In all of these test runs, the steam reformer's operating temperature and steam to carbon ratio impact the H_2, CO, and methane content in the reformate stream. As the temperature of the reformate stream at the exit of the steam reformer increase, H_2 selectivity increases while the selectivity for both CO and methane decreases. The S/C has the same effect. The lower selectivity for CO and CH4 is due to excess water in the last section of the reactor. Water-gas-shift reaction occurs and water, CO and CH_4 are converted to hydrogen and CO_2. No hydrocarbons higher than methane were detected. One shall keep in mind that at the 2g/min flow rate, the steam-fuel mixtures have about twice the residence reaction time as the 4.3g/min flow.

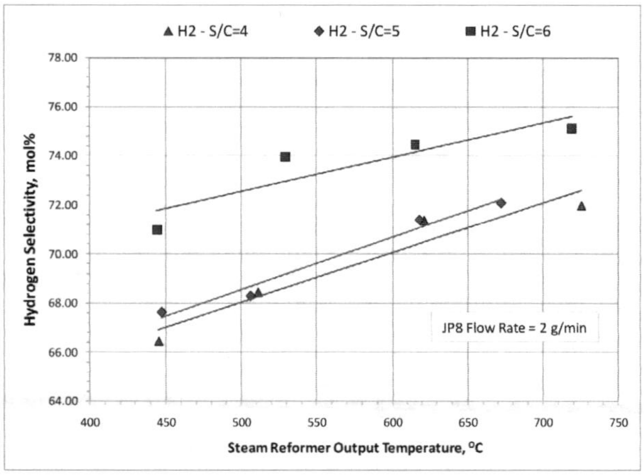

Figure 5: H_2 Selectivity as a Function of JP8 Reformate Temperature and S/C.

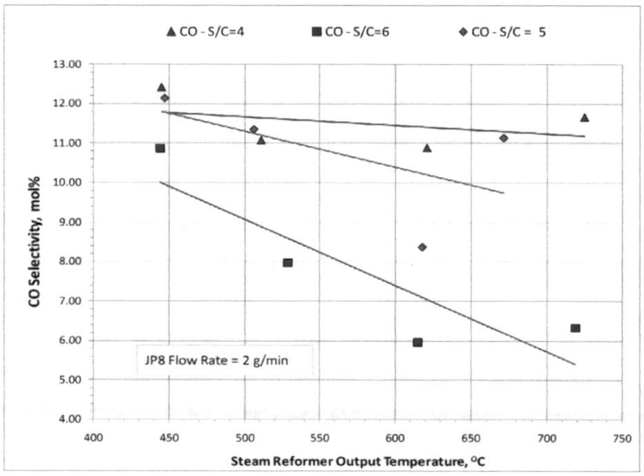

Figure 6: CO Selectivity as a Function of JP8 Reformate Temperature and S/C.

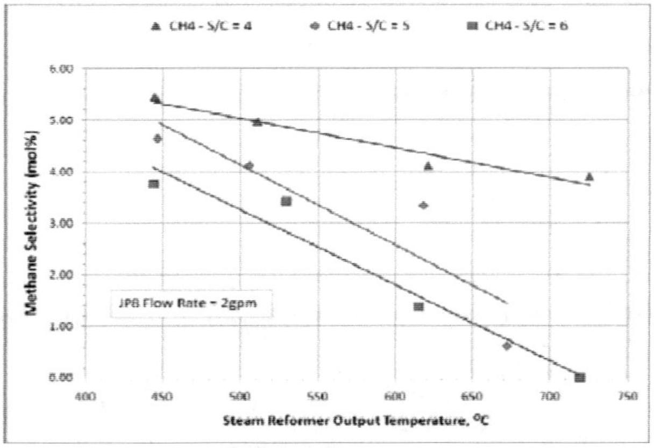

Figure 7: CH_4 Selectivity as a Function of JP8 Reformate Temperature and S/C.

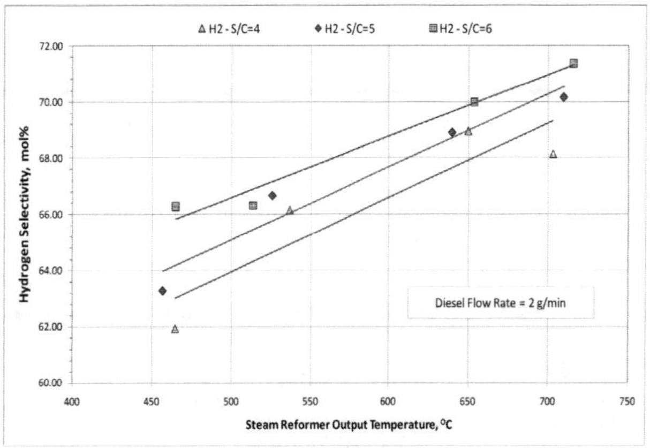

Figure 8: H_2 Selectivity as a Function of Diesel Reformate Temperature and S/C.

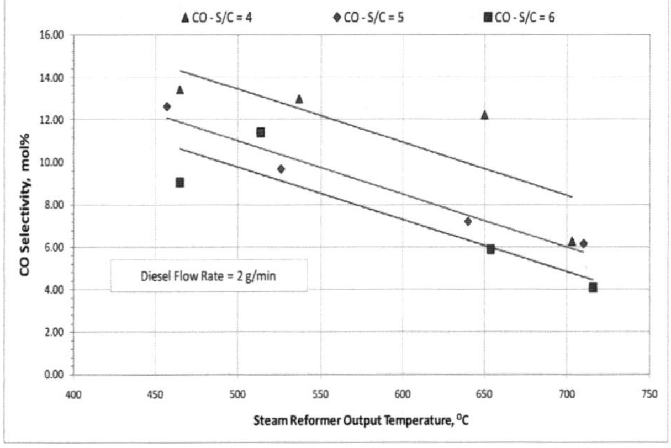

Figure 9: CO Selectivity as a Function of Diesel Reformate Temperature and S/C.

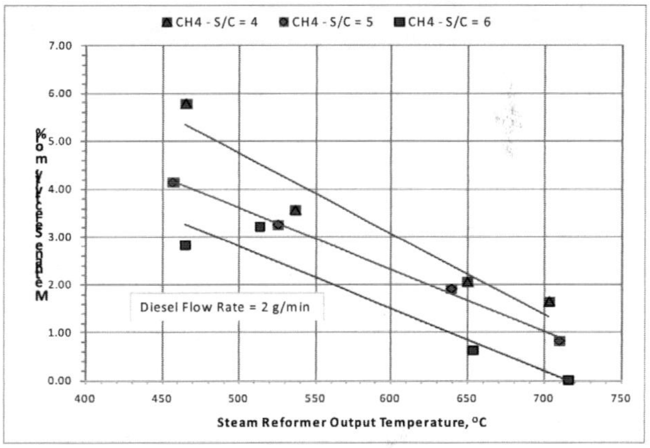

Figure 10: CH_4 Selectivity as a Function of Diesel Reformate Temperature and S/C.

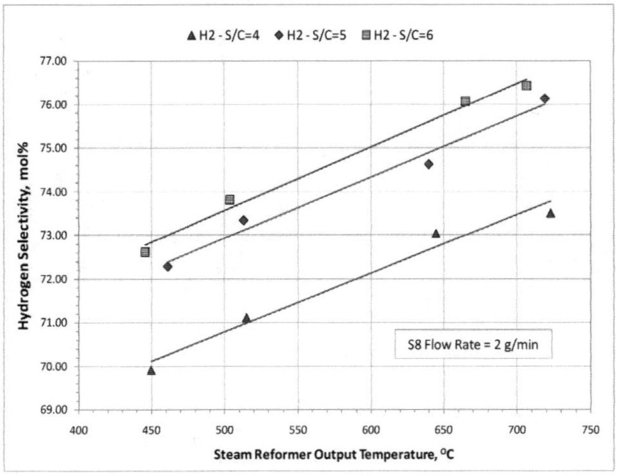

Figure 11: H_2 Selectivity as a Function of S8 Reformate Temperature and S/C.

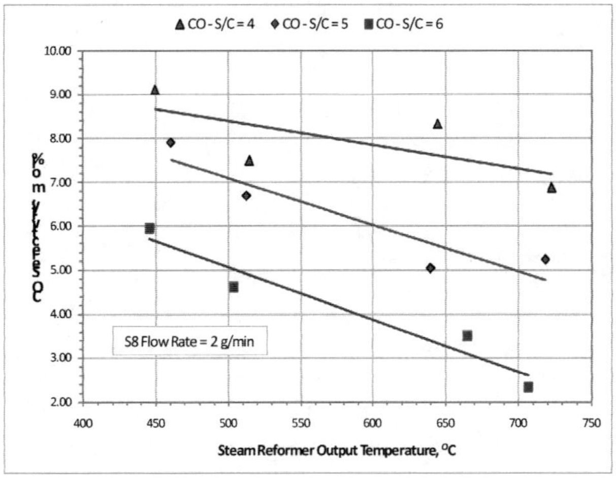

Figure 12: CO Selectivity as a Function of S8 Reformate Temperature and S/C.

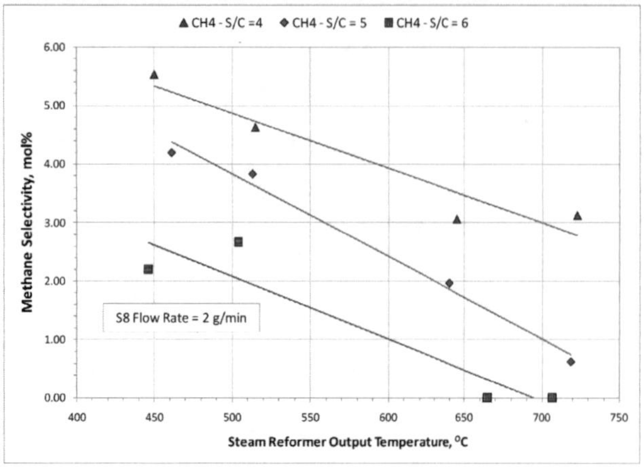

Figure 13: CH_4 Selectivity as a Function of S8 Reformate Temperature and S/C.

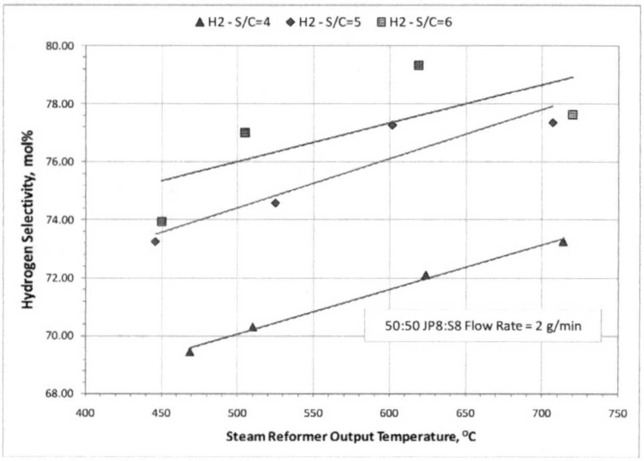

Figure 14: H_2 Selectivity as a Function of 50/50 mix Reformate Temperature and S/C.

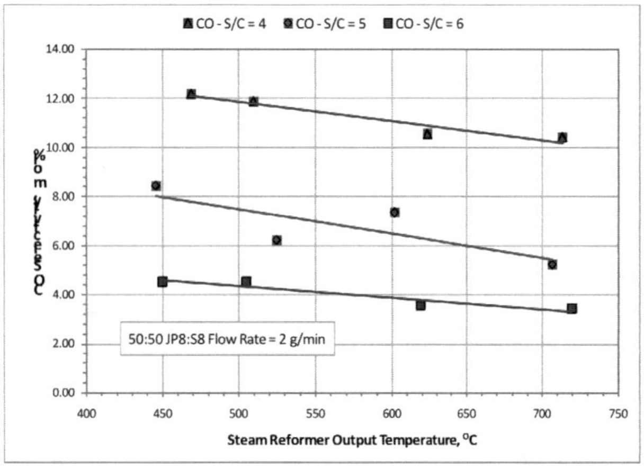

Figure 15: CO Selectivity as a Function of 50/50 mix Reformate Temperature and S/C.

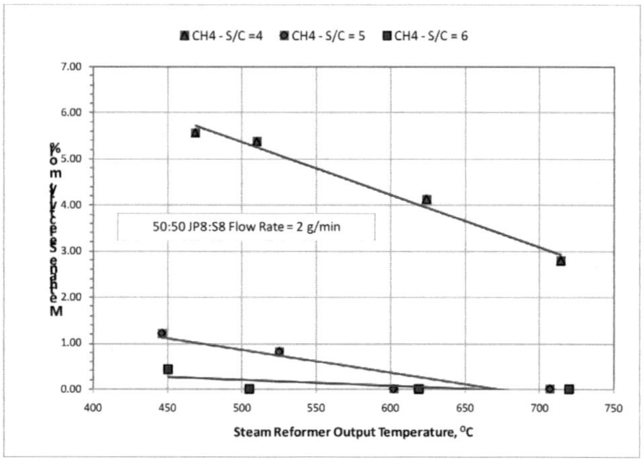

Figure 16: CH_4 Selectivity as a Function of 50/50 mix Reformate Temperature and S/C.

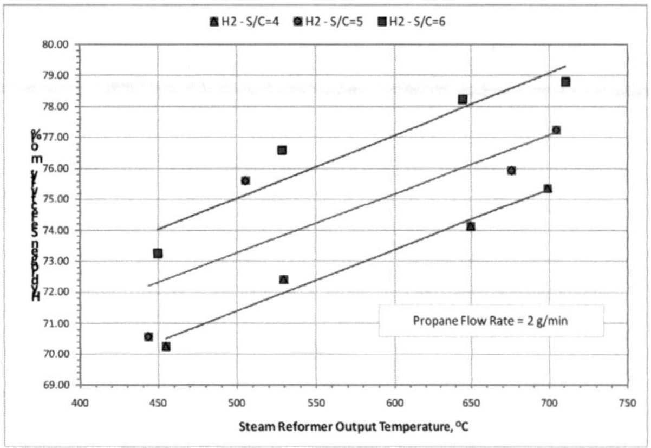

Figure 17: H_2 Selectivity as a Function of Propane Reformate Temperature and S/C.

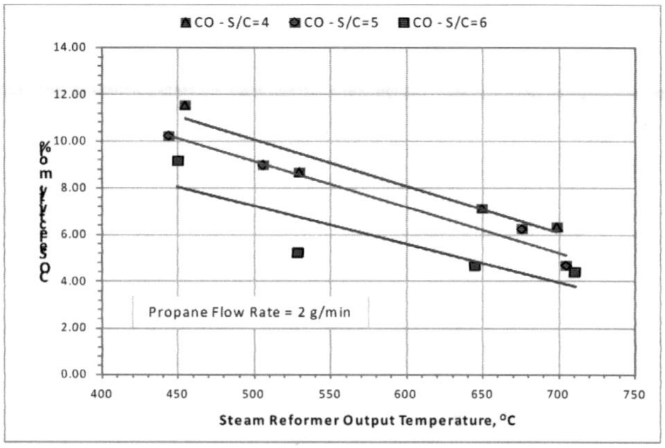

Figure 18: CO Selectivity as a Function of Propane Reformate Temperature and S/C.

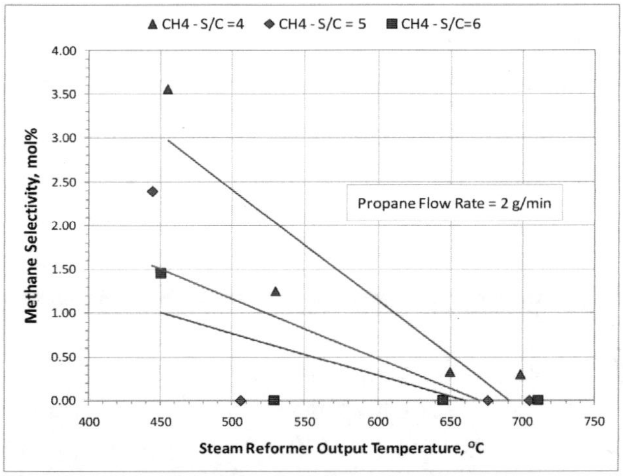

Figure 19: CH_4 Selectivity as a Function of Propane Reformate Temperature and S/C.

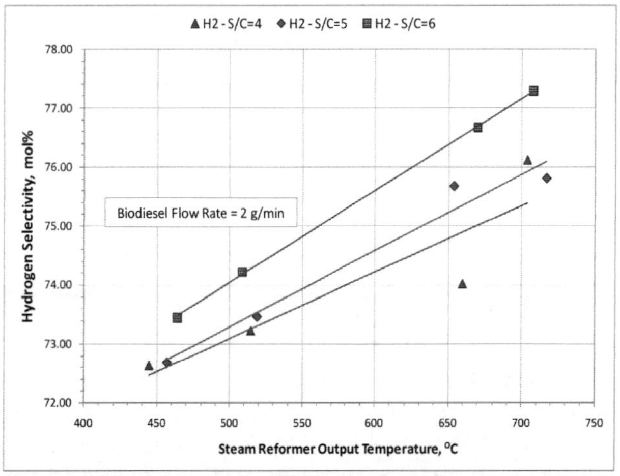

Figure 20: H_2 Selectivity as a Function of Biodiesel Reformate Temperature and S/C.

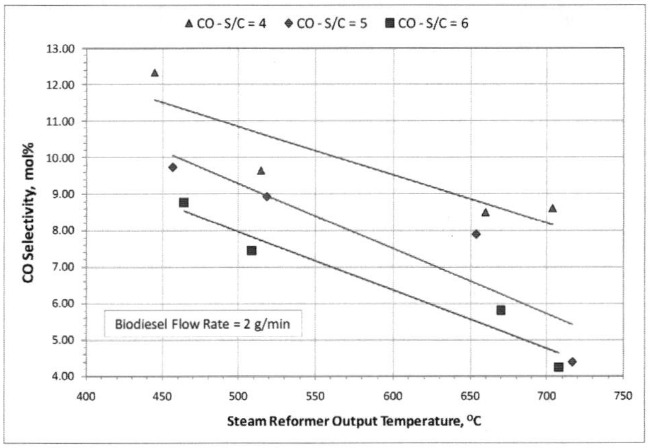

Figure 21: CO Selectivity as a Function of Biodiesel Reformate Temperature and S/C.

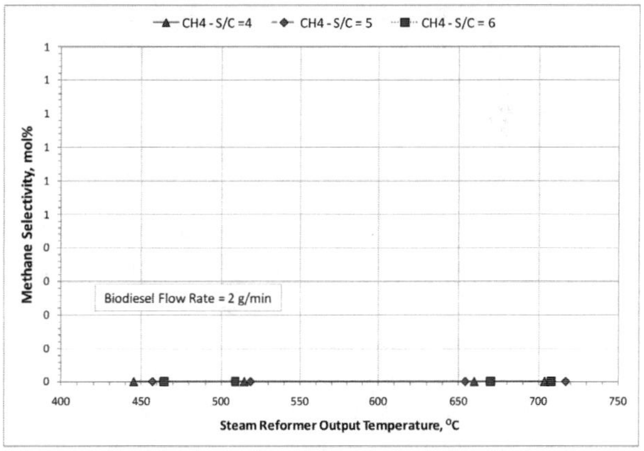

Figure 22: CH_4 Selectivity as a Function of Biodiesel Reformate Temperature and S/C.

For JP8 and diesel, methane selectivity reached zero at temperatures higher than 700°C and S/C of 6; for propane above 650°C for S/C of 5 and above 520°C for S/C of 6, for S8 no methane was detected above 650°C for S/C of 6; for the 50/50 mix no methane above 600°C for S/C of 5 and above 500°C for S/C of 6, while for biodiesel no methane was detected at all temperatures and S/C ratios tested. CO selectivities are low reaching below 10 mol% and some cases below 5 mol%. These results are important for applications using the direct steam reforming to produce hydrogen for high temperature PEM fuel cells.

Zhenq *et al* [1] used a supported Rhodium (Rh) catalyst (Rh/CeO_2-Al_2O_3) in steam reforming of a simulated jet fuel at S/C=3 and temperature of 514°C. The simulated jet fuel consists of a blend of 80 wt% dodecane (n-$C_{12}H_{26}$) and 20 wt% aromatic compounds. Strohm *et al* [2], using 2% Rh–10% Ni supported on CeO_2-modified Al_2O_3 catalyst to reform desulfurized JP8 at S/C=3 and temperature of 515°C. Current study at S/C of 4 and 511°C show higher H_2 and lower CO and CH_4 content than the both Zhenq et al and Strohm et al, Table 2. The higher S/C in current study and higher reactor capacity for the 2 g/min fuel flow rate resulted in the higher hydrogen and lower CO and methane contents.

TABLE 2: JP8 Reforming Data Comparison

	H_2 [mol%]	CO [mol%]	CO_2 [mol%]	CH_4 [mol%]
Zhenq, *et al,* 2008, [3]	60.5	5.5	20.0	14.0
Strohm *et al* 2006, [2]	61.0	3.0	23.0	13.0
Current Study	68.5	11.0	15.5	5.0

Laosiripojana *et al* [3], studied catalytic steam reforming of propane over CeO2-doped Ni/Al2O3 at high temperatures, 700-900°C. Their results show maximum hydrogen content of 80 mol% at 900°C using 14% Ce-Ni at $C_2H_6/C_3H_8/H_2O$ molar ratio of 0.65/0.35/3.0. The extrapolation of current data in Figure 23, show 80 mol% hydrogen content can be achieved at 900°C and S/C ration of 4-5. In their investigation, hydrogen content dropped drastically with reduction in operating temperatures. The hydrogen content for 800°C dropped to 60 mol%, and for 700°C dropped to 40 mol%.

Limited efforts were reported on the reforming of biodiesel. Irving *et al* [4] studied steam reforming of biodiesel using their proprietary catalyst. Their results show complete conversion of biodiesel was achieved at temperatures above 740°C with selectivities: 65 mol% hydrogen, 15 mol% CO_2, and 20 mol% CO with no CH_4 at S/C of 3 to 4. Current effort show complete conversion above 450°C and S/C of 4 with selectivities: minimum 72.6 mol% hydrogen, minimum 15.02 mol% CO_2, maximum 12.35 mol% CO, and 0 mol% CH_4. The apparent difference between the two investigations is the catalyst; however, the larger capacity of the packed-bed over that of microchannel reactor might have contributed to the difference in the two results.

The hydrogen generation mass flow rate, in g/min, was calculated for all test runs to facilitate quantitative comparison of the performance of the developed fuel processor. Figure 23 shows hydrogen generation rate as a function of steam reformer exit temperature and fuel type for the 2g/min fuel input to the steam reformer and S/C of 4.

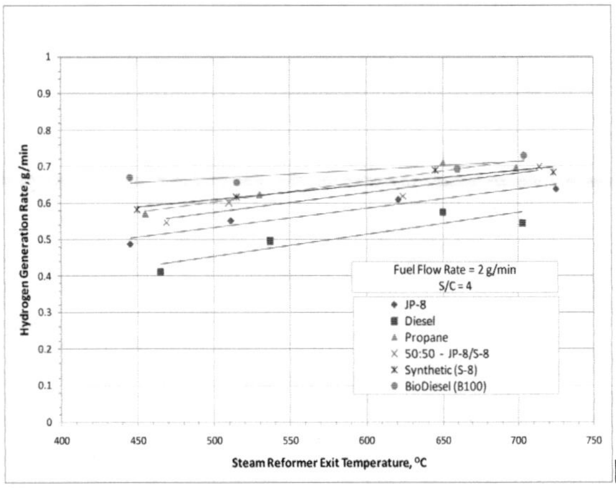

Figure 23: Hydrogen Generation Rate In Reformate Streams

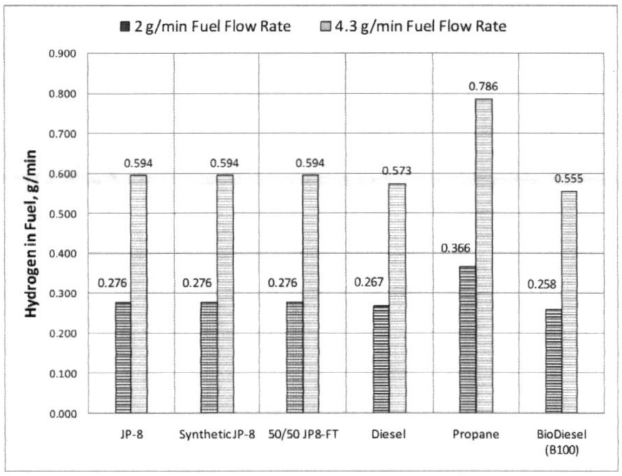

Figure 24: Hydrogen Available from Fuel for 2 and 4.3 g/min Fuel Flow Rates

Figure 24 displays the amount of hydrogen available in the tested fuels. Biodiesel contains the least amount of hydrogen per gram followed by diesel, then by the jet fuels group, while propane posses the highest amount of hydrogen per gram. However, steam reforming of diesel seems to generate the least amount of hydrogen and biodiesel the highest, Figure 23. Biodiesel is the only oxygenated fuel ($C_{20}H_{40}O_2$) among the six tested fuels. The high performance of biodiesel requires further investigation to determine whether the increase of oxygen concentration at reaction sites during fuel reforming impacts catalyst performance or whether it is due to other factors such as the C-Chain cluster structure.

CONCLUSION

Processing hydrocarbons to produce hydrogen rich stream suitable for fuel cell use has been made possible with the use of a fractionation and direct steam reforming followed by zinc oxide beds to remove hydrogen sulfide. The direct steam reforming concept is capable of reforming multiple fuels. It has been successfully used to process JP8, S8, 50/50 mix of S8 and JP8, diesel, biodiesel, and propane. Different fuels can be used intermittently without system degradation or loss of capability. The fuel processor showed superior performance over other technologies that were developed for a specific fuel. Therefore, one can conclude that catalyst selection is a major factor in the success of the reforming process.

Reforming biodiesel showed interesting results. Although biodiesel contains the least hydrogen per gram of fuel, it produced the highest hydrogen mass flow rate among the six fuels tested. Biodiesel is the only oxygenated fuel ($C_{20}H_{40}O_2$) among the six tested fuels. The high performance of biodiesel requires further investigation to determine whether the increase of oxygen concentration at reaction sites during fuel reforming impacts catalyst performance or whether it is due to other factors such as the C-chain cluster structure.

Further testing is warranted to investigate the impact of long term operations using the current six and other fuels on catalyst performance and coking deposition.

REFERENCES

[1] Zheng, J., Strohm, J.J., Song, C.,*"Steam Reforming Of Liquid Hydrocarbon Fuels For Micro-Fuel Cells," Fuel Processing* Technology 89, 2008, 440– 448.

[2] Strohm, J.J., Zheng, J., Song, C.,*" Low-Temperature Steam Reforming Of Jet Fuel In The Absence And Presence Of Sulfur Over Rh And Rh–Ni Catalysts For Fuel Cells,"* Journal of Catalysis 238, 2006, 309–320.

[3] Laosiripojana, N., Sangtongkitcharoen, W., Assabumrungrat, S.,*"Catalytic Steam Reforming Of Ethane And Propane Over Ceo2-Doped Ni/Al2O3 At SOFC Temperature: Improvement Of Resistance Toward Carbon Formation By The Redox Property Of Doping CeO2,"* Fuel 85, 2006, 323–332.

[4] Irving, P.M. and Pickles, J.S., *"Hydrogen Production From Bio-Diesel Fuel In a Microchannel Steam Reformer,"* Proceedings of the National Hydrogen Association, AHC, March 12-16, 2006, Long Beach, CA.

35d Combined Pre-Reforming-Desulfurization of High-Sulfur Diesel for Hydrogen Fueling Station Applications

Nazim Muradov[1], Karthikeyan K. Ramasamy[1], Cunping Huang[1], Ibrahim Adebiyi[1], Franklyn Smith[1], Clovis Linkous[1], Ali T-Raissi[1], and James Stevens[2]. (1) Florida Solar Energy Center, University of Central Florida, 1679 Clearlake Road, Cocoa, FL 32922, (2) Chevron Technology Ventures LLC, 3901 Briarpark, Hosuton, TX 77042

Considerable research and development efforts are being conducted, worldwide, to substitute hydrogen for gasoline and diesel in transportation sectors. A major challenge facing future Hydrogen Economy is the issue of hydrogen fuel delivery to dispensing stations. In the near term, it may be necessary to deliver diesel fuel directly to the fueling station wherein it is reformed to hydrogen, on demand. This approach has the following advantages: (1) Existing fuel delivery infrastructure can be utilized, (2) More energy is delivered per trip when the tanker is filled with diesel instead of liquid or compressed hydrogen, and (3) The fueling station would be able to service both internal combustion and fuel cell powered vehicles concurrently. Reforming high-sulfur hydrocarbon fuels (e.g., diesel, JP-8, etc.) is particularly challenging due to rapid deactivation of conventional reforming catalysts by sulfurous compounds.

Florida Solar Energy Center (FSEC), in collaboration with the Chevron Technology Ventures (CTV), has developed a new on-demand hydrogen production technology for fueling station applications. In this process, the diesel fuel is first catalytically pre-reformed to shorter chain hydrocarbons (C1-C6) before being fed to the main reformer, where it is converted to high-purity hydrogen gas. Here, we report on the catalytic pre-reforming of high-sulfur diesel to C1-C6 alkanes in a bench scale unit (in this work, 3180-5240 ppmw of thiophene was added to commercial diesel). In the pre-reformer, most sulfurous species present in the fuel were converted to predominantly hydrogen sulfide (H2S). The sulfur removal from the pre-reformate gas was required in order to avoid deactivation of the catalyst (mostly, Ni-based catalysts) in the main reformer. Desulfurization of the pre-reformate gas was carried out in a special regenerative redox system developed at FSEC, which includes a Fe-based scrubber coupled with an electrolyzer. In particular, ferrous/ferric (Fe2+/Fe3+) redox couple was used for oxidizing H2S to elemental sulfur as follows:

4Fe3+(aq) +2H2S(g) = 4Fe2+(aq) + 2S(s) + 4H+(aq)

We have shown that ferrous ion can be electrochemically oxidized back to the ferric state readily, in a closed process. Desulfurization occurred optimally at the following process conditions: 0.1 M iron (total) concentration, pH of 1.7, set by addition of sulfuric acid, and 50oC electrolyte temperature. The integrated pre-reformer and sulfur-scrubbing unit operated successfully for 100 hours meeting the required desulfurization target of 95% removal.

35e Catalytic Partial Oxidation of a Surrogate Diesel Fuel Mixture Using Pyrochlores: Effect of Reforming Metal

Daniel J. Haynes, Parsons, Inc., 3610 Collins Ferry Rd, Morgantown, WV 26507, David A. Berry, Fuel Processing R&D, National Energy Technology Laboratory, 3610 Collins Ferry Road, Morgantown, WV 26507-0880, Dushyant Shekhawat, US Dept. of Energy, 3610 Collins Ferry Rd, Morgantown, WV 26507, Mark Smith, REM Engineering Services, U.S. DOE NETL, 3610 Collins Ferry Road, Morgantown, WV 26507-0880, and James J. Spivey, Chemical Engineering, Louisiana State University, Cain Department of Chemical Engineering, South Stadium Road, Baton Rouge, LA 70803.

The formation of synthesis gas from diesel fuel for use in solid oxide fuel cells via partial oxidation is reduced by the deactivation of the reforming catalyst by polynuclear aromatics and sulfur compounds present in the fuel. Recently, it has been found that the substitution of active reforming metals into the structure of thermally stable oxides can produce a catalyst that is resistant to both sulfur poisoning and carbon formation. Of particular interest is recent work on La(2-x)SrxRhyZr(2-y)O(7- î) oxide [1, 2]. As prepared, this material showed both the pyrochlore and perovskite phases by XRD, and had superior resistance to deactivation by sulfur, aromatics, and carbon deposition compared to directly comparable formulations without Sr and/or Rh. However, the cost of Rh, and its low nominal dispersion in these recent studies (ca. 5%), suggest that other catalytically active metals should be investigated. Ni, Pt and Ru metal are less expensive than Rh, and are each well-know for their ability to convert hydrocarbon into synthesis gas. However, it is not known whether a pyrochlore substituted with either Ni, Pt or Ru will be able to retain its CPOX activity in this type of material, or how its activity and resistance to deactivation compare to that of a substituted Rh-based catalyst. Thus, the objective of the work reported here is to study the CPOX activity of directly comparable Ni, Pt, or Ru and Rh-substituted metal oxides. n-Tetradecane (TD) is used as a representative component of diesel fuel. Experiments are carried out with 50 ppmw of added sulfur (as dibenzothiophene) and 5 wt% of a representative aromatic (1-methylnaphthalene) to determine the activity/selectivity to syngas, and resistance to deactivation of these two catalysts.

145639

DRY CATALYTIC PARTIAL OXIDATION OF DIESEL-FUEL DISTILLATES INTO SYNGAS

Michael V. Mundschau, David A. Gribble, Jr., Paul N. Plassmeyer, Lee M. Henton and Nicholas R. Jentzsch, Eltron Research & Development Inc., Boulder, CO, U.S.A.

Abstract

Dry partial oxidation of volatile compounds distilled from diesel fuel is being investigated as one means for producing synthesis gas for fuel cells and for pollution control devices requiring reductants for abatement of oxides of nitrogen. Use of volatile distillates eliminates potentially complex liquid-fuel injectors and fuel-air mixers, greatly simplifying reactors. The rationale is as follows: The longer-chain *n*-alkanes have the lowest autoignition temperatures of the hydrocarbons in diesel fuel. For example, cetane (*n*-hexadecane) has an autoignition temperature of 205 °C, well below its boiling point of 286.5 °C. If attempts are made to vaporize diesel fuel without appropriate precautions, the least stable molecules may crack before they vaporize, forming alkyl free radicals that may initiate polymerization chain reactions, leading to formation of tar on fuel nozzles and cores of liquid fuel droplets. To minimize tar and to allow rapid mixing with air, commercial diesel fuel is distilled below 200 °C. In one method, vapors are condensed in a reservoir, and the distillate, containing the most volatile components of the diesel fuel, is easily re-vaporized, mixed with air, and converted into syngas in a porous membrane reactor. In order to suppress formation of carbonaceous residues as fuel is heated above autoignition temperatures in the reactor heating zone, cooler air is passed through porous walls of zirconia-based ceramic to react with alkyl radicals as they form. Hot air (>900 °C) is passed into the reactor through porous cylindrical walls surrounding the reactor hot zone containing a bed of perovskite catalyst designed to contain oxygen-anion vacancies, enhancing diffusion of oxygen and providing a reservoir of mobile oxygen to react with adsorbed carbon from beneath. For distillates formed by collecting 20% by mass of the original diesel fuel, 75 mole% of the hydrogen in the distillate is converted into molecular hydrogen; 13% into CH_4 and 12% into H_2O (the latter formed by excess oxygen used to suppress carbon); 88 mole% of the carbon in the distillate hydrocarbons is bound in CO and only 5% in CO_2; the remainder chiefly in CH_4. In runs of an arbitrary 72 hours continuous operation, carbon deposition is negligible on the fuel nozzle, walls of the heating zone, in the catalyst bed, and in the exhaust, although some graphitic carbon appears on walls just above the hot zone. No soot was seen in exhaust traps. Gas chromatography of full diesel fuel and its distillates indicates that although *n*-alkanes >C_{20} are largely eliminated, *n*-alkanes in the range C_{16}–C_{19} (still likely to crack above 200 °C) are reduced by 70–75%, and C_{13}–C_{15} reduced 50–65%, but were not eliminated due to significant vapor pressures below their boiling points and the simple means of distillation.

Introduction

Potential Applications

Considerable research effort has been made for converting the hydrocarbons in diesel fuel[1-16], JP-8[17-22] and other fuels[1,23-29] into synthesis gas, a mixture of H_2 + CO, for use in solid-oxide fuel cells. Additional world-wide interest exists for producing sufficient H_2 and CO from diesel fuel and JP-8 as reductants for NOX traps[30-32] used for abatement of the oxides of nitrogen in pollution-control devices for diesel vehicles and jet aircraft. In one version of NOX trap, NO_2, in part produced by catalytic conversion from other oxides of nitrogen, is captured by compounds of barium to form $Ba(NO_3)_2$. The NOX traps are regenerated by reaction with H_2 and CO to form $Ba(OH)_2$ and $BaCO_3$, respectively, releasing molecular nitrogen, N_2. On-board generation of syngas simplifies regeneration of NOX traps, otherwise requiring ammonia or urea as reductants. Production of synthesis gas from carbonaceous fuels is also the first step in the production of ultra-pure hydrogen required for proton exchange membrane (PEM) fuel cells, after water-gas shift[33] ($H_2O + CO = H_2 + CO_2$), partial removal of sulfur by sorbents,[34,35] and purification of hydrogen through hydrogen-transport membranes.[36-41]

Background Issues

Diesel fuel is a variable mixture, containing over 400 distinct hydrocarbons, organic compounds of sulfur and nitrogen, and oxygenated additives.[42] Combustion analysis of commercial ultra-low sulfur diesel fuel has shown hydrogen-to-carbon atomic ratios ranging from 1.87 : 1 to 1.78 : 1, and sulfur mass contents from 7 to 9 ppmw, which is below the legal sulfur mass limit of 15 ppmw.[3] Using $H_{1.8}C_1$ as an approximate empirical formula of the average molecule in diesel fuel,[42] the partial oxidation reaction is written: $H_{1.8}C_1 + ½\,O_2 = 0.9\,H_2 + CO$. The hydrogen-to-carbon ratio (along with the concentration of oxygen in the system) dictates the thermodynamic operating conditions favoring production of $H_2 + CO$ at equilibrium and suppressing formation of elemental carbon. For stoichiometric conditions of partial oxidation requiring one atom of oxygen for each atom of carbon in the fuel, operating temperature must exceed about 950 °C at one atmosphere to favor production of $H_2 + CO$ over undesired products (H_2O, CO_2, CH_4) and to completely suppress deposition of carbon onto partial oxidation catalysts designed to bring the system to equilibrium.[3] The temperature for partial oxidation of diesel fuel at stoichiometric conditions is significantly higher relative to well-established industrial steam reforming of methane, for example, performed at temperatures as low as 700–800 °C.[33] In the case of steam reforming of CH_4, the higher H-to-C atomic ratio of 4-to-1, as well as higher O-to-C atomic ratio, suppresses deposition of elemental carbon at lower temperatures.

Formation of elemental carbon is favored by both thermodynamics and reaction kinetics as diesel fuel is heated between 200–600 °C.[1,3] At autoignition temperatures,[43] the weakest chemical bonds in the system break, forming alkyl free radicals.[43-46] For diesel fuel, long-chain *n*-alkanes with carbon number $>C_{12}$ have the lowest autoignition temperatures of 204–205 °C, corresponding to breaking of the weakest carbon-carbon bonds.[5,43] The longer-chain *n*-alkanes crack below their boiling points, making them difficult to volatilize. Other hydrocarbons in diesel fuel reach autoignition temperatures below 600 °C.[43] The aromatic compounds such as naphthalene and benzene are among the most stable and have autoignition temperatures of 526 °C and 562 °C, respectively, as carbon-hydrogen bonds break, leaving the aromatic rings intact.[5,43]

Under excess oxygen and conditions favoring complete combustion, hydrocarbon-free radicals react with molecular oxygen with high probability, propagating radical chain reactions leading to deep oxidation products, $CO_2 + H_2O$.[47] However, in the absence of oxygen (or under very oxygen lean conditions), the alkyl free radicals react with other hydrocarbons initiating radical polymerization[45,48,49] (or react internally initiating cyclicization reactions). Upon loss of hydrogen, polymers lead to tar and to other carbonaceous residues.

Formation of carbonaceous residues above autoignition temperatures encountered in the diesel-fuel heating zone can be suppressed by allowing the alkyl free radicals to react with steam, for example. Diesel fuel can be vaporized by heating to 180 °C (well below cracking temperatures of the least stable *n*-alkanes) and mixing with steam at the same temperature.[5] Long-chain *n*-alkanes need not be heated to their normal boiling point to vaporize, if heated sufficiently to yield a high vapor pressure in steam. Flow of sufficient steam ensures continued evaporation.[50] The steam-fuel vapor mixture can then be safely heated to reforming temperatures without danger of carbon formation. However, the volume of water needed to raise sufficient steam exceeds the volume of diesel fuel. Handling, transport and storage of water is undesirable. Quantity of stored water can be reduced using autothermal reforming, in which both steam and oxygen are mixed,[1] or by re-circulating $H_2O + CO_2$ from fuel cell or engine exhaust.[1]

In dry partial oxidation, alkyl free radicals formed above the autoignition temperatures are reacted with molecular oxygen in the fuel heating zone to form various oxygenated organic compounds, including peroxygenated compounds, per-ethers, organic acids, esters, aldehydes and ketones.[45,47] Quantity of oxygen and temperature in the fuel heating zone must be carefully controlled to avoid excessive exothermic reactions forming thermodynamically favored elemental carbon.[7,51]

Poor mixing of liquid diesel fuel and air can lead to deposition of carbon in oxygen-lean regions of the reactor and to deep oxidation products, $H_2O + CO_2$, in oxygen-rich regions. To avoid relatively complex liquid-fuel injectors/air mixers,[52] Huang and King[18] and Huang et al.[19] have investigated distillation techniques for JP-8, in which volatile compounds distilled from JP-8 are mixed with air and reformed. Distillation also lowers sulfur concentration in the distillate.[18,19] The present study applies distillation to diesel fuel and the relatively straightforward reforming of the most volatile compounds into synthesis gas. This technique might be especially applicable for regeneration of NOX traps in diesel vehicles.

Experimental

Experimental details are published elsewhere.[3,5] Briefly, diesel fuel is distilled below cracking temperatures of 200 °C, and the most volatile components are collected and stored in a reservoir. The volatile compounds are re-vaporized (below 200 °C) and mixed with air in a porous membrane reactor (Figure 1). Cooler air (<200 °C) is effused through walls of porous zirconia-based ceramic and mixed with vaporized fuel. The balance of air required for partial oxidation is electrically heated to above 900 °C and effused through porous cylindrical ceramic walls surrounding a bed of perovskite oxidation catalyst (Figure 1).[3,53,54] To completely suppress carbon, excess oxygen, 1.2 times that required for partial oxidation, is employed.

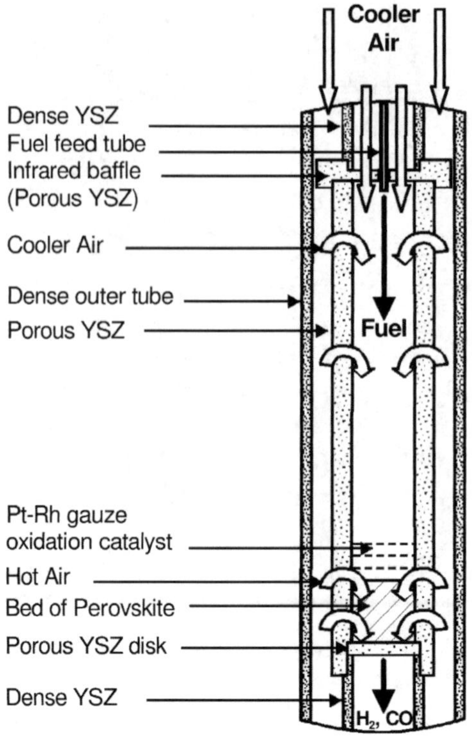

Figure 1. Laboratory porous membrane reactor for dry partial oxidation of diesel fuel. Cooler air is mixed with volatile diesel fuel distillates at top to react with alkyl free radicals and to suppress formation of carbonaceous residues in fuel heating zone. Hot air enters through porous walls of a zirconia-based ceramic cylinder surrounding a bed of perovskite oxidation catalyst.

Results

Figure 2 shows gas chromatograms of full diesel full (top) as received from the automotive pump, and a distillate formed by collecting 20% by mass of the most volatile components in a reservoir. Table 1 gives approximate relative quantitative analysis of the *n*-alkanes dominant in the fuel. Although the distillation has greatly reduced the quantity of the higher molecular weight *n*-alkanes that are most likely to crack just above 200 °C, it has by no means eliminated them, due to the simple method of distillation.

Figure 3 shows results of dry partial oxidation of the volatile distillates. Conversion of the distillates was nearly complete, with negligible deposition of carbon and naphthalene in the exhaust. Concentration of methane was higher than expected for equilibrium with the perovskite catalysts operating above 950 °C, implying that equilibrium is not achieved. The activity of the perovskite catalysts, $La_{0.5}Sr_{0.5}FeO_{3-\delta}$, was relatively stable to the sulfur in the system and to the elevated temperatures over the arbitrary 72 h of the run. Good results are attributed, in part, to ease of volatilization of the distillates, ease of mixing with air, and possible elimination of more difficult-to-reform components that remained in the non-volatilized residuum. Considerable parasitic power is consumed in externally heating the air and system to >900 °C, and the laboratory system remains far from a practical device.

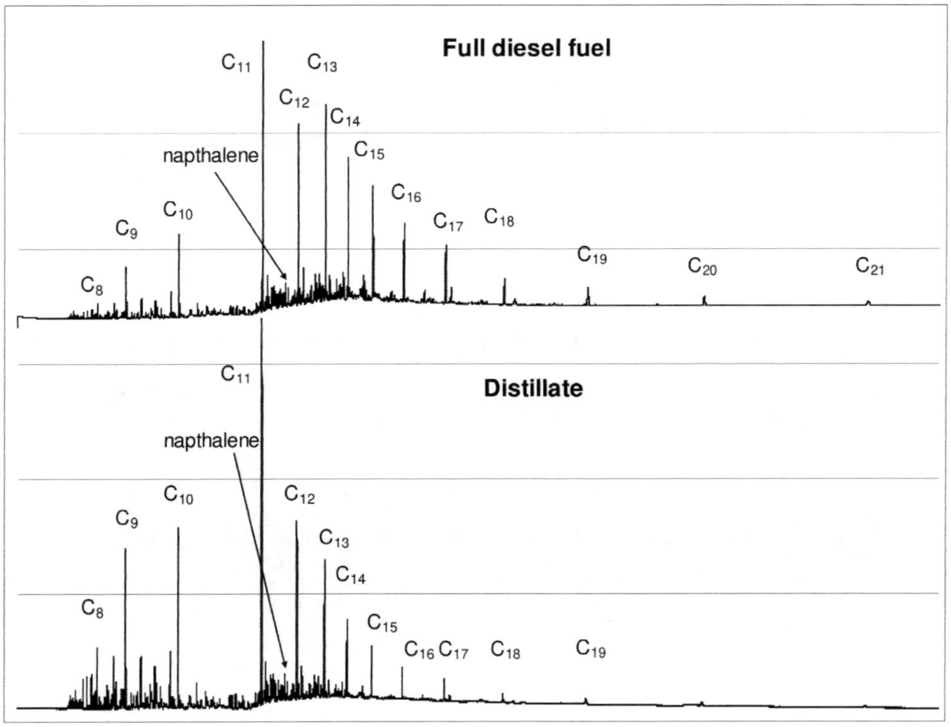

Figure 2. Gas chromatograms showing hydrocarbon distributions of full diesel fuel (top) and a distillate containing 20 mass % of the original fuel. Labels mark carbon numbers of *n*-alkanes. Scales: arbitrary intensity vs. elution time (~2 hours).

Table 1. Hydrocarbon Analysis in Diesel Fuel and Distillate

Alkane	**Diesel Fuel as Received weight %**	**Distillate Fraction ≤ 175 °C weight %**	**% change**
C_8	0.6	1.6	180.7
C_9	2.5	5.2	113.4
C_{10}	5.5	8.6	57.1
C_{11}	8.5	8.5	0.2
C_{12}	7.1	5.1	-28.4
Unidentified < C_{12}	30.0	54.5	81.4
Total < C_{12}	54.1	83.5	54.3
C_{13}	7.7	3.9	-48.6
C_{14}	6.0	2.4	-60.1
C_{15}	5.4	1.8	-66.6
C_{16}	4.6	1.4	-70.1
C_{17}	4.4	1.2	-73.0
C_{18}	2.5	0.6	-75.2
C_{19}	2.2	0.6	-73.5
C_{20}	1.5	0	-100.0
C_{21}	1.0	0	-100.0
Unidentified > C_{12}	10.4	4.5	-56.5
Total > C_{12}	45.9	16.5	-64.1

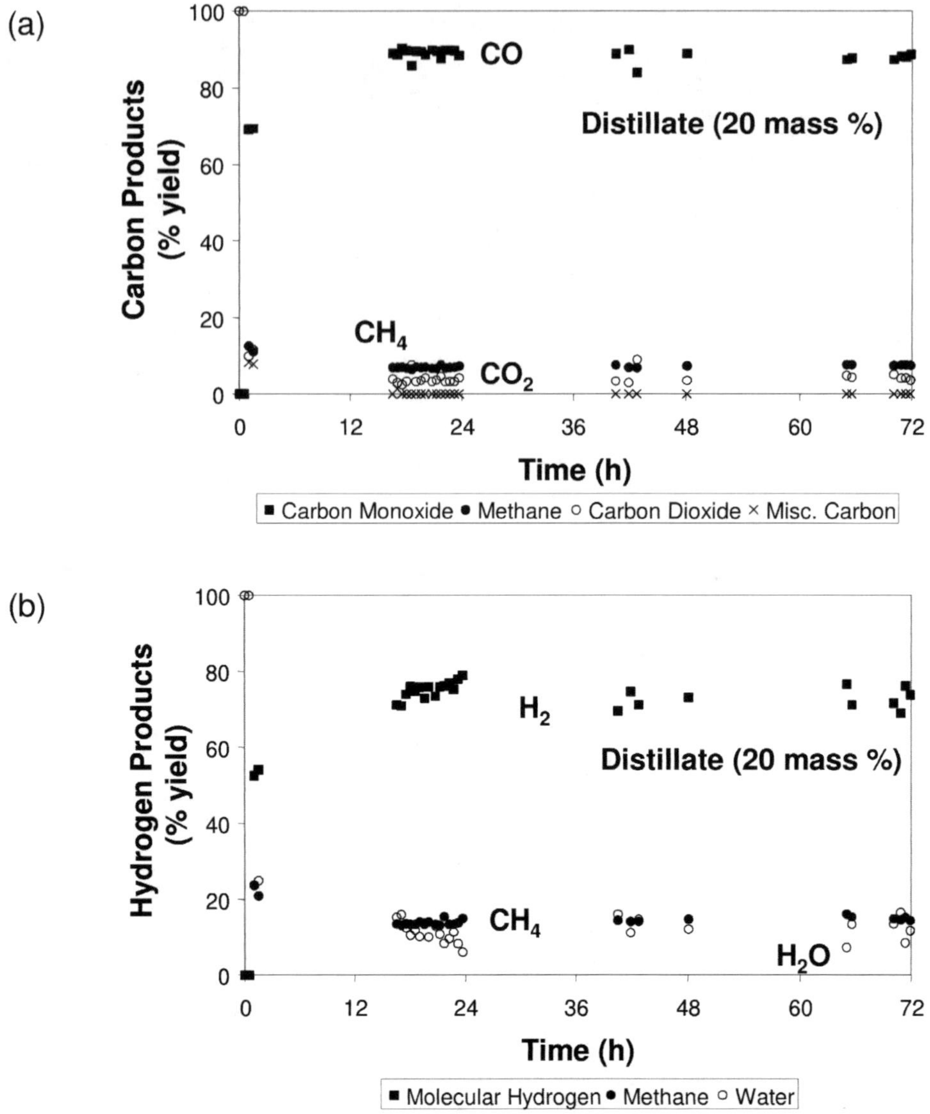

Figure 3. Percent yields. Nearly 90% of the carbon in the distillate is converted into CO and ~75% of the hydrogen in the distillate appears as molecular hydrogen.

Acknowledgements

Authors thank the U.S. Department of Energy, NETL, and the Small Business Innovative Research Program, for funding under grant DE-FG02-05ER84394, and Vivian P. Knaus of Eltron Research for preparation of figures and the manuscript.

References

1. D. Shekhawat, D.A. Berry, T.H. Gardner, J.J. Spivey. Catalytic reforming of liquid hydrocarbon fuels for fuel cell applications, *Catalysis*, **19**(2006)184–253.
2. D.J. Haynes, D.A. Berry, D. Shekhawat, J.J. Spivey. Catalytic partial oxidation of *n*-tetradecane using pyrochlores: Effect of Rh and Sr substitution. *Catal. Today*, **136**(2008)206–213. DOI: 10.1016/j.cattod.2008.02.012

3. M.V. Mundschau, C.G. Burk CG, D.A. Gribble Jr. Diesel fuel reforming using catalytic membrane reactors. *Catal. Today*, **136**(2008)190–205. DOI: 10.1016/j.cattod.2008.02.003
4. M.V. Mundschau. *Catalytic Membrane Reactor and Method for Production of Synthesis Gas*. United States Patent Application Publication US 2008/0169449 A1, Jul. 17, 2008; WO/2008/031024.
5. Michael V. Mundschau, David A. Gribble, Jr., Lee M. Henton, Nicholas R. Jentzsch and Paul N. Plassmeyer, Reforming diesel-fuel distillates with membrane reactors, *Asia-Pacific Journal of Chemical Engineering,* in press.
6. M. Krumpelt, T.R. Krause, J.D. Carter, J.P. Kopasz, S. Ahmed. Fuel processing for fuel cell systems in transportation and portable power applications. *Catal. Today*, **77**(2002) 3–16.
7. D-J. Liu, M. Krumpelt, H-T. Chien, S-H. Sheen. Critical issues in catalytic diesel reforming for solid oxide fuel cells, *Journal of Materials Engineering and Performance,* **15**(4)(2006)442–444. DOI: 10.1361/105994906XI17279
8. L. Hartmann, K. Lucka, H. Köhne. Mixture preparation by cool flames for diesel-reforming technologies, *J. Power Sources,* **118**(2003)286–297. DOI: 10.1016/S0378-7753(03)00100-9
9. T. Aicher, B. Lenz, F. Gschnell, U. Groos, F. Federici, L. Caprile, L. Parodi. Fuel processors for fuel cell APU applications, *J. Power Sources,* **154**(2006)503–508. DOI: 10.1016/j.powsour.2005.10.026
10. T. Aicher, L. Griesser. Novel process to evaporate liquid fuels and its application to the catalytic partial oxidation of diesel, *J. Power Sources,* **165**(2007)210–216. DOI: 10.1016/j.powsour.2006.12.018
11. T.H. Gardner. *Hexaaluminate Catalysts for the Partial Oxidation of Middle Distillate Fuels,* Dissertation, College of Engineering and Mineral Resources, West Virginia University, Morgantown, West Virginia, U.S.A., 2007.
12. D.J. Haynes. *The Catalytic Partial Oxidation of n-Tetradecane on Rh and Sr Substituted Pyrochlores.* Thesis, Department of Chemical Engineering, Louisiana State University, U.S.A., 2007.
13. J.J. Krummenacher, K.N. West, L.D. Schmidt. Catalytic partial oxidation of higher hydrocarbons at millisecond contact times: decane, hexadecane, and diesel fuel, *J. Catal.* **215**(2003) 332–343. DOI: 10.1016/S0021-9517(03)00011-3
14. R. Subramanian, G.J. Panuccio, J.J. Krummenacher, I.C. Lee, L.D. Schmidt, Catalytic partial oxidation of higher hydrocarbons: reactivities and selectivities of mixtures, *Chemical Engineering Science,* **59**(2004)5501–5507. DOI: 10.1016/j.ces.2004.09.012
15. P.K. Cheekatamarla, A.M. Lane. Catalytic autothermal reforming of diesel fuel for hydrogen generation in fuel cells I. Activity tests and sulfur poisoning, *J. Power Sources,* **152**(2005)256–263. DOI: 10.1016/j.powsour.2005.03.209
16. I. Kang, S. Yoon, G. Bae, J.H. Kim, J. Bae, D. Lee, Y. Song. The micro-reactor testing of catalysts and fuel delivery apparatuses for diesel autothermal reforming. *Catal. Today*, **136**(2008) 249–257. DOI: 10.1016/j.cattod.2008.01.012
17. S. Roychoudhury, M. Lyubovsky, D. Walsh, D. Chu, E. Kallio. Design and development of a diesel and JP-8 logistic fuel processor, *J. Power Sources,* **160**(2006)510–513. DOI: 10.1016/j.powsour.2005.12.101
18. X. Huang, D.L. King, Gas-phase hydrodesulfurization of JP-8 light fraction using steam reformate. *Ind. Eng. Chem. Res.*, **45**(2006)7050–7056. DOI: 10.1021/ie0607301

19. X. Huang, D.A. King, F. Zheng, V.S. Stenkamp, W.E. TeGrotenhuis, B.Q. Roberts, D.L. King. Hydrodesulfurization of JP-8 fuel and its microchannel distillate using steam reformate. *Catal. Today,* **136**(2008)291–300. DOI: 10.1016/j.cattod.2008.01.011
20. B. Lenz, T. Aicher, Catalytic autothermal reforming of jet fuel, *J. Power Sources,* **149**(2005)44–52. DOI: 10.1016/j.powsour.2005.02.010
21. P. Erri, P. Dinka, A. Varma, Novel perovskite-based catalysts for autothermal JP-8 fuel reforming, *Chemical Engineering Science,* **61**(2006)5328–5333. DOI: 10.1016/j.ces.2006.03.046
22. I.C. Lee. Rhodium supported on thermally enhanced zeolite as catalysts for fuel reformation of jet fuels, *Catal. Today,* **136**(2008)258–265. DOI: 10.1016/cattod.2008.03.027
23. Wenliang Zhu, Guoxing Xiong, Wei Han, Wenshen Yang, Catalytic partial oxidation of gasoline to syngas in a dense membrane reactor, *Catal. Today,* **93-95**(2004)257–261. DOI: 10.1016/cattod.2004.06.126
24. C. Palm, P. Cremer, R. Peters, D. Stolten, Small-scale testing of precious metal catalyst in the autothermal reforming of various hydrocarbon feeds, *J. Power Sources,* **106**(2002)231–237.
25. Kuen-Song Lin, Sujan Chowdhury, Chia-Chieh Shen, Chuin-Tih Yeh, Hydrogen generation by catalytic gasification of motor oils in an integrated fuel processor, *Catal. Today,* **136**(2008)281–290. DOI: 10.1016/cattod.2008.02.017
26. Dong Ju Moon, Jong Woo Ryu, Kye Sang Yoo, Dae Jin Sung, Sang Deuk Lee, Development of *iso*-octane fuel processor system for fuel cell applications, *Catal. Today,* **136**(2008)222–227. DOI: 10.1016/cattod.2008.02.022
27. Inyong Kang, Sangho Yoon, Gyujong Bae, Jung Hyun Kim, Joongmyeon Bae, Daehoon Lee, Younghoon Song, The micro-reactor testing of catalysts and fuel delivery apparatuses for diesel autothermal reforming, *Catal. Today,* **136**(2008)249–257. DOI: 10.1016/cattod.2008.01.012
28. S. Pengpanich, V. Meeyoo, T. Rirksomboon, J. Schwank, *iso*-Octane partial oxidation over Ni-Sn/$Ce_{0.75}Zr_{0.25}O_2$ catalysts, *Catal. Today,* **136**(2008)214–221. DOI: 10.1016/cattod.2008.01.018
29. L. Dorazio, M.J. Castaldi, Autothermal reforming of tetradecane ($C_{14}H_{30}$): A mechanistic approach, *Catal. Today,* **136**(2008)273–280. DOI: 10.1016/cattod.2008.01.017
30. M.V. Twigg, A.J.J. Wilkins. Diesel exhaust system including NOX-trap, United States Patent, US 6,857,265 B2, Feb. 22, 2005.
31. K. Liu, W.G. Wnuck, W.P. Leenhouts, Intermittent application of syngas to NOX trap and/or diesel engine, United States Patent, US 6,964,156 B2, Nov. 15, 2005.
32. K. Liu, F.A. Kocum. Generating syngas for NOX regeneration combined with fuel cell auxiliary power generation, United States Patent, US 7,090,043 B2, Aug. 15, 2006.
33. M.V. Twigg, (Ed). *Catalyst Handbook, Second Edition.* Manson Publishing: London, 1997.
34. Vikram Reddy Kalakota, *Sulfur Removal Using Regenerable Sorbents of Rare Earth/Transition Metal Oxides*, Thesis, Department of Chemical Engineering, Louisiana State University, August, 2008.
35. Li-Ping Chang, Zong-You Zhang, Xiu-Rong Ren, Fan Li, Ke-Chang Xie, Study in the stability of sorbents for removing H_2S from hot coal gas, *Energy & Fuels,* **23**(2009) 762–765. DOI: 10.1021/ef8005838
36. A. Basile. Hydrogen production using Pd-based membrane reactors for fuel cells. *Top. Catal.*, **51**(1)(2008)107–122. DOI: 10.1007/s11244-008-9128-6

37. S.N. Paglieri. Palladium Membranes in *Nonporous Inorganic Membranes,* A.F. Sammells, M.V. Mundschau, Editors, (Wiley-VCH, Weinheim, 2006) pp. 77–93.
38. S. Paglieri, J.D. Way. Innovations in palladium membrane research, *Sep. Purif. Methods*, **31**(1)(2002)1–69.
39. M.V. Mundschau, X. Xie, C.R. Evenson IV. Superpermeable Hydrogen Transport Membranes, in *Nonporous Inorganic Membranes,* A.F. Sammells, M.V. Mundschau, Editors (Wiley-VCH, Weinheim, 2006) pp. 107–138.
40. M.V. Mundschau. Hydrogen separation using dense composite membranes: Part 1 Fundamentals, in *Inorganic Membranes for Energy and Environmental Applications*, A.C. Bose, Editor (Springer, New York, 2009) pp. 125–153.
41. D.H. Anderson, C.R. Evenson IV, T.H. Harkins, D.S. Jack, R. Mackay, M.V. Mundschau, Hydrogen separation using dense composite membranes: Part 2: Process Integration and Scale-up for H_2 Production and CO_2 Sequestration, in *Inorganic Membranes for Energy and Environmental Applications*, A.C. Bose, Editor (Springer, New York, 2009) pp. 155–171.
42. G.A. Olah, Á. Molnár, *Hydrocarbon Chemistry* (Wiley, New York, 1995).
43. E.M. Goodger, *Hydrocarbon Fuels* (John Wiley & Sons, Inc, New York, 1975).
44. E.W.R. Steacie, *Atomic and Free Radical Reactions* (Reinhold Publishing Corporation, New York, 1946).
45. W.A. Waters, *The Chemistry of Free Radicals* (Oxford University Press, Oxford, 1948).
46. Carleton Ellis, *The Chemistry of Petroleum Derivatives* (The Chemical Catalog Co., Inc. New York, 1934).
47. Bernard Lewis and Guenther von Elbe, *Combustion, Flames and Explosions of Gases, Second Edition* (Academic Press, New York, 1961).
48. J.C. Bevington, *Radical Polymerization* (Academic Press, New York, 1961).
49. P.J. Flory, *Principles of Polymer Chemistry* (Cornell University Press, Ithaca, 1986).
50. M. Wei, Y. Wang, L. Reh, Experimental investigation of the prevaporized premixed (vpl) combustion process for liquid fuel lean combustion, *Chemical Engineering and Processing*, **41**(2002)157–164.
51. D. Diarra, S. Velaga, K. Lucka and H. Köhne, *CFD analysis of mixing and residence time distribution in cool flame vaporizers*, European Conference on Computational Fluid Dynamics, ECCOMAS CFD 2006, P. Wesseling, E. Ofiate, J. Périaux, Editors (TU Delft, The Netherlands, 2006).
52. S.D. Pack, *An Innovative Injection and Mixing System for Diesel Fuel Reforming.* Final Report for U.S. Department of Energy Grant Number: DE-FC26-04NT42229, February, 2008.
53. F.S. Galasso, *Structure, Properties and Preparation of Perovskite-Type Compounds* (Pergamon Press, Oxford, 1969).
54. L.G. Tejuca, J.L.G. Fierro, J.M.D. Tascón, Structure and reactivity of perovskite-type oxides, *Adv. Catal.* **36**(1989)237–328.

35g Effect of Catalyst Layer Formation and Character On Partial Oxidation of Liquid Hydrocarbons In the Presence of Oxygen-Conducting Supports

Mark Smith, REM Engineering Services, U.S. DOE NETL, 3610 Collins Ferry Road, Morgantown, WV 26507-0880, David A. Berry, Fuel Processing R&D, National Energy Technology Laboratory, 3610 Collins Ferry Road, Morgantown, WV 26507-0880, Dushyant Shekhawat, US Dept. of Energy, 3610 Collins Ferry Rd, Morgantown, WV 26507, Daniel J. Haynes, Parsons, Inc., 3610 Collins Ferry Rd, Morgantown, WV 26507, and James J. Spivey, Chemical Engineering, Louisiana State University, Cain Department of Chemical Engineering, South Stadium Road, Baton Rouge, LA 70803.

The application of oxide catalyst layers to supports with oxygen-conducting properties provides opportunities for greatly improved performance in reforming liquid hydrocarbons for solid oxide fuel cell applications. In particular, fuel feeds that contain sulfur and aromatic contaminants present the greatest challenges as these species will deactivate metal catalysts through poisoning and carbon formation. By substituting the active metal into an oxide structure that is stable at the high temperatures used in reforming, and supporting it on an oxygen-conducting support (OCS), deactivation due to thermal sintering and carbon formation can be greatly minimized. Because the active metal is not present in isolated clusters that are subject to sulfur poisoning, deactivation due poisoning is avoided.

The synthesis of the catalyst of interest here requires preparation of fine particles of the active metal oxide and dispersion of these particles on the OCS such that there is intimate contact between the particles and the OCS. Here, the catalytically active oxide layer is barium hexaaluminate substituted with 3wt% Ni; the OCS support is doped ceria. The performances of these systems for the partial oxidation of n-tetradecane, containing sulfur and aromatics, are compared to Ni metal catalyst over alumina, and bulk Ni-substituted barium hexaaluminates (BNHA), for the production of a H2 and CO rich stream.

Kinetic studies of the autothermal reforming of tetradecane over Pt/Al_2O_3 catalyst in a fixed-bed reactor

Rajesh D. Parmar[1], Arunabha Kundu[1], Christopher Thurgood[1,2], Brant A. Peppley[1], Kunal Karan[1]
[1]Queen's-RMC Fuel Cell Research Center, Kingston, Canada
[2]Royal Military College, Kingston, Canada

Abstract

Kinetics of autothermal reforming of tetradecane on $Pt-Al_2O_3$ catalyst between 750 °C and 900 °C is investigated. Experimental results obtained from NETL (US-DOE) are used for model parameter estimation and validation. Two LHHW type rate models are developed and subjected to parameter estimation and model discrimination. Fundamental LHHW model in which hydrocarbon is adsorbed on the catalyst surface as alkyl intermediate species by scission of C-H bond gave physically meaningful parameters. Parameters are estimated by using generalized reduced gradient method in spreadsheet and sequential quadratic programming in Matlab. The estimated parameters for the selected model are thermodynamically consistent. Predictions from the kinetic model are in good agreement with experimental data.

Introduction

Fossil fuels will continue to be the dominant source of energy while the demand of energy continues to increase. To minimize the impact of associated carbon emissions, it is essential to utilize efficient means of converting fossil fuel energy into electricity and/or heat. Fuel cells are such a device. Most type of fuel cells, however, requires hydrogen or hydrogen-rich stream as a fuel. Hydrogen although present abundantly on earth is obscure in its pure form. Further, distribution and storage of hydrogen, for fuel cells, pose significant challenge. On-site, on-demand production of hydrogen by reforming of high density, easily transportable fuel can address these challenges.

High energy density and developed transportation/storage infrastructure makes diesel and gasoline attractive options as fuels for reforming. Reformed-diesel-fuelled-SOFCs are also being considered as an auxiliary power units in transport trucks where they require power to heat or to cool the cabin areas and to power electrical systems for refrigeration, lighting, computers and other electronic devices.[1] Successful study and development of onboard reforming can reduce idling time, fuel consumption, emissions and provide longer engine life. Although it is said that diesel fuels are easier to reform than gasoline[2] because of their lower content of aromatic hydrocarbons, an experimental study revealed that gasoline can be reformed easily because of highly reactive olefin and oxygenate fractions in gasoline. It was also argued that aromatics in gasoline are more reactive than aromatics in diesel (generally polyaromatics)[3].

Recent work in this area is mainly focused on experimental study of commercial diesel, and diesel surrogates.[4] Studies reporting kinetics of diesel reforming in heterogeneous environment are scarce. Dorazio *et al.*[5] proposed a gas phase mechanistic study on tetradecane reforming but the proposed mechanism did not capture the experimentally observed behavior of the process. Sotelo-Boyas *et al.*[6] presented a fundamental kinetic model based on detailed description of the fundamental chemistry of transformation of each individual hydrocarbon. Berry *et al.*[7,8] developed a simple power low expression for diesel autothermal reforming (ATR). Pacheco *et al.*[9] developed and validated mathematical model for ATR of i-C_8 using published reaction network kinetic model of methane reforming proposed by Xu et al.[10]. Rostrup-Nielsel[11] initially proposed the Langmuir-Hinshelwood-Hougen-Watson (LHHW) type kinetic model for ethane steam reforming and extended the generalized reaction mechanism for higher hydrocarbon steam reforming[12]. However, no work has reported estimation of kinetic parameters using Rostrup-Nielsel mechanism for higher hydrocarbons reforming.

The objective of present work is to develop a LHHW type kinetic model for ATR of tetradecane. Experimental data of tetradecane reforming supplied by NETL is used to estimate the model parameters and to validate the model. Two approaches are used in developing surface reaction mechanisms. In the first approach, a mechanism similar to that proposed by Rostrup-Nielsen[12] is assumed. In this approach, dual site hydrocarbon adsorption and carbon-carbon breakage step is considered irreversible. In the second approach, as reported by Zaera[13], it is assumed that hydrocarbon (saturated hydrocarbons which are quite stable) is chemisorbed with scission of one of the C-H bonds. Parameters of kinetic rate expressions based on both approaches are obtained by maximizing the coefficient of determination (R-squared coefficient). The obtained parameters are then tested for thermodynamic consistency.

Model Development

Reforming has a complex reaction mechanism consisting of thousands of gas phase and surface reactions. Reforming may comprise different reactions including cracking, isomerization, dehydrogenation/hydrogenation, hydrogenolysis and hydrocyclisation; with possibility of all reactions occurring simultaneously. In reforming hydrocarbon for hydrogen generation, a hydrocarbon molecule may undergo cracking or thermolysis when it comes in contact with high temperature reactor wall before reaching to catalyst. As such the hydrocarbon molecule fed will not be the only molecule that undergoes the reforming but a whole spectrum of lighter hydrocarbons that formed due to thermolysis. To simplify such a complex problem, the presence of thermolysis product is usually ignored in the reforming analysis. It is also assumed that the hydrocarbon molecule (tetradecane) undergoes total oxidation and produced water is utilized in steam reforming along with the water fed into the reactor and the produced CO_2 takes part in the reaction through water gas shift reaction scheme. Hence dry reforming step is not taken in to account. For steam reforming it is assumed that hydrocarbon produce CO and H_2, and CO reacts with water to form CO_2 as per water gas shift reaction (WGS). Hence steam reforming forming direct CO_2 step is avoided. So major reactions used for kinetic modeling of ATR of tetradecane are:

$$C_{14}H_{30} + 21.5\ O_2 \rightarrow 14\ CO_2 + 15\ H_2O \qquad \Delta H_{298} = -8805\ kJ/mol \qquad (1)$$

$$C_{14}H_{30} + 14\ H_2O \rightarrow 14\ CO + 29\ H_2 \qquad \Delta H_{298} = 2170\ kJ/mol \qquad (2)$$

$$CO + H_2O \rightarrow CO_2 + H_2 \qquad \Delta H_{298} = -41\ kJ/mol \qquad (3)$$

Kinetic rate expression for tetradecane total oxidation

In this study, for total oxidation, rate expressions similar to that reported by Jin *et al.*[14] for methane oxidation is used. Pacheco *et al.*[9] used same activation energy as Jin et al in their simulations and only changed the pre-exponential factor to fit the model to data. However, we have estimated all kinetic parameters.

$$r_1 = k_1 p_{C_{14}H_{30}} p_{O_2} \tag{4}$$

where rate constant can be given as

$$k_1 = A_1 \exp(-E_1/RT) \tag{5}$$

where A_1 is preexponential factor in mol / g_{cat} s atm^2, E_1 is the activation energy in kJ/mol, R is the gas constant in kJ/mol K, and T is the temperature in K

Kinetic rate expression for tetradecane steam reforming and WGS reaction

In order to develop the rate expressions for steam reforming and WGS, two approaches are considered and discussed as below.

Reaction scheme 1

In the first approach, the kinetic model developed from the general mechanism provided by Rostrup-Nielsen[12] for hydrocarbon reforming is adopted. It is assumed that hydrocarbon is chemisorbed on dual sites involving dehydrogenation followed by rupture of carbon-carbon bond and formation of surface radical CH_2. Surface reaction is considered as irreversible and surface species CH_2-L_1 and O-L_1 are assumed to be most abundant surface intermediates. In case of platinum catalyst, CO can be sufficiently adsorbed on platinum catalyst, so CO adsorption on surface of active catalyst L_1 is assumed. Since CO_2 is weekly adsorbed on the platinum catalyst it is considered negligibly adsorbed. The developed rate expressions are not reported as further analysis did not give physically meaningful parameters.

Reaction scheme 2

The second approach is based on a more fundamental understanding of reaction mechanisms of hydrocarbon reforming processes as provided by Zaera[13]. It is considered that the hydrocarbon adsorption is accompanied by C-H bond scission followed by subsequent elementary reactions of adsorbed alkyl intermediate to form different products. It is assumed that the surface species CH_2-L, O-L, H-L, and CO-L represents the major adsorbed species. CH_2-L species is assumed to react with O-L species to form CH_2O-L which is similar to the Xu *et al.*'s[10] methane reforming mechanism. By adopting certain simplifying assumptions, the following rate expressions for steam reforming and water gas shift reaction are derived.

$$r_2 = k_2 \frac{p_{C_{14}H_{30}}}{\sqrt{p_{H_2}}} \bigg/ DEN_1^2 \tag{6}$$

$$r_3 = \frac{k_3}{P_{H_2}} \left(p_{CO} p_{H_2O} - \frac{P_{H_2} P_{CO_2}}{K_{eq,WGS}} \right) \bigg/ DEN_1^2 \tag{7}$$

where

$$DEN_1 = 1 + a \frac{p_{H_2}^2 p_{CO}}{P_{H_2O}} + K_{H_2O} \frac{p_{H_2O}}{P_{H_2}} + K_{CO} p_{CO} + \sqrt{K_{H_2} p_{H_2}} \tag{8}$$

where K_i can be given as $K_i = \exp\left(\Delta S_i^o / R\right)\exp\left(-\Delta H_i^o / RT\right)$, ΔH^o and ΔS^o is the standard enthalpy change in kJ/mol and standard entropy change in kJ/mol K respectively. $K_{eq,WGS}$ is the equilibrium constant for water gas shift reaction.

Experimental data

For parameter estimation of the kinetic model, experimental data[4] supplied by the NETL (US-DOE),. In the NETL study, tetradecane was used as a surrogate for diesel and ATR experiments were carried out in a fixed bed reactor system. The reactor was operated continuously at steady state. 0.9 g of Pt/alumina catalyst was used for these runs. Pt loading was 0.611 wt% and surface area was 103 m^2/g. The temperature was varied between 750 and 900 °C and GHSV was in the range of 43,000 – 200,000 hr^{-1}. There were in total 13 experiments with O_2/C ratio 0.3 and steam/carbon ratio 1.5. Prior to gas analysis, the reactor effluent gas was passed through condenser to remove water and any other unreacted or heavy hydrocarbons. Dry effluent analysis included H_2, CO, CO_2, CH_4, C_2H_6, C_2H_4, C_3H_8, C_4H_{10}, C_5H_{10}, C_5H_{12}, C_6H_{12}, C_6H_{14}, C_7H_{14}, C_6H_6, and unconverted O_2. Concentration of water is obtained by O atom balance.

Model Discrimination and Parameter Estimation

Experimental data are used to obtain the parameters of developed kinetic models and discriminate between the models. The packed bed reactor represents a distributed system, an integral analysis of the reactor data was necessary for parameter estimation, which is based on minimization of SSE (squared sum of residuals) of outlet flow of components. Assuming isothermal and isobaric operations, the following set of differential equations are solved

$$\frac{dF_i}{dW} = \frac{dF_i}{\rho A dZ} = \sum_{j=1}^{j=M} \upsilon_{i,j} r_j \quad for\, i = 1 \ldots n \tag{9}$$

where ρ is bulk density of catalyst, which is 0.83 gm/cc, and A is the cross section area, which is 0.866 cm^2. The outlet calculated flow of H_2 and O_2 for all 13 experiments is compared with experimental flow rates. Ratio of SSE/TSS (total squared sum) is minimized, that is coefficient of determination (R^2) is maximized.

In order to get initial estimate, an Euler method is used in spreadsheet. The error in Euler method is minimized by using small step size (i.e. 1.25 cm length of the catalyst bed is divided into 200 equal parts). A solver optimizer, which employs the simplex, generalized reduced gradient, and branch and bound methods to find an optimal solution and sensitivity information, is used to maximize the coefficient of determination (R^2). The obtained parameters are then fed as an initial guess to minimize SSE/TSS ratio using standard fmincon Matlab optimization function, which uses sequential quadratic programming (SQP) method. ODE45 Matlab function, which uses 4th and 5th order Runge-Kutta algorithm, is used to solve the set of simultaneous differential equations. For solver and fmincon optimization functions, physically meaningful constraint on some of the parameters (for example all pre-exponential factors and all activation energies should be greater than zero etc.) is applied so that the obtained parameters are thermodynamically consistent.

In case of the first kinetic model, which is based on Rostrup-Nielsen[12] mechanism, one of the constants in denominator is always obtained negative and accordingly the model was abandoned. The second kinetic model, which is based on a more fundamental mechanism, is found to give thermodynamically consistent and physically meaningful estimated parameters. For obtained outlet molar flow of H_2, CO, CO_2, H_2O, and O_2, mean square regression ratio test is applied.

$$Test\ Ratio = MSR/MSE \tag{10}$$

where MSR is mean square regression and MSE is mean square error.

Predictions for 13 experimental runs in the data set each comprising molar flow of 5 species (H_2, CO, CO_2, H_2O, and O_2) were carried out. That is, a total of 65 data points (i.e. n=65) were used. Thirteen parameters are estimated (i.e. p=13). The calculated ratio of MSR/MSE is 358.7, which is much greater than $F_{(p-1),(n-p),0.05}$. Hence we reject the Null hypothesis that MSR and MSE are equal and conclude that model has picked up significant trend.

Results and Discussion

Over the range of experimental conditions, the developed kinetic model seems to predict the experimental outlet composition very much closely. It can be seen from Fig. 1 that the model has not only captured the general trend but also predicts the outlet composition over a wide range of GHSV or residence time. In case of CO_2, at high GHSV and high temperature model predicts high amount of CO_2. However, experimental data shows nearly same composition at high GHSV and all temperatures. For such a low concentration, discrepancy can be attributed, at least partially, to the experimental/measurement error. Model has captured the outlet oxygen composition very well (Fig. 2(a)). At 900 °C, experimental data shows oxygen conversion increases and then decreases with lowering GHSV, which indicates some experimental or measurement error. Model has predicted smooth decease in oxygen concentration (i.e. high conversion) with a decrease in GHSV and the predicted composition is close to that of experimental ones. The outlet C_{14} concentration is calculated from given conversion based on carbon, it is observed (not shown here) that total experimental hydrocarbon mole fraction (including C_{14}) is higher than the predicted C_{14} mole fraction and experimental C_{14} mole fraction is lower than the predicted C_{14} mole fraction. This result Is not surprising and simply points to the deficiency of not considering the formation of all hydrocarbons (for example CH_4 and higher hydrocarbons) in the overall mechanism. Inclusion of all hydrocarbon would result into high moles of hydrocarbons (as per mechanism one mole of C_{14} forms more moles of hydrocarbons) and would have predicted higher mole fraction than the current values. Even inclusion of formation of other hydrocarbons resulted into lower value of C_{14} predicted as more amount of C_{14} would have consumed. This is further confirmed by comparing the experimental and predicted volumetric flow as shown in Fig. 2(b). The predicted volumetric flow is lower than the experimental volumetric flow. This indicates the effect of not considering the other hydrocarbons in the calculation. Hence, inclusion of all hydrocarbons by providing some rate expression which describe formation of all hydrocarbons from C_{14} could result into better prediction for all hydrocarbons. However, it should be emphasized that including all hydrocarbons could alter the predicted mole fraction of all components as shown in Fig. 1 and 2(a). But since the mole fraction is very low, the change in predicted more fraction would be minor. It is also observed that hydrocarbon molecule undergoes oxidation reaction and produces water and CO_2 for initial 15 to 20 % bed length and then steam reforming and water gas shift reaction controls the product composition. Around 50 % of the oxygen is consumed in initial 15 to 20 % of the reactor length.

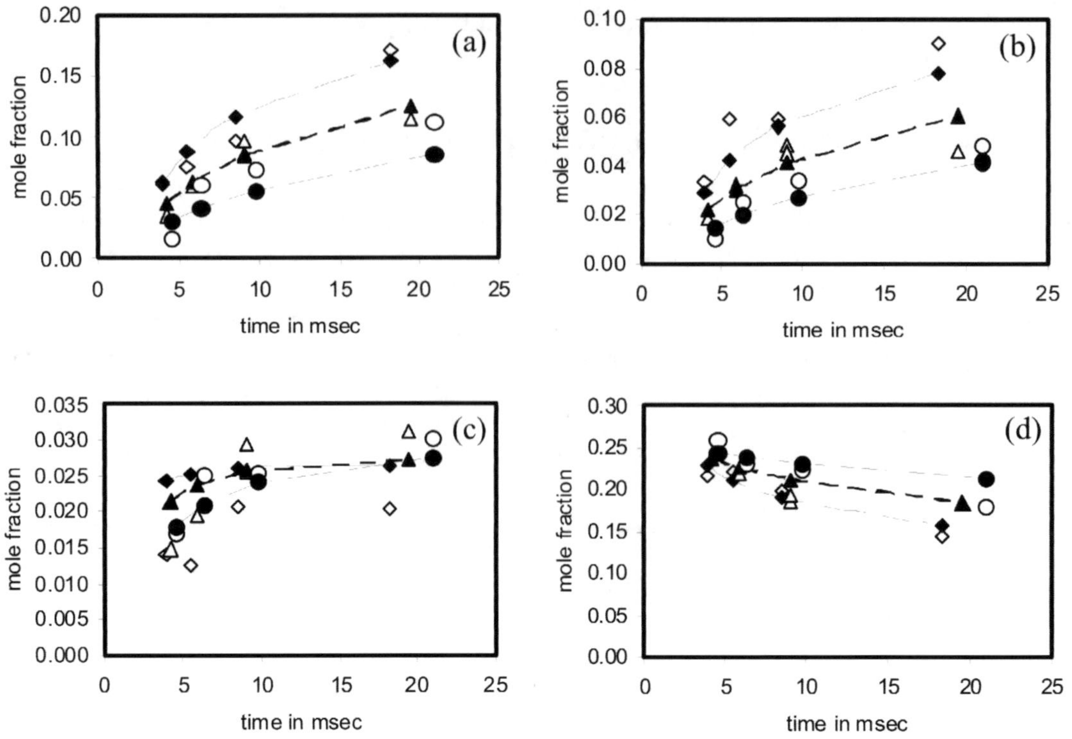

Figure 1. Experimental vs. predicted mole fraction of (a) H_2; (b) CO; (c) CO_2; (d) H_2O. ○, 750 °C; Δ, 825 °C; ◊, 900 °C; experimental and -●-, 750 °C; -▲-, 825 °C; -♦-, 900 °C; predicted.

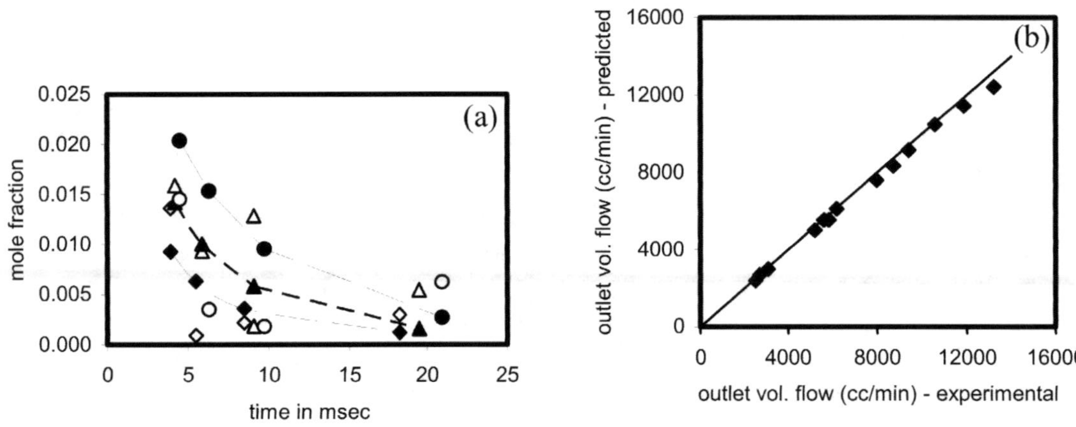

Figure 2. (a) Experimental vs. predicted mole fraction of O_2; ○, 750 °C; Δ, 825 °C; ◊, 900 °C; experimental and -●-, 750 °C; -▲-, 825 °C; -♦-, 900 °C; predicted. (b) Experimental vs. predicted outlet volumetric flow.

Thermodynamic consistency of estimated parameters.

The estimated parameters are reported in Table 1. Jin *et al.*[14] reported the activation energy for total oxidation of methane as 162 kJ/mol. The same value of activation energy is used by Pacheco *et al.*[9] for isooctane oxidation. However, the current analysis estimated lower activation energy (47.28 kJ/mol) for tetradecane. This could be attributed to the lower auto-ignition temperature of tetradecane (200 °C) than methane (630 °C).[15] For steam reforming of tetradecane the activation energy is 102.66 kJ/mol. In general, it is said that the activation energy should be greater than heat of reaction (i.e. E > ΔH). In case of steam reforming of tetradecane, this argument is not applied as rate constant is lumped parameter.

Table 1. Estimated parameters

k_1		k_2		k_3	
A_1 (mol / g_{cat} s atm^2)	E_1 (kJ/mol)	A_2 (mol / $atm^{1/2}$ g_{cat} s)	E_2 (kJ/mol)	A_3 (mol / atm g_{cat} s)	E_3 (kJ/mol)
$1x10^2$	47.28	$5.50x10^2$	102.66	$2.73x10^2$	141.14
K_{H_2O}		K_{CO}		K_{H_2}	
$\Delta S^o_{H_2O}$ (kJ/molK)	$\Delta H^o_{H_2O}$ (kJ/mol)	ΔS^o_{CO} (kJ/molK)	ΔH^o_{CO} (kJ/mol)	$\Delta S^o_{H_2}$ (kJ/molK)	$\Delta H^o_{H_2}$ (kJ/mol)
0.0105	63.62	-0.151	-90	-0.125	-80.16
a	0.5572				

Since water adsorption step ($H_2O + L \rightleftarrows O\text{-}L + H_2$) is not actually any dissociative or non-dissociative chemical or physical adsorption step, its estimated parameters are not subjected to thermodynamic consistency. However, Xu *et al.*[10] estimated $\Delta S^o_{H_2O}$ and $\Delta H^o_{H_2O}$ for nickel catalyst 0.100 kJ/mol K and 88.68 kJ/mol, respectively. In this study, parameter estimation for Pt catalyst resulted in 0.01058 kJ/mol K and 63.62 kJ/mol respectively.

Estimated parameters for dissociative chemisorption of H_2 and chemisorption of CO are subjected to following consistency test.[16]

$$\Delta H^o_{ad} < 0 \tag{11}$$

$$0 < -\Delta S^o_{ad} < S^o_g \tag{12}$$

$$10 \le -\Delta S^o_{ad} \le 12.2 - 0.0014\Delta H^o_{ad} \quad \text{(in cal/mol)} \tag{13}$$

For both CO chemisorption and H_2 dissociative chemisorption, the estimated heats of adsorption is less than zero satisfying the first consistency criterion. Theoretically H-H bond energy is 436 kJ/mol and Pt-H bond energy[17] is 261.5 kJ/mol. This gives -87 kJ/mol heat of dissociative chemisorption of H_2 on Pt metal atom. The estimated value (-80.16 kJ/mol) is close to the theoretical value. Guerrero-Ruiz[18] reported that heat of chemisoption of CO (approximately 140 kJ/mol on Pt/activated carbon catalyst) is higher than that of hydrogen. Estimated heat of chemisorption of CO (-90 kJ/mol) on Pt/alumina catalyst is also higher than that of hydrogen but it is lower than that reported for Pt/activated carbon catalyst.

For CO and H_2, the S^o_g values at 298 K are 0.198 and 0.13 kJ/mol respectively. The estimated $-\Delta S^o_{ad}$ values for CO and H_2 are 0.151 kJ/mol and 0.125 kJ/mol respectively, which are less than S^o_g. Hence, a second consistency criterion (equation (12)) is also satisfied.

Third criterion suggests that the minimum entropy change during dissociative or non-dissociative adsorption should be a minimum of 10 e.u. (cal/mol.K) and should be less or equal to $12.2-0.0014\Delta H^{o}_{ad}$, where ΔH^{o}_{ad} is in cal/mol. For CO and H_2, estimated $-\Delta S^{o}_{ad}$ values are 36.09 cal/mol K and 29.88 cal/mol K respectively and the value of the term $12.2-0.0014\Delta H^{o}_{ad}$ are 42.12 and 35.19 respectively. Hence, the third consistency criterion (equation (13)) is also satisfied.

Conclusions

Kinetics of tetradecane reforming on $Pt\text{-}Al_2O_3$ catalyst is investigated. Two LHHW type rate models are developed and subjected to parameter estimation and model discrimination. The second fundamental LHHW model in which hydrocarbon is adsorbed on the catalyst surface as alkyl intermediate species by scission of C-H bond is found to give physically meaningful parameters. The kinetic parameters for the model are estimated by using generalized reduced gradient method and sequential quadratic programming. The estimated parameters for the selected model are thermodynamically consistent. A good agreement between the predicted values from the model and those of the experimental is observed with derived kinetic parameters. Some discrepancy for C_{14} concentration is observed due to exclusion of high hydrocarbons in the mechanism. The derived model also shows that, for the given operating conditions, the total oxidation reaction is dominant at initial 15 to 20 % of the total reactor length.

References

(1) Argonne National laboratories (2007), http://www.transportation.anl.gov/engines/idling.html

(2) Thormann, J., P. Pfeifer, U. Kunz and K. Schubert (2008), "Reforming of Diesel Fuel in a Micro Reactor," *International Journal of Chemical Reactor Engineering,* 06, pp. 1-17.

(3) Nilsson M., X. Karatzas, B. Lindstrom and L.J. Pettersson (2008), "Assessing the Adaptability to Varying Fuel Supply of an Autothermal Reformer," *Chemical Engineering Journal*, 142, pp. 309-317.

(4) Berry D.A., D. Shekhawat and T.H. Gardner (2003), "U.S.DOE Progress Report for Development of Reaction Kinetics for Diesel-Based Fuel Reformer," pp. 1-6.

(5) Dorazio L. andM. Castaldi (2008), "Autothermal Reforming of Tetradecane ($C_{14}H_{30}$): A Mechanistic Approach,"*Catalysis Today,* 136, pp. 273-280.

(6) Sotelo-Boyas R. andG. Froment (2009), "Fundamental Kinetic Modeling of Catalytic Reforming," *Ind. Eng. Chem. Res.*, *48*, pp. 1107–1119.

(7) Berry, D.A., D. Shekhawat and T.H. Gardner (2003), "U.S.DOE Progress Report for Hydrogen, Fuel Cells, and Infrastructure Technologies Program," pp. 147-152.

(8) Shekhawat, D., D.A. Berry, T.H. Gardner andJ.J. Spivey (2006), "Catalytic Reforming of Liquid Hydrocarbon Fuels for Fuel Cell Applications," *The Royal Society of Chemistry*, 19, pp. 184-253.

(9) Pacheco, M. J. Sira and J. Kopasz (2003), "Reaction Kinetics and Reactor Modeling for Fuel Processing of Liquid Hydrocarbons to Produce Hydrogen: Isooctane Reforming," *Applied Catalysis A: General*, 250, pp. 161-175.

(10) Xu, J. and G.F. Froment, (1989), "Methane Steam Reforming, Methanation and Water-Gas Shift: I, Intrinsic Kinetics," *AIChE Journal*, 35, pp. 88-96.

(11) Rostrup-Nielsen, J.R. (1973), "Activity of Nickel Catalyst for Steam Reforming of Hydrocarbons," *Journal of Catalysis*, 31, pp. 173-199.
(12) Rostrup-Nielsen, J.R. (1984), "Catalytic Steam Reforming. *Catalysis, Science and Technology*, " ed. J.R.Anderson and M.Boudart, Springer-Verlag, Berlin, 5, pp. 1-118.
(13) Zaera, F. (2002), "Selectivity in Hydrocarbon Catalytic Reforming: a surface Chemistry Perspective," *Applied Catalysis A: General*, 229, pp. 75-91.
(14) Jin, W., S. Gu, S. Li, P. Huang, N. Xu and J. Shi (2000), "Experimental and Simulation Study of Catalyst Packed Tubular Dense Membrane Reactor for Partial Oxidation of Methane to Syngas," *Chemical Engineering Science*, 55, pp. 2617-2625.
(15) http://www.faqs.org/faqs/sci/chem-faq/part5/section-1.html
(16) Vannice, M.A. (2005), "Kinetics of Catalytic Reactions," Springer: New York.
(17) Miyazaki, E. (1978), "The Heat of Chemisorption of H_2 on Transition d-Metals," *Surface Science*, 71, pp. 741-747.
(18) Guerrero-Ruiz, A. (1993), "Some Aspects of Reversible Chemisorption on Supported Platinum Catalysts," *Reaction Kinetics and Catalysis Letter*, 49, pp. 53-60.

NATURAL GAS DIRECT CARBON FUEL CELL FOR HYDROGEN AND ELECTRICTY PRODUCTION

Meyer Steinberg
HCE, LLC
February 10, 2008
HCEI-2-08-001

Introduction

Over the past several years, the Direct Carbon Fuel Cell (DCFC) has been under development for converting carbon directly to electricity.[1] The DCFC is an electrochemical fuel cell that may be fueled with pure carbon or with impure carbon as in coal. The DCFC has never been fueled with natural gas (methane) or gaseous and liquid carbonaceous fuels. The invention described herein uses methane or natural gas as fuel in a direct carbon fuel cell to produce coproducts hydrogen and electricity.

The DCFC consists of an electrochemical cell containing molten salt electrolyte, usually alkali (Na, Li, K) carbonate at 600°C to 1000°C within an anode compartment and a cathode compartment. The compartments are separated by an ion permeable membrane (zirconia or other permeable high-temperature membrane).

In the cathode compartment, oxygen from air reacts with recycled CO_2 to produce carbonate ion ($CO_3^{=}$), which travels to the anode compartment through the permeable membrane. In the anode compartment, the carbonate ion reacts with the carbon fuel to produce CO_2 and releasing electrons which produces electricity for use in an external circuit. The reactions are as follows:

at the cathode $\quad 4e^- + 2CO_2 + O_2 = 2CO_3^{=}$

at the anode $\quad C + 2CO_3^{=} = 3CO_2 + 4e^-$

overall reaction $\quad C + O_2 = CO_2$

The theoretical thermodynamic efficiency in the DCFC is 100% for conversion of the enthalpy (ΔH) of oxidation of carbon to CO_2 to free energy (ΔF) of reaction, which generates the electromagnetic electricity force. Since the entropy change for the oxidation of carbon to CO_2 is zero, $\Delta H = \Delta F$.

The molten salt fuel cell is well known for converting hydrogen to electricity.[(2)] However, the theoretical thermodynamic efficiency of hydrogen as fuel is only 70%. Thus, the molten carbonate fuel cell with carbon fuel is inherently more efficient than the molten carbonate fuel cell with hydrogen fuel. In the laboratory, thermal efficiencies for the DCFC with carbon fuel have been obtained reaching 90%. For a practical current density fuel cell for electrical power production, the DCFC is projected to be 80% thermally efficient.[(1)]

It is also known that natural gas (methane) can be cracked and decomposed to carbon and hydrogen at temperatures above about 500°C. Much work has been performed on cracking methane to carbon and hydrogen in various reactors, such as a furnace heated by gas or by electricity, a high temperature plasma or in a bath of molten metal (tin)[(3, 4)] or in molten salt (sulfate or carbonate). The reaction is endothermic requiring 18 kcal/g-mol of CH_4 to drive the reaction decomposition and is as follows:

$$CH_4 = C + 2H_2$$

<u>The Invention (NGDCFC)</u>

This invention uses methane (natural gas) as a feed stock to produce coproducts hydrogen gas and electricity by combining molten-salt methane decomposition with the direct carbon fuel cell (DCFC) in one unit, known as Natural Gas Direct Carbon Fuel Cell (NGDCFC).

The anode compartment of the DCFC acts as the methane decomposer to form particulate solid carbon, which remains in the molten salt, and hydrogen gas is evolved from the reaction. The carbon in the molten salt combines with the carbonate ion ($CO_3^{=}$) and produces electricity for the power circuit and the CO_2 evolves from the anode compartment of the cell as a gas in a concentrated stream. To prevent gaseous H_2 from reacting with the highly-concentrated gaseous CO_2 to produce H_2O and CO in the anode compartment, the gases are quickly cooled as they are emitted from the anode compartment.

An alternative embodiment divides the anode compartment into two sections: an ante-chamber CH_4 decomposer section and a main anode section. The two sections are created by the addition of a louvered baffle in the anode compartment. This baffle provides added efficiency because it creates a separate ante-chamber for methane decomposition to take place within the anode compartment allowing removal of hydrogen separately from the molten salt and minimizing possible reaction with the CO_2 evolving in the main anode section.

Solid particulate carbon produced in the ante-chamber CH_4 decomposer section is retained in the molten carbonate salt. The carbon flows through the louvered baffle to the main anode section where the anodic electrode is situated, allowing the carbon to react with the carbonate ion to form CO_2. In this manner, the hydrogen evolves separately in the ante-chamber CH_4 decomposer section and the CO_2 evolves separately in the main anode section. The methane gas feed and the hydrogen gas released in the ante-chamber CH_4 decomposer section provide enough convection and turbulence for mixing the molten carbonate salt to flow the carbon through the louvers to the main anode section. The positioning of the louvers, top in one direction and the bottom in the opposite direction, promotes circulation of the molten salt between the two anode sections.

CO_2 is recycled from the main anode section to the cathode compartment by both back diffusion through the membrane and by recycle of CO_2 gas from the anode to form the carbonate ion with oxygen from the air.

A schematic of this invention is shown in FIG. 1.

The thermal energy to decompose the methane to carbon and hydrogen in the ante-chamber CH_4 decomposer section is obtained from the 20% inefficiency of the DCFC carbon oxidation reaction. Thus, the DCFC operates autothermally, requiring no additional energy input. Preheating the natural gas feed is accomplished by gas to gas heat exchangers transferring the heat from the hot H_2 and CO_2 effluent from the cell to the natural gas input to the cell.

This invention is especially amenable to producing and supplying hydrogen at vehicle gas filling stations. A natural gas fed DCFC is much more efficient and economical than conventional steam reforming of natural gas now being deployed for supplying hydrogen at gas filling stations. It also can do so with considerably reduced greenhouse gas CO_2 emission compared with the conventional steam reforming process. The CO_2 is emitted at 100% concentration, so it can be easily collected and either sold or sequestered. Furthermore, the present invention simultaneously produces electricity, which can be sold to power companies and put back into the grid or can be dispensed at the gas station for electrically powered vehicles, which some think will eventually overtake hydrogen-fueled vehicles. In any case, this system could independently supply hydrogen fueled vehicles and electrically driven vehicles.

Another decided advantage of this invention is that hydrogen and electricity are coproduced in one unit. This simplifies operations and considerably reduces capital investment that would otherwise be required using the known method involving steam reforming of natural gas.

References

(1) J.F. Cooper, et al. "Turning Carbon into Electricity," Science and Technology Review, pp-4-12, Lawrence Livermore National Laboratory (June 2001).

(2) V.R. Silimase, J. Power Sources, 160, 852-7 (2006).

(3) M. Steinberg and Y. Dong, U.S. Patent 5, 767,165 (June 16, 1998).

(4) M. Seehan, et al., "Hydrogen Production by Direct Contact Pyrolysis of Natural Gas, Energy and Fuels, 17, 705–13 (2003).

FIG. 1
Natural Gas Direct Carbon Fuel Cell (NGDCFC) for Hydrogen and Electricity Production

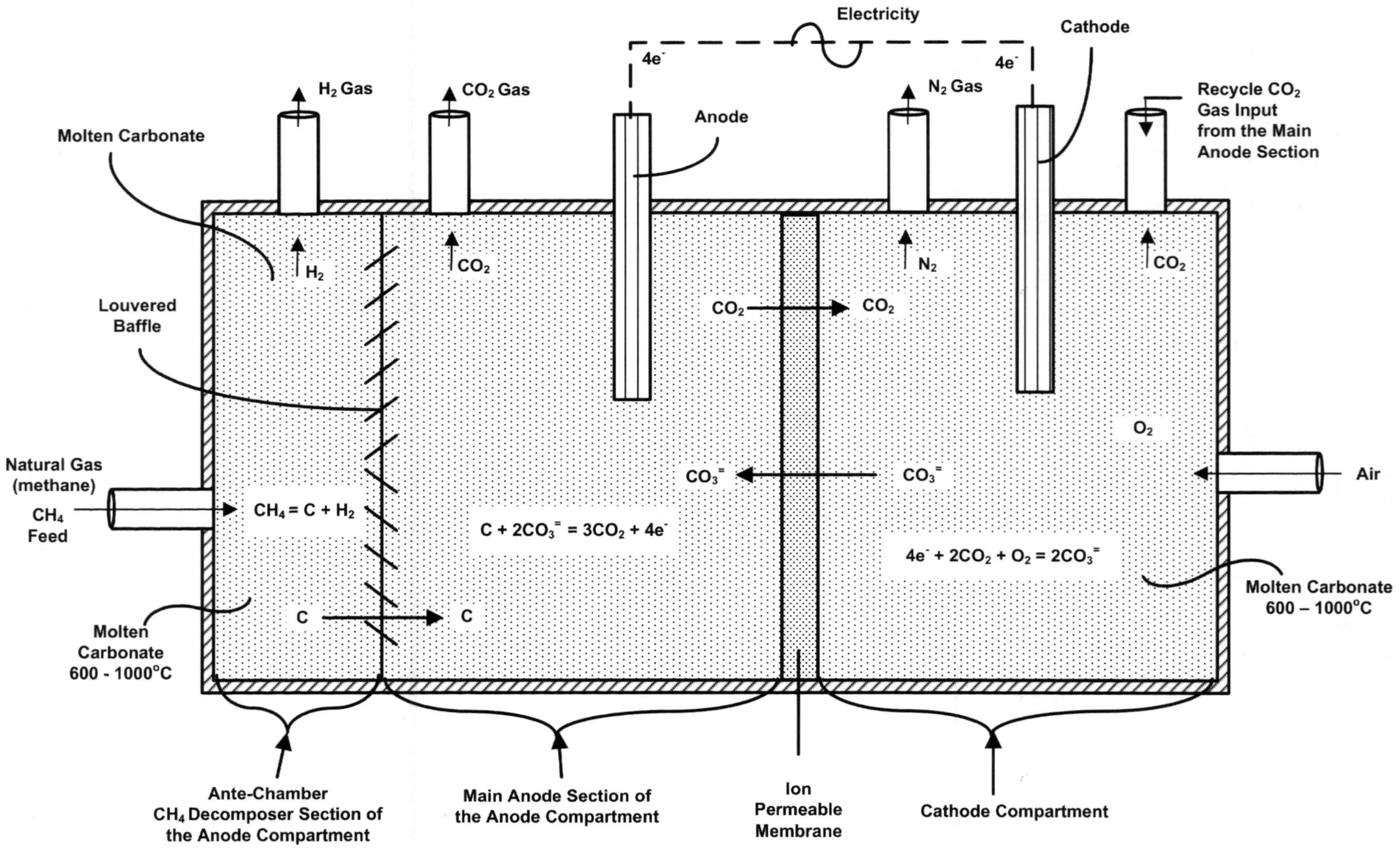

45a Dry Reforming of Biogas to Syngas And Hydrogen for Fuel Cell Applications

Nazim Muradov, Franklyn Smith, and Nathaniel Garceau. Florida Solar Energy Center, University of Central Florida, 1679 Clearlake Road, Cocoa, FL 32922

Biogas (BG) is an important renewable resource for H2 production (here, the term BG also includes landfill gas, LFG, and digester gas). The advantages of using BG as a feedstock for H2 production are four-fold: (i) it is a domestic energy resource, (ii) it is a local resource, thus, there would be no need for its delivery by expensive pipelines, (iii) it can be obtained at little or no cost, and (iv) if not utilized, BG could potentially create significant environmental damage (since methane is more potent greenhouse gas than CO2). Although the resources of BG are vast and widely available throughout the country, it remains mostly unused. Direct catalytic reforming of BG is complicated by two factors: (i) the presence of sulfur-containing compounds that could easily poison catalysts used in the reforming process and (ii) a feedstock non-uniformity. Due to the fact that methane is a predominant component in BG (typically, [CH4]= 50-70 vol.%), an undesirable methane decomposition reaction may occur at the operating temperatures of the reforming process, which would result in catalyst deactivation.

In this paper, the authors investigate syngas and H2 production via catalytic dry reforming of a biogas-mimicking gaseous mixture containing CH4:CO2=1.3:1 (molar) and 0.1 vol.% H2S. The objectives are to determine the catalytic activity, selectivity and stability of a number of transition metal catalysts for CO2 reforming of CH4 at the conditions favorable for carbon lay down. H2S was removed from the gaseous mixture by a selective adsorbent prior to the dry reforming reaction. Two types of catalysts were tested: noble metal (NM) catalysts (e.g., alumina-supported Ru, Pt, Rh, Ir, Pd) and Ni-catalysts. NN catalysts, especially, Ru, Ir and Pt catalysts, demonstrated high activity and selectivity for CO2 reforming of methane and good stability (i.e., no methane decomposition and carbon deposition occurred during the reforming reaction). The H2:CO molar ratio in the syngas was •1, and it contained about 7 vol.% of unreacted CH4, according to the following reaction:

(1.3CH4) + CO2 = 2H2 + 2CO + (0.3CH4)

The resulting syngas is suitable as a fuel for molten carbonate and solid oxide fuel cells. Ni-based catalysts demonstrated poor selectivity since they also catalyzed methane decomposition reaction (leading to the catalyst deactivation). The deposition of carbon on the Ni catalyst surface could be prevented by adding relatively small quantities of an oxidant (CO2, O2 or steam) to the feedstock. The effect of different oxidants on the Ni-catalyst stability is reported. The resulting syngas was free of unconverted methane, and it was further processed via water gas shift reaction to H2-CO2 mixtures. Hydrogen was subsequently recovered from these mixtures by cryogenic adsorption method using an activated carbon based adsorbent. The use of a pressure-swing adsorption system for the separation of the H2-CO2 mixture would produce H2 with the purity of >99.999 % that is suitable for the operation of a polymer membrane electrolyte (PEM) fuel cell.

45b Non-Catalytic, Partial Oxidation for Syngas Production

Janet L. Ellzey, Ingmar Schoegl, and Colin H. Smith. Mechanical Engineering, University of Texas at Austin, 1 University Station C2200, Austin, TX 78712

Although hydrogen is abundant in nature, it is primarily bound into other molecules such as hydrocarbons. The conversion, or reforming, of a hydrocarbon into a hydrogen-rich syngas may be accomplished through either catalytic or non-catalytic processes. We investigated non-catalytic syngas production with two different reactors: a reactor consisting of multiple parallel channels with alternating flow directions and a reactor consisting of porous inert media. In both devices, heat transfer from the hot products to the cold reactants promotes reactions of fuel-rich, syngas-producing mixtures that lie beyond the conventional flammability limit. For the parallel channel reactor, experimental results with methane/air and propane/air mixtures show that stable reaction fronts are obtained at a wide range of equivalence ratios and inlet velocities. A comparison of the reactor performance for methane/air and propane/air mixtures shows significant differences, which are attributed to the more complex reaction chemistry of rich propane combustion. With the porous media reactor, we investigated syngas production from wet ethanol, which is ethanol that has not been fully distilled or dehydrated. Ethanol is often promoted as the biofuel of the future, yet it is hindered by its overall production energy requirements, the most significant of which is water removal through distillation and dehydration. Through experiments and modeling, we studied the reforming of wet ethanol over a range of equivalence ratios, inlet velocities, and wet ethanol water fractions. Experimental results, which include species concentrations in the exhaust and gas temperatures, showed that syngas can produced as efficiently with wet ethanol as with dry ethanol. Computational results included gas and porous media temperatures, reaction front propagation speeds, and spatial and temporal species production and destruction. We also include experimental and modeling results from our previous investigations of methane and heptane fuel reforming with a porous media reactor.

Novel High stability Hydrogen Production from Ethanol Steam Reforming over Catalyst of Nano-NiO particles Based on Flowerlike CeO_2 Microspheres for EH Hybrid Vehicle Application

Jie Sun [*1,2], Yuguang Wang [1], Jigang Li[1], Tingyun Shen[1], Xinping Qiu[2], Liquan Chen[3]

1, Lab of Renewable Energy and Energy Safety, P.O.Box 1048, Beijing 102205, China;

2, Department of Chemistry, Tsinghua University, Beijing 100084, China;

3, Beijing National Laboratory for Condensed Matter Physics, Institute of Physics, Chinese Academy of Sciences, Beijing 100080, China.

Abstract

Catalysts of nickle oxide based on flowerlike ceria microspheres with high disperse and stable structure were made by hydrothermal method to achieve simultaneous dehydrogenation of ethanol and water molecules on multi-active sites. This special morphology catalyst represented novel stability more than 2000 hours with H_2 average selectivity value of 70mol.%, and >99.9% ethanol conversion from low temperature (350°C) ethanol steam re-reforming, Which were very good for the EHV application. And the discontinuous stability test with 13 times restarts was more than 600 hours. Catalytic parameters with respect to activitry, selectivity towards hydrogen production and stability with time on stream were determined and optimized.

Key words: Hydrogen production, high stability, ethanol steam reforming, EHV.

1. Introduction

Biomass ethanol as a source fuel of hydrogen has advantages of that it is renewable, it operates in a closed carbon loop, thus it involves no net CO_2 emissions into the atmosphere. Ethanol is safe and non-toxic as well as environmentally friendly. Furthermore, the infrastructure of petroleum transportation can be used by ethanol fuel [1,2].

Bio-ethanol had to be one of the replacers of gasolene to vehicles in China. That ethanol-H and petrolic hybrid vehicle (EHV) was a kind of vehicel to use the waste heat from vent-pipe to reform ethanol water mixture to produce H_2, CH_4 and CO by a micro-ethanol steam reformer which was setted in the vent- pipe to promote the combustion of gasolene and to decrease the use of gasolene obviously (about 30 vol.%). So, EHV could be a transitional products before the industrialization of fuel cell and pure H_2 engine vehicles in China.

Many reactions of ethanol over metals and metal oxides have been investigated recent years. Reactions of ethanol dehydrogenation over noble metal membrane [3], and reactions of ethanol over M/CeO_2 catalysts [4] have been studied. The catalytic properties of the supported transition metal catalysts to one of the steam reforming reactions ($C_2H_5OH + H_2O \rightarrow 2CO + 4H_2$, $\Delta H^0_{298} = +256$ kJ mol^{-1}) show that the selectivity of H_2 for the reforming reaction is in the order: Co much greater than Ni>Rh>Pt, Ru, Cu [5]. The studies of ethanol steam reforming in a molten carbonate fuel cell (MCFC) utilized the catalysts from the noble metals (Pt, Rh, etc.) to $CuO/ZnO/Al_2O_3$, $NiO/CuO/SiO_2$, Ni/MgO and Co/MgO, they got good results: the selectivity of hydrogen of 5%Rh/Al_2O_3 can reach to 30.1vol. % [6]. The steam reforming of ethanol at about 300°C over catalyst of $Cu/Ni/K/\gamma\text{-}Al_2O_3$ was also studied [7].

In recent researches, nickel catalysts were widely studied in ethanol steam reforming for

* To whom correspondence should be addressed. Tel: 8610-66757029-82140, Fax: 8610-52565168,
E-mail: magnsun@gmail.com; magnsun@mail.tsinghua.edu.cn

their good activity and low cost [1, 8-12]. Ni/Al_2O_3 catalysts had been reported to be the most effective catalysts for reforming of methane with carbon dioxide, but they suffer a serious problem of carbon deposition, nickel sintering and phase transformation [13-15]. It was well known that rare earth metal oxides was of high alkaline and was favorable for dehydrogenation of alcohols. Ni/La_2O_3 catalyst had been previously found to exhibit good performance under conditions (800°C) of reforming of methane with carbon dioxide, and to be very active and stable for the steam reforming of ethanol under certain operating conditions (above 300°C) [8, 16], While Y_2O_3 was more active for dehydrogenation than oxides of other elements in La system, so nano sized Ni/Y_2O_3 was very good for ethanol reforming to H_2 production [17]. CeO_2 is widely used as a promotor in three-way catalysts (TWC) for the elimination of toxic auto-exhaust gases, low-temperature water-gas shift reaction, fuel cells [18-20]. Nanocrystalline CeO_2 has showed improved and size-dependent properties. Actually, mesoporous ceria has shown great potential as versatile catalysts and catalyst supports due to its high surface area and increased dispersion of active secondary components[21-25].

Thermodynamics claimed when T > 200∘C, reaction ($C_2H_5OH \rightarrow CO + CH_4 + H_2$, $\Delta H^0_{298} = +49$ kJ mol^{-1}) will be favored. At higher temperatures, the methane yield is reduced via the CH_4 steam reforming reaction ($CH_4 + 2H_2O \rightarrow CO_2 + 4H_2$, $\Delta H^0_{298} = +166$ kJ mol^{-1}). But the high temperature will caused the sinter of catalysts and coke formation. There was one couple contradiction. Water-gas shift reaction ($CO + H_2O \rightarrow CO_2 + H_2$, $\Delta H^0_{298} = -41$ kJ mol^{-1}) could decrease CO and increase H_2, it was favored at temperature range 200~400∘C. So, we hope to combine the WGS reaction with the reforming reactions at an appropriate low temperature to eliminate CO, try to avoid sinter and coke formation, and improve the stability finally.

In most of our work, the choice of supports was focused on the rare earth metal oxides for their high alkaline and favor of dehydrogenation of alcohols. We ever used nickel with supports (yttria, lanthanum oxide, ceria, zirconia, low-area alumina, etc.) [10,11,17,26-28] as catalysts for ethanol steam reforming reactions. In this work, active supports of CeO_2 and La_2O_3 with special flowerlike morphology was synthesized and modified through hydrothermal method[26], nickle oxide were deposited from salt solutions on the active supporsts[10], then all were dipped on low-area cordierite ceramic carrier. We show here results from Ni-ceria and Ni/ La_2O_3 because they were more stable and gave greater WGS activity than Rh-ceria [2], and had good low temperature activity for ethanol steam reforming [9, 26].

2. Experimental

X-ray powder diffraction (XRD) was carried out on a Holland X'Pert Pro MPD X-ray diffractometer equipped with a monochromatized Cu-Kα radiation at 40 kV and 20mA.

Field Emission Scan electron microscope (FESEM) was carried out on XL30s-FEG equipment, at 10 kV observed the morphology of the products.

Steam reforming test was carried out on a self-built system. Ethanol-water mixtures were injected into a tube through an electronically controlled sprayer-slide that sprayed small droplets onto the walls of the tube, heated to 200°C. This produced rapid vaporization, so that reactants went from liquids at room temperature to products exiting the catalyst in less than 36 ms.

Products were analyzed by a gas chromatography Angilent 7890A with 2 TCDs, with mass balances closing to within 0.1%. All experiments were run for up to 80 hours on a given catalyst, and most experiments were repeated on several catalysts. No significant deactivation or variations between catalysts were noted in these experiments.

3. Results and discussions

3.1 Catalysts Characteristics

Figure 1 shows the XRD patterns of flowerlike supports NiO/CeO_2 catalysts .The peaks labeled with the symbol " * "are corresponding to the diffraction of NiO, and peaks with acclivitous labels of Miller indexes are corresponding to the diffractions of CeO_2. There is no new peaks appeared in the XRD patterns of NiO/CeO_2, meaning that the catalysts were just the mixture of nickel oxide and supports and no new phase was formed.From the diffraction peaks of NiO (111), CeO_2(200), the crystal grain sizes of nickel oxide and support materials of the catalysts, calculated with Scherrer formula, were NiO 11.8 nm and the flowerlike supports CeO_2 7.9nm.

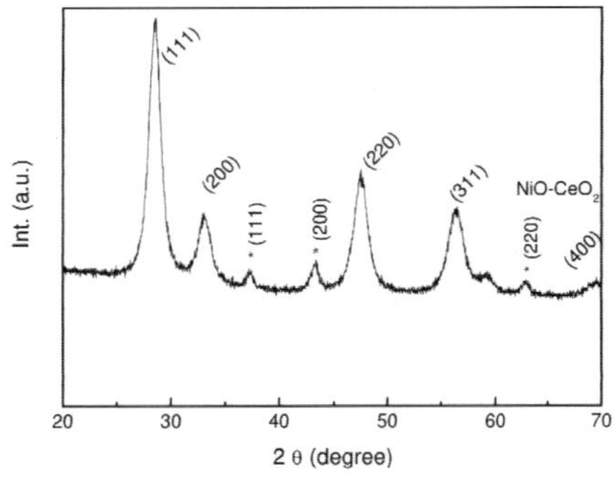

Fig.1 XRD of the samples of NiO loaded on special flowelike microspheres supports of CeO_2 .

As Figure 2 showed, the size and morphology of the supports and supported catalysts products were examined by field-emission scanning electron microscopy (FESEM) measurements. Figures 2a showed CeO_2 particles are nearly monodisperse spherical particles with flowerlike texture, respectively. The average diameter of the microspheres is 2-4 μm, it can be seen that these flowerlike microspheres are composed of many nanosheets as the petals with an average thickness of about 20-30 nm and these nanosheets interweave together forming an open porous structure. Figure 2b showed NiO particles are loaded dispersively on the flowerlike microspheres supports.

3.2 Catalysts activity and selectivity

Ethanol conversion (C_E) and product selectivity (S_P) were defined as [10].

Typical results of ethanol reforming on Ni-ceria catalysts with flowerlike active supports were shown in Figs.3, in which the selectivity of each products and the conversion of ethanol were shown as a function of reaction temperature in the temperature range of 275–650°C. These experiments were done at a total flow rate of 6.8 standard liters/min [gas hourly space velocity (GHSV)~ 1.0×10^5 hour $^{-1}$], with preheating to 200°C.

(a) (b)

Fig.2 Representative FESEM images of the flowerlike CeO_2 microspheres, and nickle oxide particles loaded on the support : (a) high-magnification FESEM image of an individual microspheres, revealing the constituent details of the microspheres; (b) high-magnification FESEM image of many NiO particles supported on an individual CeO_2 microspheres.

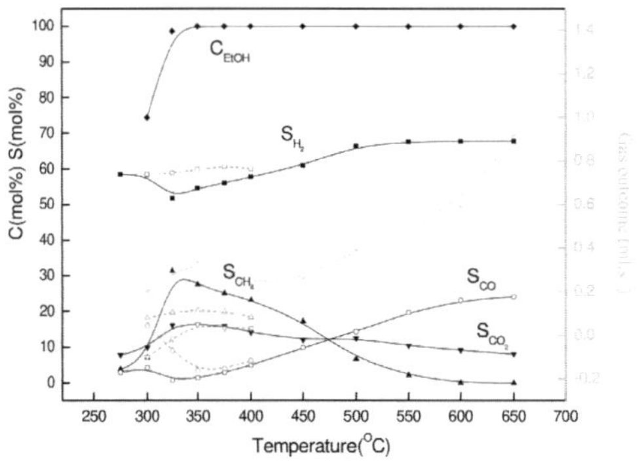

Fig.3 Effect of reaction temperature ***T*** on conversion of ethanol (C_{EtOH}) and on selectivity of hydrogen (S_{H2}), carbon monoxide (S_{CO}), carbon dioxide (S_{CO2}), and methane (S_{CH4}) obtained over the special morphology catalyst NiO/CeO_2. Experimental conditions: mass of catalyst 0.55g, H_2O/EtOH mol ratio ***R*** = 3:1, liquid flow rate 0.05cm^3 min^{-1}, total flow rate of 6.8 standard liters/min gas hourly space velocity (GHSV)~ 1.0×10^5 hour $^{-1}$, ***P*** = 1atm.

Fig. 3 showed that at a temperature of 300°C, steam reforming of ethanol occurred obviously over the special morphology catalyst Ni/CeO_2, the ethanol conversion reached 74.4mol.%, and the selectivity of hydrogen, methane, carbon dioxide and carbon monoxide was 58.3mol.%, 7.1mol.%, 10.0mol.% and 4.6mol.%, respectively. When the temperature increased to 325°C, the conversion of ethanol increased to 98.6 mol.%, the selectivity of methane increased to the maximal value 31.4 mol.%, the selectivity of CO_2 increased to 16.1mol.%, the selectivity of CO and H_2 decreased to the minimum 0.7mol.% and 51.8mol.%, indicating that reaction of $C_2H_5OH \rightarrow CO + CH_4 + H_2$, $\Delta H^0_{298} = +49$ kJ mol^{-1} and the WGS reaction had occurred evidently. In the temperature range of 325–650°C, the selectivity of H_2 and CO kept increasing to 67.8mol.% and 24.1mol.%, respectively, while the selectivity of CH_4 kept decreasing to 0, the selectivity of carbon dioxide increased slightly to 16.4mol.% from 325°C to 350°C and then kept decreasing to 8.1mol.% from 350°C to 650°C, which indicated that CH_4 steam reforming reaction ($CH_4 + 2H_2O \rightarrow CO_2 + 4H_2$, $\Delta H^0_{298} = +166$ kJ mol $^{-1}$) occurred determinately. The conversion of ethanol reached to 100% in the temperature range of 350-650°C . The outcome velocity of products gas reached a peak value of 0.34ml s^{-1} at 350°C and a valley value of 0.23 ml s^{-1} at 400°C, indicating that coke formitiaon occurred during this temperature. But with temperature increasing, water gas reaction ($C + H_2O \rightarrow CO + H_2$, $\Delta H^0_{298} = +131$ kJ mol $^{-1}$) contributed to the recover of outcome gas velocity, which kept increasing to 0.90 ml s^{-1} with the temperature increasing to 650°C.

After the first cycle activity and selectivity tests at temperature range of 275-650°C, the reformer was shut and cooled down to room temperature for 12hours and then restarted to work to show the very good repeated activity of this special catalyst in a lower temperature range of 300-400°C (red dash lines as shown in Fig.3). The results showed that the second cycle almost had the same direction to the first cycle but was more smooth. And the selectivity of hydrogen was higher than the first cycle at the same temperature, the selectivity of valley value of CO and peak value of CO_2 were both at 350°C.

3.3 Catalysts stability

The stability of the two special morphology catalysts Ni/CeO_2 with time-on-stream was examined at temperature of 350°C considering to combine the WGS reaction with the steam reforming reactions to increase H_2 and eliminate CO selectivity. It was really excited that the stability test of these two special morphology catalysts exceed 2000 hours and the discontinuous stability test with 13 times restarts was more than 600 hours.. The relationships of selectivity of each product and conversion of ethanol with time-on-stream over catalyst Ni/CeO_2 were shown in Fig.4. Under the experimental conditions, the first 100h conversion of ethanol was of 100mol.%, the first 1400h of 80mol.%, 1600-2000h of 70mol.%.The selectivity of H_2 was average above 70mol.% in the whole 2000h test. The selectivity of CO was much balanced and averaged at 2.1mol% during the whole stability test. The outcome velocity of products gas kept slowly increasing during the first 100 hours and down to a valley value at 1450 hours, then become recover in the last 100 hours, the average value of the gas outcome is about 0.38ml s^{-1}. So the yield of H_2 could be calculated to be 50.1mol%.

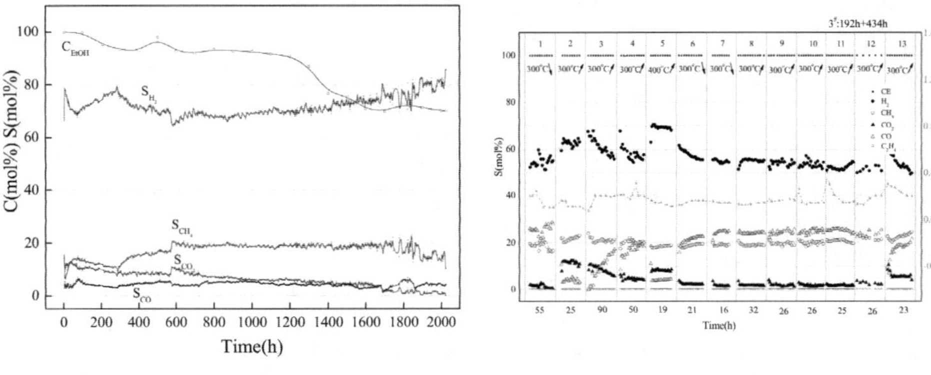

(a) (b)

Fig.4 Conversion of ethanol and selectivity of products as function of time-on-stream, obtained over the special morphology NiO/CeO_2 catalyst.(a) one time test of 2000h, (b) 13 times tests of about 600h.Experimental conditions: mass of catalyst 0.55g, H_2O/EtOH mol ratio ***R***=3:1, liquid flow rate 0.05cm^3 min^{-1}, total flow rate of 6.8 standard liters/min gas hourly space velocity (GHSV)~ 1.0×10^5 hour $^{-1}$,***T*** = 350°C, ***P*** = 1atm.

The above results means very good activity of Ni/CeO_2 catalysts with flowerlike active supports for ethanol steam reforming, which were much better than Ni/Al_2O_3 and common morphology nickle-lanthumum oxide [10] in low temperature domain, and stabilities were also much better than the nano-sized balllike nickle-cerium[28]. One of the reasons that enhanced catalytic activity of the flowerlike Ni/CeO_2 a catalysts for ethanol steam reforming should be the flowerlike supports beneficial for gas transport due to its open 3D porous structure.

Cordierite ($2MgO\ 2Al_2O_3\ 5SiO_2$) ceramic carrier might had the most active cites caused coke formation. The temperature range of coke formation to Ni/CeO_2 was at about 350~425°C. So, it could be estimated that this catalysts with flowerlike active supports would have a higher H_2 yield and longer stability if its stability had been tested at 300°C avoiding the coke formation, but a little CO elimination was sacrifized.

4. Conclusions

The modified catalysts of NiO loaded on flowerlike microspheres CeO_2 had novel low ttemperature activation to catalyze ethanol steam reforming reactions with high H_2 selectivity, CO elimination, and very good stability more than 2000 hours at 350°C. For this special morphology catalysts, the low temperature (275-350°C) activity, H_2 selectivity and CO elimination were better than traditional morphology nickle based catalysts. Slight coke were still formed in the low temperature range, cordierite ($2MgO\ 2Al_2O_3\ 5SiO_2$) ceramic carrier might had the most active cites caused coke formation. The temperature range of coke formation to Ni/CeO_2 was at about 350~425°C. So, it could be estimated that catalysts with flowerlike active supports would have a longer stablity if its stability tests had been carried out at 300°C.

This special morphology catalyst of Ni/CeO_2 had to be practical catalysts to be used in ethanol steam reforming processors for fuel cell and EH (ethanol hydrogen) vehicle applications for their low temperature activity, high H_2 selectivity and CO elimination, especially the very long stability. Fast and efficient renewable fuel reforming is one of the critical steps in producing H_2 for the "hydrogen economy," and non-corn biomass ethanol is now the most available and economical renewable fuel. This process just realizes the ideal of "Energy from nature, make the usage of energy more concordant with nature".

Acknowledgements

We acknowledge the financial support by ICD projects (Renewable energy No.1) and China NSFC No.50606020.

References

[1] S. Freni, G. Maggio, S. Cavallaro, Ethanol steam reforming in a molten carbonate fuel cell: A thermodynamic approach . J. Power Sources 1996; 62: 67-73.

[2] Deluga GA, Salge JR, Schmidt LD, Verykios XE. Renewable hydrogen from ethanol by autothermal reforming. Science 2004 ;303 : 993-997.

[3] Amandusson H, Ekedahl LG, Dannetun H. Isotopic study of ethanol dehydrogenation over a palladium membrane. J. Catal. 2000; 195: 376-382.

[4] Yee A, Morrison SJ, Idriss H. A study of the reactions of ethanol on CeO_2 and Pd/CeO_2 by steady state reactions, temperature programmed desorption, and in situ FT-IR. J. Catal. 1999; 186; 279-295.

[5] Haga F, Nakajima T, Yamashita K, Mishina S. Effect of crystallite size on the catalysis of alumina-supported cobalt catalyst for steam reforming of ethanol. Reaction Kinetics and Catal. Lett. 1998; 63: 253-259.

[6] Freni S. Rh based catalysts for indirect internal reforming ethanol applications in molten carbonate fuel cells. J. Power Sources 2001; 94: 14-19.

[7] Marino FJ, Cerrella EG, Duhalde S. Hydrogen from steam reforming of ethanol. Characterization and performance of copper-nickel suppored catalysts. Int. J. Hydrogen Energy 1998 ; 23 : 1095-1101.

[8] Fatsikoatas AN, Kondarides DI, Verykios XE. Steam reforming of biomass -derived ethanol for the production of hydrogen for fuel cell applications. Chem. Commun. 2001; 851-852.

[9] Aupretre F, Descorme C, Duprez D. Bio-ethanol catalytic steam reforming over supported metal catalysts.Catal. Commun. 2002; 3: 263-267.

[10] Sun J, Qiu XP, Wu F, Zhu WT, Wang J. H_2 fromsteamreform ing of ethanol at lowtemperature over Ni/Y_2O_3, Ni/La_2O_3 and Ni/Al_2O_3 catalysts for fuel-cell application. Int. J. Hydrogen Energy 2005; 30: 437 – 445.

[11] Sun J, Qiu XP, Wu F, Zhu WT, Wang WT, Hao SJ. Hydrogen from steam reforming of ethanol in low and middle temperature range for fuel cell application. Int. J. Hydrogen Energy 2004; 29: 1075-1081.

[12] Fatsikostas AN, Verykios XE. Reaction network of steam reforming of ethanol over Ni-based catalysts. J. Catal. 2004; 225: 439-452.

[13] Wang SB, Lu GQ. Reforming of methane with carbon dioxide over Ni/Al_2O_3 catalysts: Effect of nickel precursor. Appl. Catal. 1998; 169: 271-280.

[14] Ruckenstein E, Hu YH. Carbon dioxide reforming of methane over nickel/alkaline earth metal oxide catalysts. Appl. Catal. A 1995; 133: 149-161.

[15] Wang JB, Tai YL, Dow WP, Huang TJ. Study of ceria-supported nickel catalyst and effect of yttria doping on carbon dioxide reforming of methane. Appl. Catal. A 2001;218: 69-79.

[16] Zhang ZL, Verykios XE. A stable and active nickel-based catalyst for carbon dioxide reforming of methane to synthesis gas. J. Chem. Soc. Chem. Commun. 1995 : 71-72.

[17] Sun J, Luo DF, Xiao P, Li JG, Yu SS. High yield hydrogen production from low CO selectivity ethanol steam reforming over modified Ni/Y_2O_3 catalysts at low temperature for fuel cell application. J. Power Sources 2008; which is in press.

[18] Trovarelli A. Catal. Catalytic properties of ceria and CeO2-containing materials . ReV. Sci. Eng. 1996; 38: 439-520.

[19] Carrettin S, Concepcion P, Corma A, Nieto JML, PuntesVF. Nanocrystalline CeO2 increases the activity of an for CO oxidation by two orders of magnitude. Angew. Chem., Int. Ed. 2004; 43: 2538-2540.

[20] Park SD, Vohs JM, Gorte RJ. Direct oxidation of hydrocarbons in a solid-oxide fuel cell . Nature 2000; 464: 265-267.

[21] Zhan ZL, Barnett SA. An octane-fueled olid oxide fuel cell .Science 2005; 308: 844-847.

[22] Patil S, Kuiry SC, Seal S. Nanocrystalline ceria imparts better high- temperature protection. Proc. R. Soc. London, Ser. A 2004; 460: 3569-3587.

[23] Gupta S, Brouwer P, Bandyopadhyay S, Patil S, Briggs R, Jain J, Seal SJ. Nanosci. Nanotechnol. 2005, 5: 1101.

[24] Wang JA, Dominguez JM, Montoya A, Castillo S,Navarrete J, Moran-Pineda M, Reyes-Gasca J, Bokhimi X. New insights into the defective structure and catalytic activity of Pd/ceria .Chem. Mater. 2002; 14: 4676-4683.

[25] Sun J, Wu F, Qiu XP, Wang F, Hao S, Liu Y. Hydrogen Production from ethanol steam reforming over Ni/ Al_2O_3 and Ni/ La_2O_3 catalysts at low temperature. Chin. J. Catal. 2004; 25: 551-555.

[26] Sun CW, Sun J, Xiao GL, Zhang HR, Qiu XP, Li H, Chen LQ. Mesoscale Organization of Nearly Monodisperse Flowerlike Ceria Microspheres. J. Phys. Chem. B 2006;110: 13445-13452.

[27] Deng XZ, Sun J, Yu SS, Xi JY Zhu WT, Qiu XP.Steam reforming of ethanol for hydrogen production overNiO/ZnO/ZrO_2 catalysts. Int. J. Hydrogen Energy 2008; 33: 1008-1016.

[28] Wang J, Sun J, Qiu XP, Wu ZH, Chen LQ.Hydrogen production from steam reforming of ethanol using Ni/CeO_2 and Ni-Cu/CeO_2 catalysts. Chin. J. Power Sources 2006; 30: 234-237.

45d Ni-Fe Binary Metal Catalysts for Hydrogen Production Via Auto-Thermal Reforming of Ethanol

Lihong Huang[1], Jian Xie[1], Rongrong Chen[1], Deryn Chu[2], and Andrew T. Hsu[1]. (1) Richard G. Lugar Center for Renewable Energy, Indiana University - Purdue University Indianapolis, 799 West Michigan Street, ET 114, Indianapolis, IN 46202, (2) U.S. Army Research Laboratory, 2800 Powder Mill Rd., Adelphi, MD 20854

Auto-thermal reforming (ATR) of bio-ethanol is a promising process for hydrogen production, because of its self-sustainment for reaction heat. Ni-based alumina supported catalysts has been extensively studied in ATR of ethanol; Hydrogen selectivity, coke deposition, as well as carbon-containing product distribution are main challenges to be addressed.

Ni-Fe binary catalysts have been studied, and show a higher activity and stability in ATR of ethanol for hydrogen production. Over an optimized Ni-Fe/Al2O3 catalyst, hydrogen selectivity has been increased by a factor of 70.52 %, 81.47 %, and 47.44 % at 400 „aC, 500 „aC and 600 „aC, respectively, compared with the un-promoted catalyst. Moreover, the optimized catalyst shows a significantly improved stability without obvious loss in catalytic performance during a durability test at 600 „aC, while the catalyst without iron promoter performs poorly with remarkable decrease (about 60 %) in hydrogen selectivity within 20 hours. For carbon selectivity, meanwhile, higher carbon dioxide yield has been achieved over iron promoted catalysts, i.e. 54.3 % (promoted sample) vs. 9.6 % (un-promoted sample). For the ratio of dioxide to monoxide, iron promotes a higher selectivity of dioxide than that of monoxide, with an approximately double gain in the ratio from 0.93 to 1.82, which can be attributed to the high activity of iron in the water-gas shift reaction (WGSR). Furthermore, over the un-promoted catalyst, the dehydration of ethanol is dominant in ATR of ethanol; as a result, ethylene becomes the main carbonaceous product and reached 76.7 % on carbon basis, which can result in carbon deposit on the catalyst surface via polymerization and dehydrogenation. By comparison, only about 2.5 % of ethylene selectivity has been found over the promoted sample, and coke deposition is constrained. The zirconia and ceria contained Ni-Fe binary catalysts also show a comparable performance, and the lower selectivity to carbon monoxide and ethylene are observed, which shows the improvement in WGSR and the restrain on dehydration of ethanol.

The characterization results indicate that the active component in Ni-based catalysts is the partially reduced spinel phase NiAl2O4, which favors the ethanol dehydration reaction route, and results in a high ethylene selectivity and a low hydrogen selectivity at low temperatures. Within the Ni-Fe binary catalysts, the mixed crystals in spinel phase as NiAl2O4-FeAl2O4 are formed, where the catalysts' structure and electronic properties are modified: NiAl2O4 crystal is distorted because of heterogeneous iron atoms coordination, and more defect sites are formed, which results in changes in reduction as well as reactants adsorption and activation, and in an improvement of activity. Meanwhile, FeAl2O4 phase in mixed crystals promotes the ethanol dehydrogenation reaction. Furthermore, more acid sites on Al2O3 surface are also covered by these new species. Consequently, the synergistic effect of the NiAl2O4-FeAl2O4 mixed crystal induces the reaction network into the dehydrogenation route, which is coupled with a high activity of partially reduced NiAl2O4 for acetaldehyde transformation through steam reforming, partial oxidation, and water-gas shift reaction processes. This route preference explains the high selectivity to hydrogen and C1 products of Ni-Fe binary metal catalysts.

DESIGN AND PILOT SCALE TESTS OF MONOLITHIC CATALYSTS FOR AUTOTERMAL/STEAM REFORMING OF NATURAL GAS AND BIOFUELS

Sergey Beloshapkin[a], *Julian Ross*[b], *Vladislav Sadykov*[c], *Natalia Mezentseva*[c], *Galina Alikina*[c], *Zakhar Vostrikov*[c], *Yulia Fedorova*[c], *Vladimir Pelipenko*[c], *Vladimir Usoltsev*[c], *Sergey Tikhov*[c], *Oleg Smorygo*[d], *Vladimir Ulyanitskii*[e]

[a]*Materials and Surface Science Institute, University of Limerick, Ireland*

[b]*Centre of Environmental Research, University of Limerick, Limerick, Ireland*

[c]*Boreskov Institute of Catalysis, Novosibirsk, Russia*

[d]*Powder Metallurgy Institute, Minsk, Belarus*

[e]*Lavrentiev Institute of Hydrodynamics, Novosibirsk, Russia*

Introduction

Transformation of fuels (fossil fuel, biofuels) into syngas or hydrogen is one of the most important tasks of catalysis in the energy-related fields. Catalysts comprised of precious metals and/or Ni supported on complex oxides with a high lattice oxygen mobility are known to be very efficient and stable to coking in autothermal/steam reforming of a variety of fuels including oxygenates. Monolithic substrates with a good thermal conductivity are promising for providing an efficient heat transfer within the reactor to prevent emergence of hot spots/cool zones deteriorating performance. This work presents results of research aimed on design of such catalysts and characterization of their performance parameters at the pilot-scale level.

Experimental

Nanocomposite active components were comprised of complex fluorite-like oxides $Ln_x(Ce_{0.5}Zr_{0.5})_{1-x}O_{2-y}$ (Ln = La, Pr, or Sm, x=0-0.3) or perovskite –like oxides $La_{0.8}Pr_{0.2}Mn_{0.2}Cr_{0.8}O_3$ prepared via modified Pechini route [1-5]. In some cases, Y-doped zirconia (YSZ) was mixed with fluorite-like or perovskite-like oxides to form composites with improved sintering resistance [2, 4]. Precious metals (up to 1 wt.% of Pt, Pd, Ru), NiO or $LaNiO_3$ (up to 10 wt. %) were supported by the incipient wetness impregnation. Details of these active components synthesis and characterization are given elsewhere [1-5].

Several types of heat-conducting substrates including compressed Ni-Al foam, fechraloy foil or gauze (Fig. 1) protected by corundum layer (either segregated by controlled oxidation or supported via blast dusting) and Cr-Al-O microchannel cermets were used [5-8]. Stacked flat and corrugated bands of foil were winded into the Arkhimed spiral forming triangular channels (Fig. 1). After spark-welding of tungsten rods as electrical current leads, such a monolithic substrate was also used for evaporation of liquid fuels sprayed via a nozzle ensuring also a good mixing of fuel vapors with steam and air (Fig. 2). To prepare structured/monolithic catalysts, different substrates were washcoated with a slurry of mixed oxides with addition of peptizers and surfactants [5-8]. Basic characteristics of structured catalysts studied in this work are presented in Table 1.

Monolithic catalysts were tested in the reactions of autothermal (ATR)/steam (SR) reforming of natural gas, acetone, ethanol, anisol and sunflower oil using flow kinetic installations and specially designed reactors including a pilot tubular stainless steel reactor equipped with the electric current heated fuel/water evaporation unit and heating coils to change the catalyst temperature up to 1000 ^{o}C.

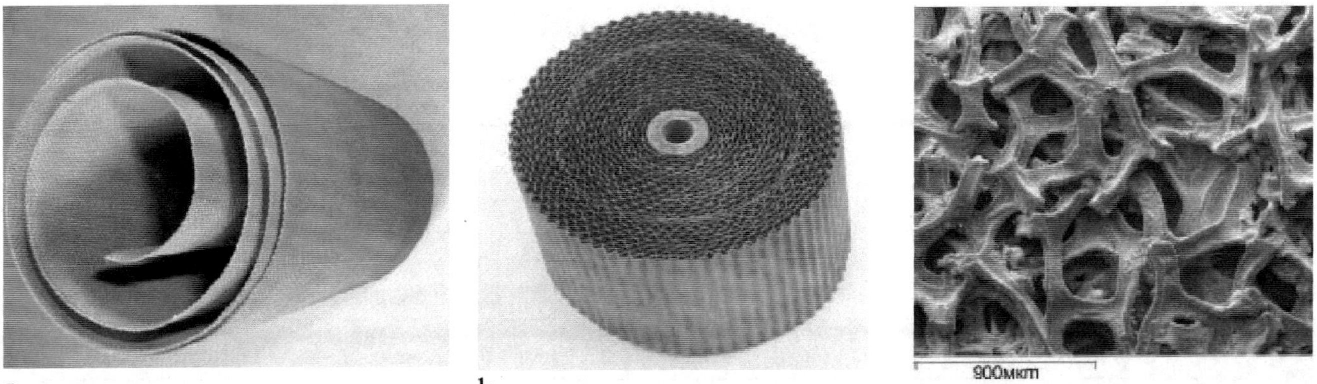

a b c

Fig. 1. Typical images of heat-conducting substrates: a-Fechraloy gauze, b- Fechraloy thin foil monolithic substrate, c-Ni-Al compressed foam substrate [6-8].

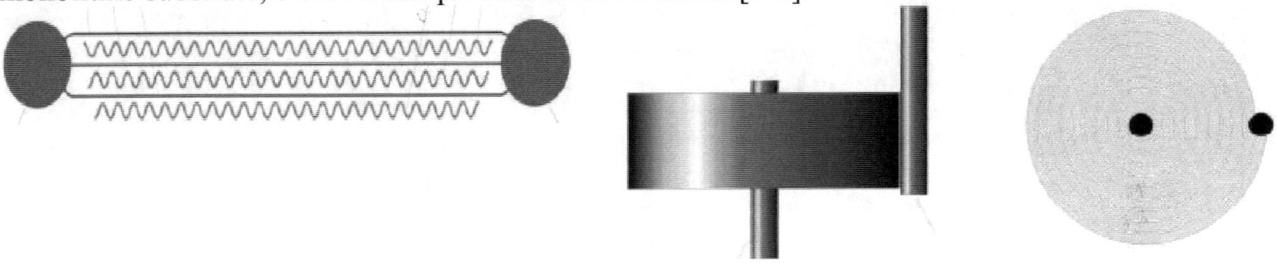

a b c

Fig. 2. Monolithic Fechraloy foil-based evaporation/mixing monolithic unit heated by passing the electric current. a- a general scheme with two tungsten rods as current leads, b-side view, c-top view.

The molar ratios of H_2O/C and O_2/C in the feeds were varied in the range of 1-7 and 0-0.5, respectively; the gas flow rate was up to 2 m^3/h.

Results and discussion

Methane

For fluorite-like complex oxides supported on corundum-protected Fechraloy foil or gauze substrates and promoted by precious metals and/or lanthanum nickelate, stable performance without deterioration due to coking even in stoichiometric concentrated feed for days-long tests was observed. At the same type of active component, a higher performance was observed for catalysts on gauze substrates (Fig. 3) due to more developed surface area. Addition of Ru to Pt only decreases performance (cf. catalysts 1 and 3). $LaNiO_3$ promoted by Pt demonstrates a good performance in the case of its combination both with Pr-Ce-Zr-O and La-Ce-Zr-O fluorite-like oxides (cf. catalysts 2 and 6, Fig. 3). The increase of contact time improves syngas yield (Fig. 4).

Nanocomposite active component comprised of complex perovskite-like oxide combined with YSZ and promoted by Ni also demonstrates a high activity and performance stability in CH_4 SR at short contact time in the intermediate temperature (IT) range (Fig. 5) which is important for such application as internal reforming in IT solid oxide fuel cells [8].

Ethanol

In the reaction of ethanol steam reforming, the highest performance was demonstrated by Ni-containing catalysts (Fig. 6), ensuring a high yield of syngas even at short contact time. Though syngas

Table 1. Basic types of monolithic catalysts

N°	Sample Description	
	Type of substrate	Active component (wt.%)
1	FeCrAl gauze (cylinder D 15 mm, L 26 mm)	5.9 Pr-Ce-Zr-O +0.92 Pt
2	The same as 1	5.9 Pr-Ce-Zr-O +7 $LaNi(Pt)O_3$
3	The same as 1	6.6 Pr-Ce-Zr-O+1.3 Pt+0.3 Ru
4	FeCrAl foil 20 μm (cylinder D 15 mm, L 26 mm)	5.5 Pr-Ce-Zr-O +1.2 Pt
5	The same as 4	6.3 Pr-Ce-Zr +1.0 Pt
6	FeCrAl gauze (the same as 1)	5.4 La-Ce-Zr-O + 10 $LaNi(Pt)O_3$
7	Ni-Al compressed foam (plate 10x20x1 mm)	4% (50 $La_{0.8}Pr_{0.2}Mn_{0.2}Cr_{0.8}O_3$ + 30 NiO + 20 YSZ +0.7Ru)
8	$CrAlO_y/CrAl_x$ cermet (cylinder D 19 mm, L 21 mm, 20 channels).	3.3 ($Pr_{0.15}Ce_{0.35}Sm_{0.15}Zr_{0.35}O_2$ + 1.5 Ru)
9	FeCrAl foil 200 μm (D 53 mm, L 12.5 mm)	2 Ce-Zr-O + 1 $LaNi(Pt)O_3$
10	FeCrAl foil 20 μm (D 50 mm, L 50 mm).	10 Ce-Zr-O + 7 $LaNi(Pt)O_3$
11	Corundum honeycomb monolith, (D 50 mm, L 50 mm)	10 Ce-Zr-O + 7 $LaNi(Pt)O_3$

content increases with the steam excess (Fig. 6, 7), performance was stable even for feeds with a small excess of steam. The excess of steam primarily increases the hydrogen content due to effect of water gas shift reaction (Fig. 7). Complex oxide active component promoted by Ru demonstrates also a high performance when supported on cermet substrate (Fig. 8, 9) without performance deterioration. This suggests that Cr present in the cermet is not involved in any side reactions leading to coke deposition. Similar to the case of active components supported on FeCrAl substrates, main by-product here is CH_4 formed via cracking of ethanol. A weak variation of CH_4 content with the steam excess implies that at short contact times CH_4 steam reforming proceeds slower than ethanol transformation. The increase of ethanol content in the feed provides a high yield of syngas even at short contact times (Fig. 9). At long contact times and high temperatures (Fig. 10), complete transformation of ethanol into syngas is apparently approached due to steam reforming of residual methane as well. For oxy-steam reforming of ethanol which is attractive from the heat management point of view, catalysts on monolithic heat-conducting substrates with nanocomposite active components also provide a high yield of syngas at short contact time (Fig. 11). In this case, nearly constant H_2/CO ratio ~ 1 can be achieved at a broad variation of H_2O/ethanol ratio.

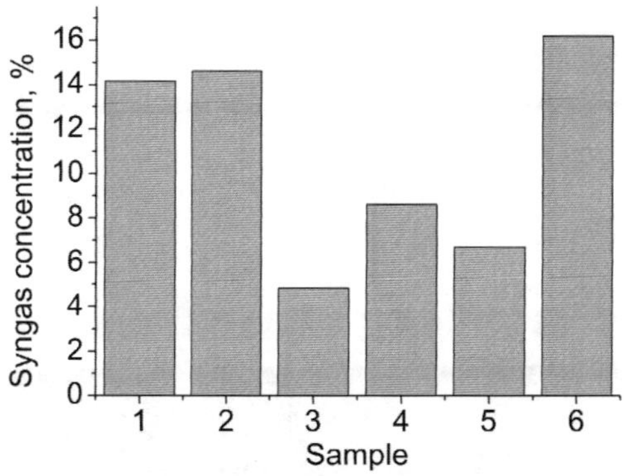

Fig. 3. Comparison of performance of catalysts (numbering as in Table 1) in CH_4 SR. Feed 50% CH_4 +50% H_2O, contact time 0.06 s, T_{inlet} 750 °C.

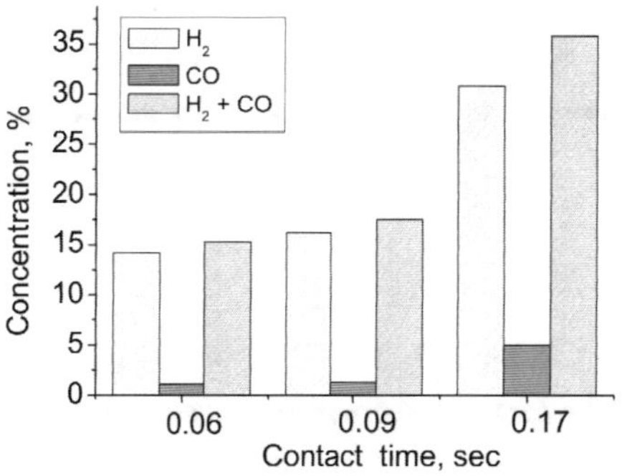

Fig. 4. Effect of contact time on products content in CH_4 SR on catalyst 2. Feed CH_4 + H_2O (CH_4/H_2O = 0.8), T_{inlet} 750 °C.

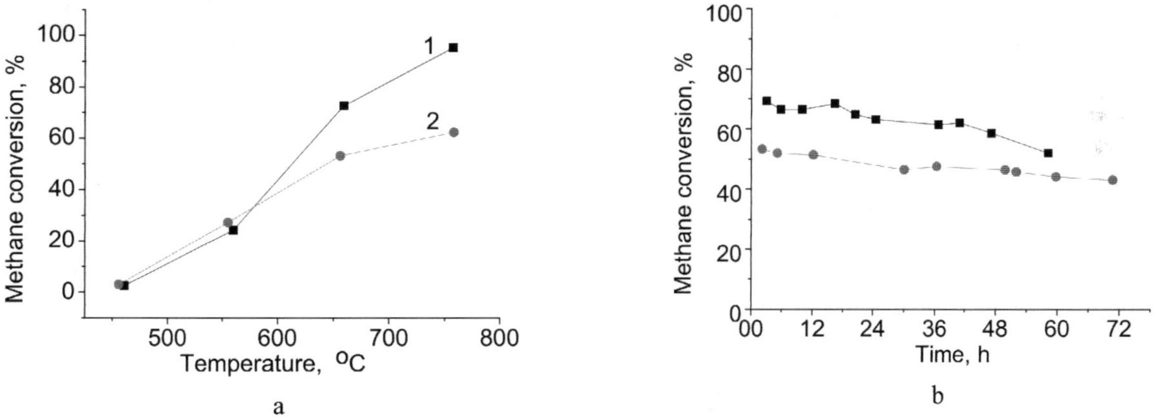

Fig. 5. Temperature dependence of CH_4 conversion (a) and its variation with time-on-stream at 650 °C (b) for the 0.25 mm fraction of nanocomposite active component 50% $La_{0.8}Pr_{0.2}Mn_{0.2}Cr_{0.8}O_3$ + 30% NiO + 20% YSZ +0.7Ru (1) and catalyst 7 (Table 1) with the same active component (2). Contact time 10 ms for (1) and 20 ms for (2), feed composition 20% CH_4 +40% H_2O, Ar balance.

Acetone

Monolithic catalysts designed in this work were found to be also very efficient and stable to coking in the reaction of acetone steam reforming (Fig. 12). This is very important since acetone as by-product of bio-oil steam reforming is known to be responsible for coking of traditional catalysts. In this reaction, $LaNiO_3$-containing active components also provide the highest performance. Syngas yield increases with the contact time achieving a high level already at τ~0.2 s, which is certainly attractive for the practical application. The same types of nanocomposite active components are efficient and stable in acetone transformation into syngas even when supported on monolithic corundum substrate (Fig. 13). Similar to the case of ethanol steam reforming, main by-product here is also CH_4. Oxygen addition into the feed does not affect syngas yield but helps to improve the heat balance and ensures performance stability even at moderate excess of steam which is attractive from the process economy point of view.

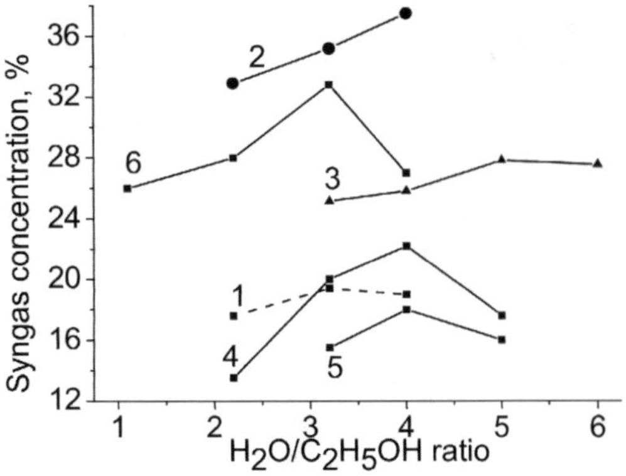

Fig. 6. Effect of the H_2O/C_2H_5OH ratio in the feed on exit syngas (H_2+CO) concentration in ethanol SR on monolithic catalysts (Table 1). Feed 10% EtOH + H_2O + N_2 balance, 750 °C, contact time 0.085 sec.

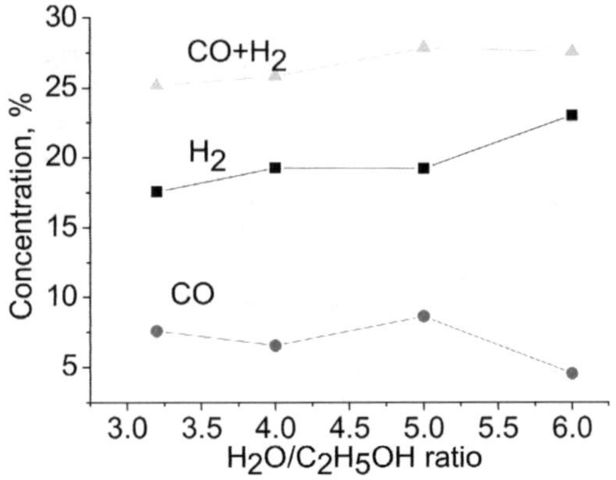

Fig. 7. Effect of the H_2O/C_2H_5OH ratio in the feed on product concentrations in ethanol SR on catalyst 3 (Table 1). 750 °C, contact time 0.085 s, feed composition 10% EtOH + H_2O + N_2 balance.

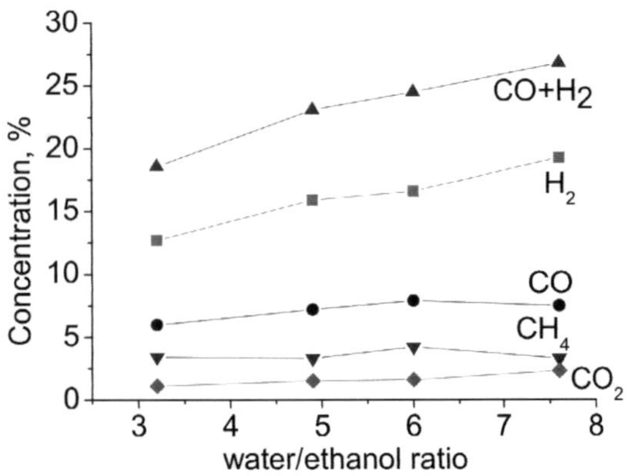

Fig. 8. Effect of the H_2O/C_2H_5OH ratio in the feed on product concentrations in ethanol SR on catalyst 8 (Table 1). 750 °C, contact time 0.1 s, feed composition 10% EtOH + H_2O + N_2 balance

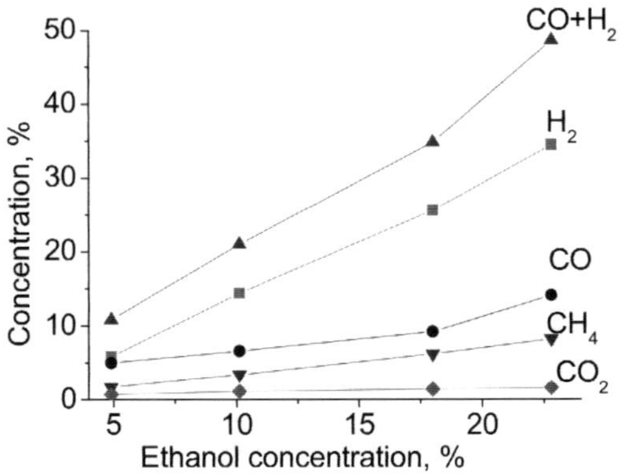

Fig. 9. Effect of C_2H_5OH content in the feed on product concentrations in ethanol SR on catalyst 8 (Table 1). 750 °C, contact time 0.1 s, feed composition EtOH + H_2O + N_2 balance, H_2O/EtOH =3.2, 750 °C, contact time 0.1 s.

Anisole, sunflower oil

For these feeds a high and stable performance of developed catalysts was demonstrated as well. Some typical results are shown in Figs. 14 and 15. Due to well-known high coking ability of these fuels, their stable performance was obtained only in the case of oxygen addition to the feed (oxy-steam reforming). Moreover, only unique design of evaporation and mixing unit used in this work allowed to obtain stable and reproducible results preventing cracking of fuels in supplying lines. The most efficient performance in transformation of these heavy fuels was provided by combination of catalysts based on heat-conducting metallic substrates (placed in the front part of the layer, efficiently transfer heat generated due to fast oxygen consumption) and those based on corundum monolithic substrates (placed in the rear part, provide a high conversion level due to developed surface area).

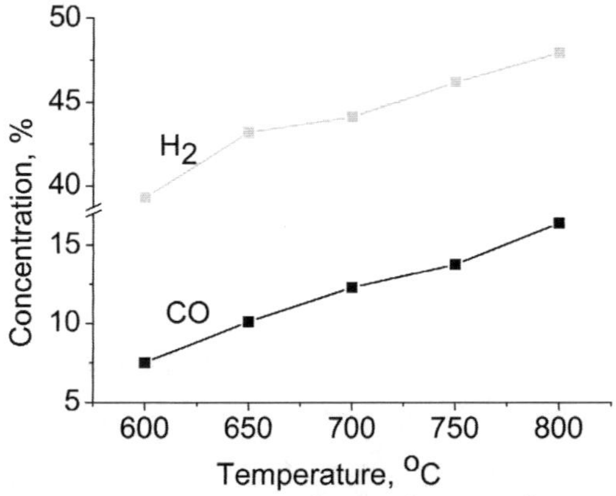

Fig. 10. Temperature dependence of H_2 and CO concentrations in products of ethanol SR for catalyst 4 (Table 1). Feed composition 10 % C_2H_5OH + 40% H_2O +50% N_2, contact time 6.5 s.

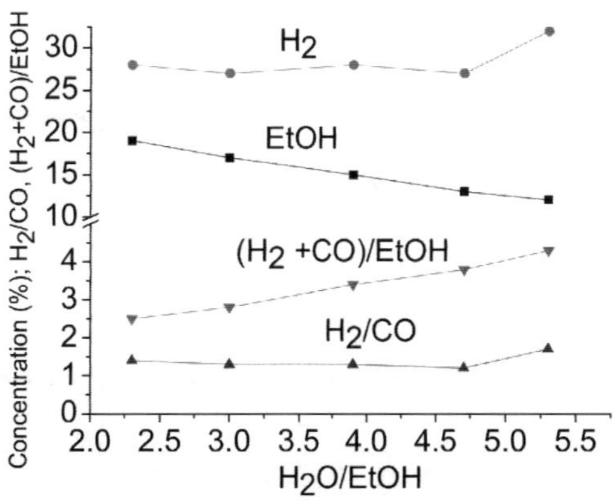

Fig. 11. Effect of H_2O/C_2H_5OH ratio on product concentrations in oxysteam reforming on catalyst 10 (Table 1). Inlet feed: 7% O_2+ H_2O + EtOH+N_2 balance, T inlet 700 °C, contact time 0.3 s.

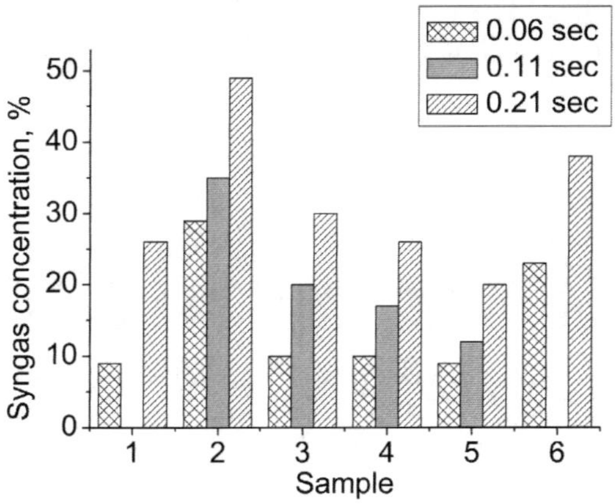

Fig. 12. Effect of contact time on syngas concentration in converted feed for acetone steam reforming on monolithic catalysts (Table 1). Feed composition 10% acetone +50% H_2O+ 40% N_2, inlet temperature 750 °C.

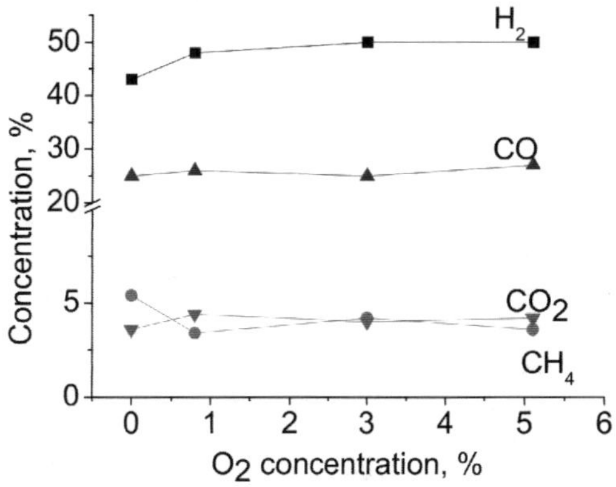

Fig. 13. Effect of oxygen content in the feed on products concentration in acetone oxysteam reforming on catalyst 11 (Table 1). Contact time 0.5 s, inlet feed 24% acetone + 48% H_2O + O_2, N_2 balance, T inlet 600 °C, T outlet 700 °C

Conclusions

Structured catalysts comprised of complex fluorite-like/perovskite- like oxides supported on heat-conducting metal/cermet substrates and promoted by precious metals and/or Ni possess a high and stable performance in steam reforming of CH_4 and biofuels. This provides new possibilities for transformation of a broad range of a variety of fossil and biofuels into syngas for the energy field.

Acknowledgments

This Work is supported by INTAS 05-1000005-7663, INTAS YSF 06-1000014-5773, SOFC 600 FP6 EC Projects and Integration project 57 of SB RAS –Belarus National Academy of Sciences.

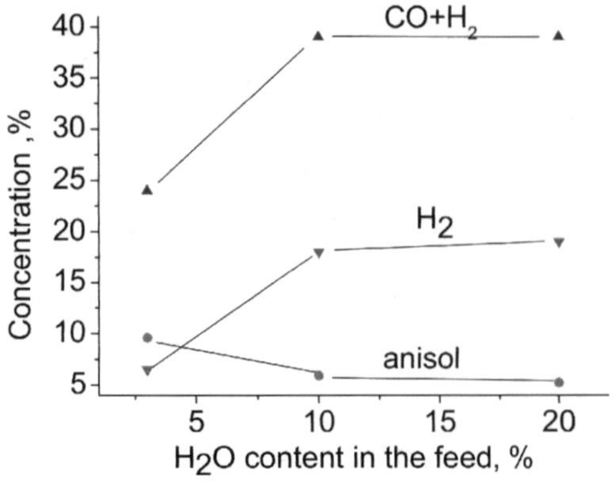

Fig. 14. Effect of H_2O content in the feed on product concentrations in the oxysteam reforming of anisole on catalyst 10 (Table 1). Contact time 0.06 s, T exit 850 ^{0}C, feed composition 18% O_2 + anisole + H_2O +N_2 balance.

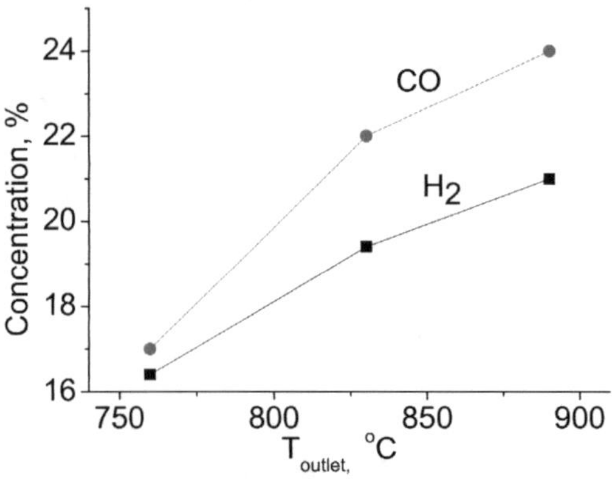

Fig. 15. Effect of catalyst temperature on the product concentration in the oxysteam reforming of sunflower oil on stacked layer of catalysts 10 (front part) and 11 (rear part). Feed composition 0.7% of sunflower oil + 15% H_2O +20% O_2, N_2 balance, contact time 0.2 s.

References

1. P. Yaseneva, S. Pavlova, V. Sadykov, G. Alikina, A. Lukashevich, S. Belochapkine, J. Ross, (2008), "Combinatorial approach to the preparation and characterization of catalysts for biomass steam reforming into syngas", *Catalysis Today*, 137, pp. 23-28
2. V. Sadykov, N. Mezentseva, A. Smirnova, J. Irvine, O. Vasylyev (2008), "Effect of complex oxide promoters and Pd on activity and stability of Ni/YSZ (ScSZ) cermets as anode materials for IT SOFC", *Catalysis Today,* 131, pp. 226–237.
3. V. Sadykov, N. Mezentseva, A. Smirnova, J. Irvine (2007), "Doped Nanocrystalline Pt-Promoted Ceria-Zirconia as Anode Catalysts for IT SOFC:Synthesis and Properties", *Mater. Res. Soc. Symp. Proc.,* 1023, pp. JJ02-07.1-6.
4. V. Sadykov, N. Mezentseva, J. Ross, (2007), "Pt-supported nanocrystalline ceria-zirconia doped with La, Pr or Gd: factors controlling syngas generation in partial oxidation/autothermal reforming of methane or oxygenates", *Solid State Phenomena*, 128, pp. 239-248.
5. V. Sadykov, S. Pavlova, S.Tikhov, V. Usol'tsev, V. Parmon (2005), "Selective Oxidation of Hydrocarbons into Synthesis Gas at Short Contact Times: Design of Monolith Catalysts and Main Process Parameters",. *Kinetics and Catalysis,* 46, pp. 227–250.
6. S. Tikhov, V. Usoltsev, V. Sadykov (2006), "CrAl alloy –based cermet monolith with polymodal pore structure for partial oxidation of methane to synthesis gas", *Studies in Surface Science and Catalysis,* 162, pp. 641-648.
7. V. Sadykov, S. Pavlova, S. Tikhov (2007), "Design of structured catalysts based on metallic monoliths for syngas production via partial oxidation of natural gas", *Studies in Surface Science and Catalysis* 172, pp. 241-244.
8. V. Sadykov, N. Mezentseva, V. Pelipenko, O. Smorygo, B. Rietveld (2008), "Anode Materials for IT SOFC Based on NIO/YSZ Doped with Complex Oxides and Promoted by Pt, Ru or Pd: Properties and Catalytic Activity in the Steam Reforming of CH_4", *Proc. 8th European SOFC Forum, 30 June – 4 July 2008 – Lucerne/Switzerland, A0526 .1-6 (CD).*

Thermodynamic and Kinetic Analysis of the Thermal Partial Oxidation Process of Ethanol

Z. Al-Hamamre
University of Jordan, Faculty of Engineering and Technology
Chemical Engineering Department
11942 Amman, Jordan
Tel. 00962/(0)6 5355000/ Ext: 22895

Email: z.hamamre@ju.edu.jo

Abstract

In this study thermodynamic and kinetic analysis of the Thermal partial oxidation of ethanol with the aim of producing syngas (a CO - H_2 rich mixtures) is performed. Equilibrium and kinetic calculations are performed in order to find the limiting parameters for the thermal partial oxidation. The effect of air ratio and mixture inlet temperatures on the reforming efficiency (syngas yield), the reaction progress and the equilibrium time are investigated. The optimum practical operating conditions of the partial oxidation process of ethanol are identified.

It is found that the reforming efficiency the reforming efficiency within the O/C range from 0.3 to 1.65 increases with increasing the process temperature and reach a maximum value at $(O/C)_{max}$. beyond this value, the efficiency decreases with increasing the O/C ration and after O/C = 0.9, it become independent of the mixture inlet temperature. The value of $(O/C)_{max}$ is shifted to lower C/O value at higher reaction temperature. The kinetic simulations suggest that three different stages exist during the partial oxidation process of ethanol: oxidation region, water gas shit reaction- reforming region and reforming region. The reforming reactions in the 3rd region are the reaction process limiting step. Additionally, it is found that increasing the pressure reduces the oxidation region and retired the progress of hydrogen and CO formation. Due to preheating limitations (self ignition and reactor material stability) and the kinetic behavior of the TPOX process of ethanol, practical operating conditions could be bounded in the O/C range of 1.2 to 1.35 and mixture inlet temperature range of 650K to 850K.

Key words: Ethanol, Kinetic, Thermodynamic, Thermal Partial Oxidation (TPOX), Syngas

Introduction

The consumption of fossil fuel is continuously growing and fuel prices are sharply fluctuating. These prices are directly related not only to the political situation in oil producing countries but also to the political relation between oil producing countries and oil consuming countries. In addition, because of the exhaust gases associated with the use of such fuels, which threaten the environment and mankind, and due to the limited resources which may approach their end, new alternatives in the field of power supply are needed that make less use of the world's fossil energy resources and convert energy in a cleaner way.

The use of biofuels can significantly reduce both the dependency on fossil fuels and the amount of greenhouse gases emitted to the atmosphere. The potential for biofuels to be considered to be "carbon neutral" depends upon the carbon that is emitted being reused by further plant growth. Ethanol, as a biofuel, is a clear, colorless alcohol fuel produced from sugars found in grains, such as corn, sorghum, and wheat, as well as potato skins, rice, and yard clippings [1]. The use of ethanol as a source for syngas can not only reduce the emissions to the atmosphere and the dependency on fossil fuels, but also increase the efficiency of the energy utilization process [2,3].

Syngas can be produced from hydrocarbon fuel by several processes: Steam reforming, partial oxidation and auto-thermal reforming. Compared to other reforming technologies, the exothermic partial oxidation process has several advantageous characteristics [4, 5]. There is no need for external heat sources and additional fluids like water. It has a good dynamic response and it can handle light as well as heavy hydrocarbon fuels. Also, the use of inert porous material to run the soot free process and the high flame stability that can be achieved utilizing this technology increases the interest in investigating the thermal partial oxidation process of hydrocarbon fuels [6, 7, 8, 9, 10]. Most of the work that have been reported in the literature concerned with hydrogen production from ethanol using catalytic processes, [2, 11-17]. However, no work has been done concerned with ethanol reforming by thermal partial oxidation.

In this work, thermodynamic and kinetic analysis of the thermal partial oxidation process of ethanol is performed. The reaction temperature, the reformat composition and the reforming efficiency are estimated for different ethanol-air mixture temperatures and compositions (different C/O ratios). Additionally, the reaction progress with time and the equilibrium time are also predicted in the kinetic analysis part of this work. The simulations in this work are performed using a limited test license CHEMKIN-PRO program. The program is the newest version of CHEMKIN [18, 19, 20]. The kinetic scheme and the thermodynamic data of [21] are used in the analysis.

THERMODYNAMIC AND KINETIC ANALYSIS

By means of equilibrium calculations the conditions can be investigated, under which high syngas yield can be reached. Kinetic calculations deliver the information, how fast the reactions proceed under different conditions, e.g. the C/O rations and the operating temperature. By the coupling of the thermodynamic and kinetic examinations an optimum set of parameters can be found. In these calculations the mixing process is not considered, i.e. a perfect gas mixture is assumed.

Thermodynamic Analysis of Ethanol Thermal Partial Oxidation (TPOX)

In order to determine if the partial oxidation of ethanol for syngas production is possible and to what extent these reactions can progress, a thermodynamic study is performed. The analysis is also used to identify the appropriate operating parameters and to investigate the product composition. The thermodynamic analysis is performed using the EQUIL code of CHEMKIN PRO. For combustion systems, EQUIL can be used to determine adiabatic flame temperatures.

For ethanol, the global partial oxidation reaction is described by the following equation:

$$C_2H_5OH + 0.5(O_2 + 3.77N_2) \rightarrow 2CO + 3H_2 + 1.88N_2 \tag{1}$$

The maximum syngas (H_2 and CO) concentrations can be achieved at $\lambda = 0.15$ at which y_{H2}= 43.6 % and y_{CO} = 29.1 %. However, other factors like the possible operating temperature, pressure and the time needed for the process play a major role in determining the practical operating conditions

Effect of mixture inlet temperature and oxygen concentration

Fig. 1 shows the effect of the preheating temperature on the adiabatic flame temperature of methane/air mixture.

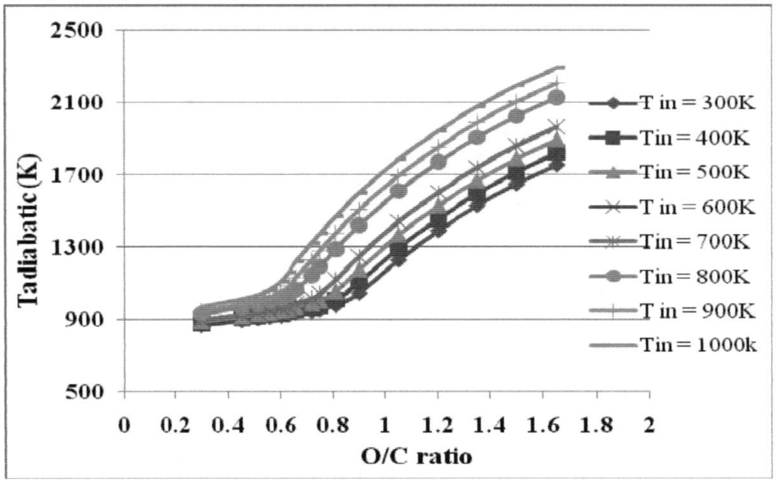

Fig. 1: The adiabatic flame temperature for ethanol/air mixture as a function of the O/C ratios and different mixture inlet temperature.

For an O/C ratio of 0.3, an air preheating temperature of 1000 K will result in an adiabatic flame temperature of 1011 K. Higher adiabatic flame temperatures require higher preheating temperatures at which the risk of self ignition in the mixing volume increases. For O/C = 1.2 to 1.35, a 800 K air temperature results in adiabatic flame temperatures of 1760 K and 1900 K, respectively.

On the other hand, with higher O/C values the hydrogen yield and the reforming efficiency decrease as can be seen in Fig. 2, which shows the reforming efficiency as a function of the O/C ratio for different mixture inlet temperatures. The reforming efficiency can be expressed in several ways, depending on the complexity of the process. It could be expressed interims of the heating value of the produced reformate gas including the enthalpy of the process streams. More details are available in [22]. Since this woke concerned with TPOX process of ethanol to produce syngas (a mixture of CO and H_2), the reforming efficiency is defined as the ratio between the heating values of the gas mixture which is produced from the reforming process to the heating value of the original fuel, as shown in Equ. (2),

$$\eta = \frac{(n_{H_2} * LHV_{H_2} + n_{CO} * LHV_{CO})}{n_{fuel} * LHV_{fuel}} \qquad (2)$$

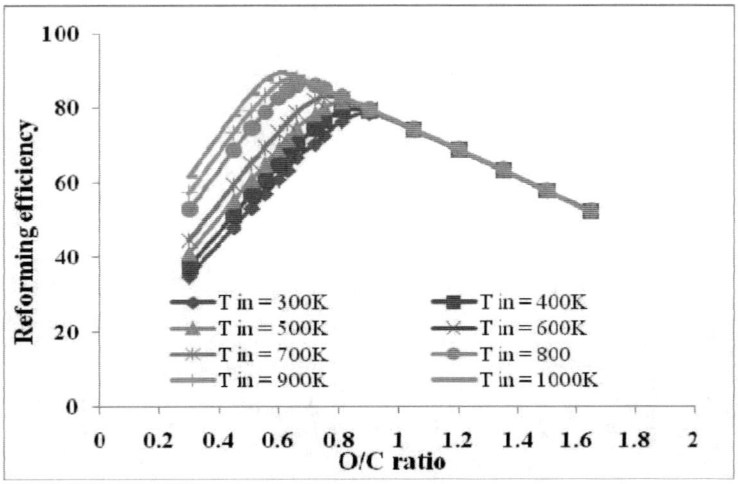

Fig. 2: The effect of the oxygen/carbon ratio on the reforming efficiency for different mixture inlet temperature.

It can be seen that O/C ratio less than 0.9, the reforming efficiency increases with increasing the process temperature and reach a maximum value at $(O/C)_{max}$. after this value, the reforming efficiency decreases with increasing the O/C ratio. For O/C values higher than or equal to 0.9, the reforming efficiency is nearly independent on temperature and has approximately constant value at a given O/C ratio.

This behavior can be explained by considering Fig. 3 (left) and 3 (right) which shows the effect of oxygen concentration on the mole number of H_2 and CO, respectively, and for different mixture inlet temperatures. The mole number is defined as the mole of species produced divided by the initial mole of fuel used. The mole number of H_2 and CO increases with increasing the O/C ratio and reach a maximum at $(O/C)_{max}$, after which they decrease with increasing the O/C ratio. The $(O/C)_{max}$ depends on the mixture inlet temperature. At a mixture inlet temperature of 300K, the $(O/C)_{max} = 0.9$ at which the mole number of H_2 = 2.52 mole H_2 /mole fuel and that of CO is 1.64 mole CO /mole fuel. The $(O/C)_{max}$ ratio decreases to 0.66 at mixture inlet temperature of 800K and the corresponding mole number of H_2 is 2.74 (mole/mole fuel) and that of CO is 1.83 (mole/mole fuel). Moreover, for O/C ratio higher than $(O/C)_{max}$, the mole number of H_2 and CO become independent on the mixture inlet temperature. Additionally, for O/C ratio < $(O/C)_{max}$, the mole number of H_2 and CO increase with increasing the mixture inlet temperature, while for O/C ratio > $(O/C)_{max}$, the mole numbers are independent on the mixture inlet temperature.

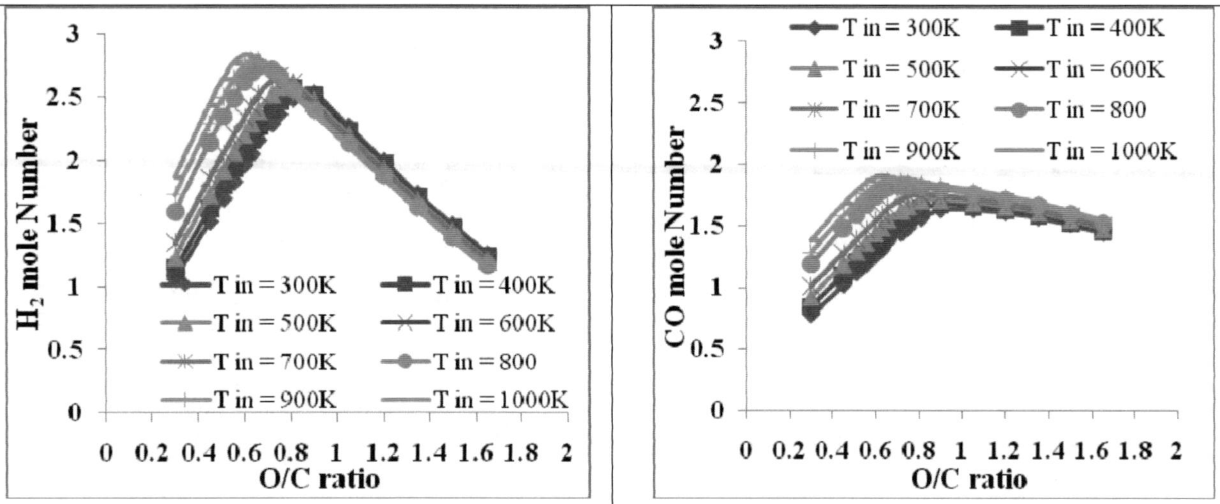

Fig. 3: The effect of the (O/C ratio) on the H_2 (left) and CO (right) mole numbers.

The behavior of the mole number of H_2 and CO presented in Fig. 3 shows the importance of the inlet mixture temperature in enhancing the conversion of ethanol to produce syngas. Additionally, the high inlet mixture temperature reduces the formation of hydrocarbons such as methane, this is shown in Fig. 4 which shows the mole number of CH_4 as a function of the O/C ratios at different inlet mixture temperature.

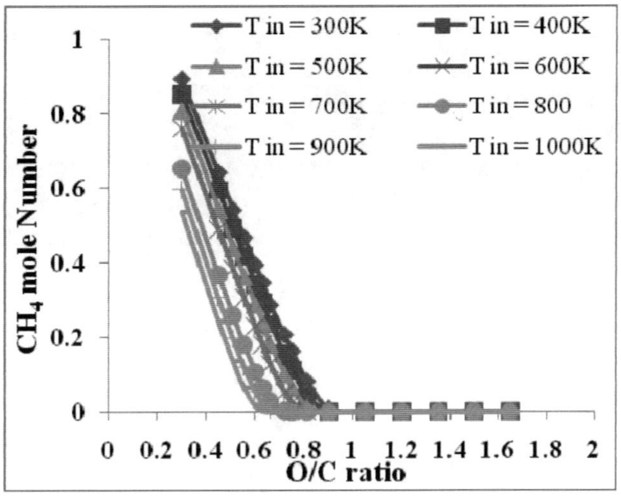

Fig. 4: The effect of the mixture inlet temperature on the CH_4 mole number.

It can be concluded that high mixture inlet temperature is necessary to favorer the formation of CO and H_2 rich reformate and reduce the CO_2 and H_2O fractions.

Because the equilibrium analysis cannot offer any insights into the reaction process, i.e. reaction progress with time, a kinetic analysis was necessary to determine the practical conditions of the reforming process.

Kinetic Analysis of Ethanol Thermal Partial Oxidation (TPOX)

The kinetic analysis was performed to investigate the time required for the reactions to complete and the process controlling step. In order to estimate the process time, in Fig. 5 the time dependent hydrogen concentrations for several temperatures are plotted for O/C = 1.2 and 1.35. It can be seen that for a temperature of 1300 K and O/C = 1.2, more than 1000 sec is required to achieve equilibrium. This means that a large reactor volume is required for the process. increasing the reaction temperature accelerates the reaction kinetic and thus reduces the reaction time. With increasing the O/C ratio, the reaction becomes faster due to the fact that more oxygen available for the oxidation process. From the figure, the equilibrium times are 0.226 sec, 0.373 sec and 0.466 sec at S/C = 1.2 and 1.35, respectively. However, the low hydrogen mole fraction is achieved as it can be seen in Fig. 6. For lower O//C ratios, to achieve such high reaction temperatures, a preheating of the air over 1000K is needed as shown in Fig. 1. At this temperature level there is a risk of self ignition in the fuel-air mixture preparation unit the TPOX reformer. Therefore, a lower preheating temperature is preferable although the syngas yield (reforming efficiency) is lower. Also, the material specification of the reaction zone and the maximum temperature that can be sustained limit the degree of preheating.

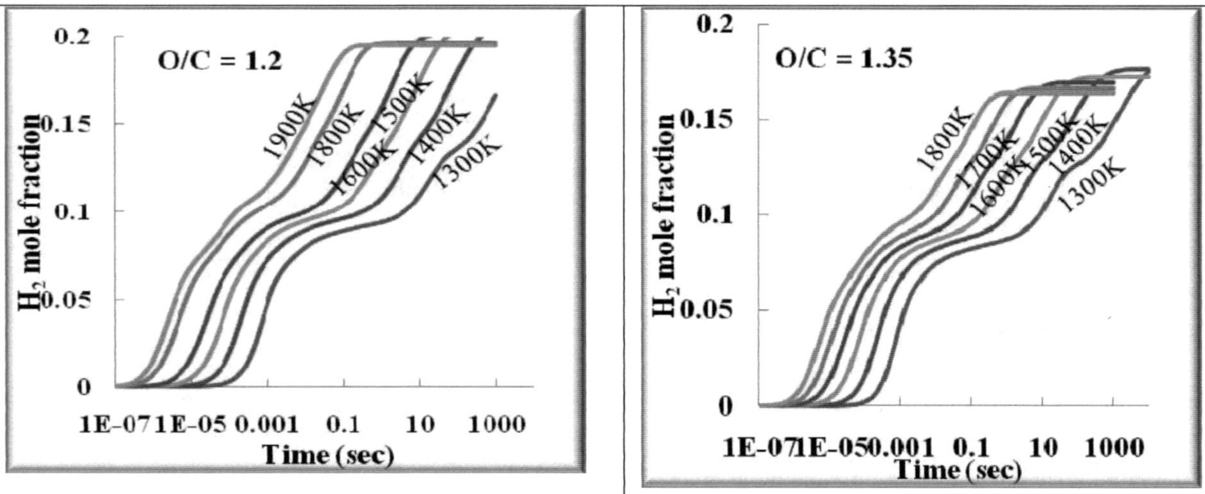

Fig. 5: The time dependent hydrogen concentrations for O/C = 1.2 (left) and O/C = 1.35 (right) at different reaction temperatures.

To understand the kinetic behavior of partial oxidation processes, a representative example of major concentrations of the species produced from the TPOX is presented in Fig. 6 for O/C = 1.2 and a process temperature of 1600K (which is equal to the adiabatic flame temperature at preheating temperature of 600K). A subdivision of the partial oxidation into three major time dependent zones can be observed.

For different O/C ratios and reaction temperature the behavior of the chemistry remains the same, only the concentration levels and time scales change. In the first reaction zone the fastest reactions, the partial and complete oxidation reactions occur. These reactions lead to the consumption of Ethanol and oxygen in this zone and the formation of H_2O, CO_2, H_2, CO. and CH_4. C_2H_2 which is supposed to be a soot precursor, is also formed in this region. In the second zone, slower reactions such as the water gas shift reaction and steam reforming reactions of CH_4 and C_2H_2 occur. The 3rd region represent a slow reactions region in which steam reforming reactions and dry reforming with CO_2 occur.

It can be observed that in the partial oxidation process of ethanol the water gas shift and the reforming reactions are the process limiting step. In Fig. 6, the oxidation time is 0.00164 sec (reaction time for the first region to complete) and the time for the reforming reactions in the 2nd and 3rd region is about 10 sec.

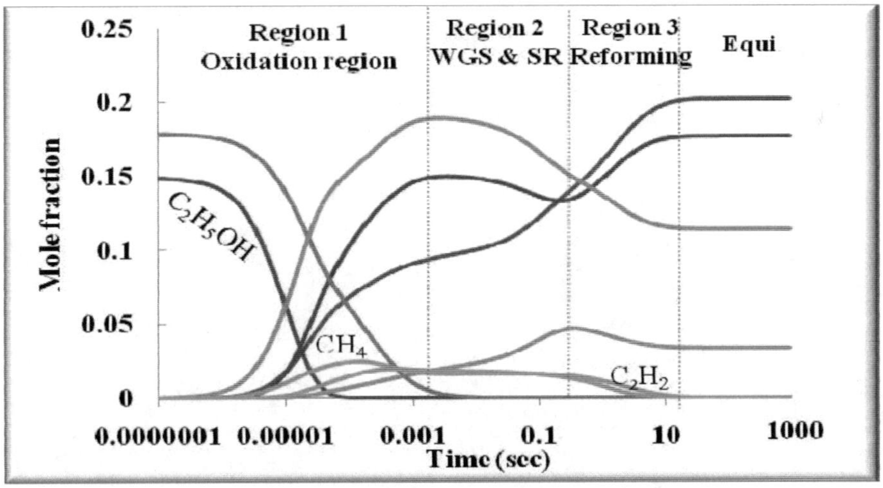

Fig. 6: Kinetic reaction zones in the partial oxidation of ethanol for O/C = 1.2 and process temperature of 1600 K.

Summary and Conclusion

In this work thermodynamic and kinetic investigations for the thermal partial oxidation of ethanol are performed. Thermodynamic predictions, within the O/C range from 0.3 to 1.65 and for temperature range 300k to 1000, show that for O/C ratio less than 0.9, the reforming efficiency increases with increasing the process temperature and reach a maximum value at $(O/C)_{max}$. After this value, the reforming efficiency decreases with increasing the O/C ratio. For O/C values higher than or equal to 0.9, the reforming efficiency is nearly independent on temperature and has approximately constant value at a given O/C ratio. The thermodynamic analysis shows that mole number of H_2 and CO increases with increasing the O/C ratio and reach a maximum at $(O/C)_{max}$, after which H_2 and CO concentrations decrease with increasing the O/C ratio. The $(O/C)_{max}$ depends on the mixture inlet temperature. The value of $(O/C)_{max}$ is shifted to lower C/O value at higher reaction temperature.

The kinetic simulations suggest that three different stages exist during the partial oxidation process of ethanol: oxidation region, water gas shit reaction- reforming region and reforming region. The reforming reactions in the 3rd region are the reaction process limiting step. The reaction temperature affects significantly the equilibrium time, the higher reaction temperature the faster is the reaction progress. Increasing the pressure reduces the oxidation region and retired the progress of hydrogen and CO production. From thermodynamic and kinetic simulations, the practical operating conditions could be bounded in the air ratio range of 1.2 to 1.35 and mixture inlet temperature range of 650K to 850K. lower O/C ratios might result in reforming efficiency, however, higher preheating is necessary to accelerate the reaction kinetics and this increases the risk of self ignition.

REFERENCES

[1] Pimentel David and Patzek Tad W., Ethanol Production Using Corn, Switchgrass, and Wood; Biodiesel Production Using Soybean and Sunflower, Natural Resources Research, 2005; Vol. 14, No. 1.

[2] Haga, F., T. Nakajima, K.Yamashita, and S. Mishima, Effect of crystallite size on the catalysis of Alumina-supported cobalt catalyst for steam reforming of ethanol, Reaction kinectic. catalysis letter, 1997; 63: 2: 253-259.

[3] Athanasios, N.F.and D.I. Kondaridesm, Production of hydrogen for fuel cells by reformation of biomass-derived ethanol, Catalysis Today, 2002; 75: 145-155

[4] Kumar R.; Ahmed S.; Krumpelt M., The Low-Temperature Partial-Oxidation Reforming of Fuels for Transportation Fuel Cell Systems, Fuel Cell Seminar, Orlando.

[5] Tindall B.; Crews M., Alternative technologies to steam-methane reforming, Hydrocarbon Processing, August, 1996; P. 85

[6] Bingue JP, Saveliev AV, Kennedy LA. Optimization of hydrogen production by filtration combustion of methane by oxygen enrichment and depletion. Int J Hydrogen Energy 2004; 29:1365–70.

[7] Hall MJ, Peroutka XN. A porous media burner for reforming methanol for fuel cell powered electric vehicles. Society of Automotive Engineers Special Publications SP-1108; 1995. p. 101-109.

[8] Pedersen-Mjaanes H, Chan L, Mastorakos E. Hydrogen production from rich combustion in porous media. Int J Hydrogen Energy 2005;30:579–92.

[9] Z. Al-Hamamre, D. Trimis, K. Wawrzinek, Thermal partial oxidation of methane in porous burners for hydrogen production, 7th International Conference on Technologies and Combustion for a Clean Environment (Clean Air VII), Lisbon, Portugal, 2003.

[10] Z. Al-Hamamre, S. Voß, D. Trimis, Hydrogen Production by Thermal Partial Oxidation of Hydrocarbon Fuels in Porous Media Based Reformer, International Journal of Hydrogen Energy, article in Press. http://dx.doi.org/10.1016/j.ijhydene.2008.10.085

[11] Athanasio, N.F. and X.E Verykios, Reaction network of steam reforming of ethanol over Ni-based catalysts, Journal of Catalysis, 2004; 225: 439-452

[12] Breen, J.P., R. Burch and H. M. Coleman, (2002), Metal-catalyzed steam reforming of ethanol in the production of hydrogen for fuel cell applications, Applied catalysis B: Environmental, 2002; 39: 65-74.

[13] Cavallaro, S., V. Chiodo, S. Freni, N. Mondello and F. Frusteri, Performance of Rh/Al2O3 catalyst in the steam reforming of ethanol: H2 production for MCFC, Applied catalysis A: General, 2003; 249: 119-128.

[14] Freni, S., S. Cavallaro, N. Mondello and L. Spadaro and F. Frusteri, Steam reforming of ethanol on Ni/MgO catalyst: H_2 production for MCFC, Journal of power sources, 2002;108: 53-57.

[15] Freni, S., G. Maggio, and S.Cavallaro, Ethanol steam reforming in a molten carbonate fuel cell: A thermodynamic approach. Journal of Power Sources,1996; 62: 67-73.

[16] Galvita, A.A., G.L. Semin, V.D. Belyaev, V.A. Semikolenov, P.Tsiakaras and V.A. Sobyanin, Synthesis gas production by steam reforming of ethanol, Applied catalysis A: General,2001; 220: 123-127.

[17] Abayomi John Akande, Production of Hydrogen by Reforming of Crude Ethanol, M.Sc thesis, Department of Chemical Engineering,University of Saskatchewan, Saskatoon, Saskatchewan, 2005.

[18] Kee R.J., Rupley F.M., Miller J.A.: Sandia National Laboratories Report, SAND89-8009, 1990a.

[19] Kee R.J., Grcar J.F., Smooke M.D., Miller J.A.: Sandia National Laboratories Report, SAND85-8240, 1990b.

[20] Lutz, A.E., Kee R.J., Miller, J.A.: Sandia National Laboratories Report, SAND87-8248, 1990.

[21] Marinov, N. M, A Detailed Chemical Kinetic Model for High Temperature Ethanol Oxidation, Int. J. Chem. Kinet. 1999; 31:183-220

[22] Schäfer J., Entwicklung eines hochintegrierten Dampfreformers für mobile PEM-Brennstoffzellensysteme, PhD thesis, Fluid Mechanics Institute, University Erlangen-Nuernberg, 2006.

PURE HYDROGEN AND PURE CARBON DIOXIDE FROM GASEOUS HYDROCARBONS BY CHEMICAL LOOPING REFORMING

Tobias Pröll, Johannes Bolhàr-Nordenkampf, Philipp Kolbitsch, Klemens Marx and Hermann Hofbauer
Vienna University of Technology, Austria

Abstract

A dual circulating fluidized bed pilot plant was operated in chemical looping reforming conditions at a scale of 140 kW fuel power with natural gas as fuel. A nickel based oxygen carrier was used as bed material. The results show that the oxygen in the air reactor can be completely absorbed by the solids and that the fuel reactor exhaust gas is a syngas in thermodynamic equilibrium with no methane left at atmospheric pressure. Even though no steam was added to the natural gas feed no carbon formation was found for global excess air ratios larger than 0.4. A process configuration is proposed for production of pure H_2 and CO_2 from chemical looping reforming. Two separation steps where the pure gases can be obtained are arranged consecutively and the final off gas is largely recycled into the CLR fuel reactor. The performance of the system will depend on the necessary amount of steam to prevent carbon deposition on the particles at low global air/fuel ratios.

Introduction

Today, hydrogen is produced predominantly by reforming of methane or other light hydrocarbons of fossil origin. Thus, natural gas is the main primary resource for hydrogen. Hydrocarbons are transformed to synthesis gas in basically two types of catalytic processes: methane steam reforming (MSR), where CH_4 and H_2O are contacted to a fixed bed catalyst in externally heated, usually tubular reformers [1] and autothermal reforming (ATR), where hydrocarbons, typically refinery by-products, are contacted to catalysts and free oxygen is added to release the necessary heat for heating up the feedstock and for the reforming reactions. In both cases, the catalysts are normally based on nickel as the active component. These technologies were established as a standard long ago and have been optimized with respect to performance and economics over the last decades. Therefore, it can be assumed that, unless the boundary conditions change significantly, these technologies are the optimum choice for hydrogen production on the large scale.

However, there is increasing concern about CO_2 emissions from fossil fuels. Chemical looping combustion (CLC), a technology where metal oxides are used to transport oxygen from one reaction zone into another where the scope is full oxidation of a hydrocarbon fuel, has been intensively investigated in the recent years [2,3]. Hundreds of materials have been developed and systematically screened [2,4] and the technology has recently been successfully demonstrated for more than 1000 hours [5] and at scales of up to 150 kW [6]. A variant of the process, where simply less oxygen is provided than needed for full conversion of the fuel, allows production of synthesis gas. The process can then be called chemical looping reforming (CLR). In this respect the technology has been first proposed in several patent applications by Standard Oil Development Company in the early 1950-ies (e.g. [7]). It has been re-discovered as a variant of CLC [8] and recently investigated in theory and at laboratory scale at Chalmers University of Technology [9]. In CLR, the catalyst is provided as loose particles and is circulated between the reforming reactor, also called fuel reactor in chemical looping vocabulary, and the so-called air reactor by means of fluidization (Figure 1). The particles fulfill three main tasks:

- heat transport
- selective oxygen transport
- reforming catalyst

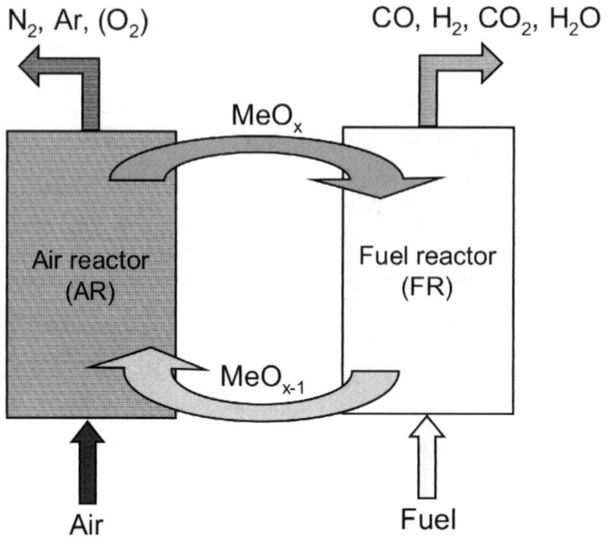

Figure 1. Chemical looping reforming (CLR).

Because of the second functionality above, the particles are called the "oxygen carrier" of the chemical looping system. Apart from thermodynamic and catalytic properties, such oxygen carriers need to show sufficient mechanical strength for their use in fluidized bed reactor systems. Especially for nickel-based oxygen carriers, some very suitable candidates have been designed and provided at relevant quantities recently [5]. Carbon deposition in CLR has recently been studied by de Diego et al. [10]. At Vienna University of Technology, a pilot plant consisting of two interconnected circulating fluidized bed (CFB) reactors has been designed, built and operated. Nickel-based oxygen carriers produced from commercially available raw materials [5] have been used as bed material. The unit is dimensioned for CLC of natural gas at a fuel power of 120 kW [11]. In a recent series of experiments, the amount of air to the air reactor was reduced step-wise to investigate CLR operation. In the following, the main characteristics of these experiments are reported. Based on the most relevant results, the potential of the technology for coupled production of pure H_2, pure CO_2 and an oxygen depleted air stream is discussed.

Experimental

The 100-150 kW pilot plant

The chemical looping unit is a dual circulating fluidized bed (DCFB) system for gaseous fuels. It has been designed with a direct focus on scalability to larger size. Steam-fluidized loop seals between the reactors avoid mixing of the gas phases. The flow regime in the air reactor is fast fluidization and in the fuel reactor turbulent fluidization. Downstream of each reactor, gas and solids are separated in cyclone separators. The gases are cooled separately. For determination of fuel reactor exhaust gas concentrations a Rosemount NGA 2000 (CO: 0–100%, CO_2: 0–100% , O_2: 0–25%, H_2: 0–100%, CH_4: 0–100%) is used and, additionally, an online gas chromatograph Syntech Spectras GC 955, which allows cross-checking of carbon species and determination of the N_2 content for evaluation of possible gas leakages from the air reactor to the fuel reactor. The air reactor exhaust stream is analyzed using a Rosemount NGA 2000 (CO: 0–100%, CO_2: 0–100%, O_2: 0–25%). After gas analysis sampling, the gas streams pass a common post combustion chamber equipped with a support burner. The exhaust of the post combustor is cooled, cleaned in a bag filter and sent to the chimney. A sketch of the pilot plant is shown in Figure 2.

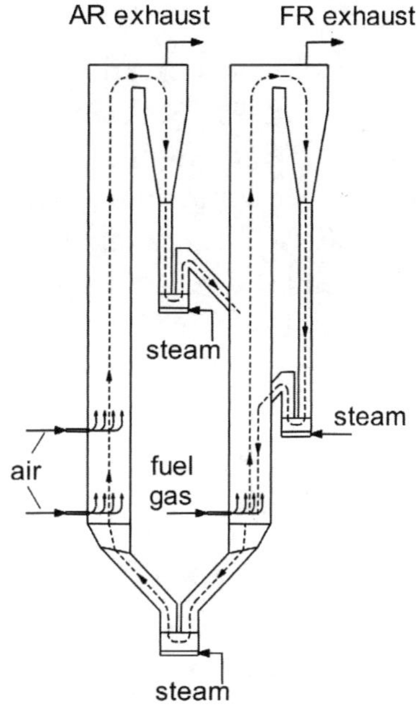

Figure 2. Chemical looping pilot plant.

A variable cooling system, consisting of three double jackets, is attached to the air reactor. It allows controlled heat extraction for varying operating conditions. The properties of the pilot rig, the design and the variable cooling, enable operation points from combustion to reforming (global stoichiometric air/fuel ratio range 0.4 – 1.5). A detailed description of the reactor system, the cooling system and the other auxiliary units can be found elsewhere [11].

CLR experiments

A first series of experiments was performed to investigate system performance under chemical looping reforming conditions. Natural gas from the Viennese grid (98.7 vol.% CH_4) was used as fuel. The fuel power was kept constant at 140 kW while the global stoichiometric air/fuel ratio was decreased step-wise starting from an air/fuel ratio of 1. The operating temperature in the fuel reactor was kept constant at 1176 K in a first run and at 1020 K in a second run. This was accomplished by proper adjustment of the cooling duty of the cooling jackets attached to the air reactor. In each operating point shown below, the system has been operated for at least 30 minutes to achieve steady state with respect to the gas phase concentrations. Data are collected on-line and evaluated using a data reconciliation method for fulfillment of mass and energy balances [12].

Operating results

The composition of the gas streams exiting the reactor system is shown versus the global air to fuel ratio in Figure 3 for the experimental run where the fuel reactor temperature was kept at approximately 1020 K. The broken lines in Figure 3a indicate the concentrations assuming thermodynamic equilibrium. Apart from global air ratios higher than 0.7, where oxidation of the particles decreases the reforming activity, equilibrium is established in the fuel reactor. Above this value, equilibrium is not reached mainly because of insufficient CH_4 conversion. This is clearly a limitation by reactivity. At lower air ratios, the particles are present in a less oxidized form showing higher reforming activity. The equilibrium concentration of CH_4 is practically zero at atmospheric pressure and at this temperature over the whole air/fuel ratio range.

On the air reactor side, the oxygen uptake by the particles clearly improves with decreasing air/fuel ratio. H_2O is introduced into the air reactor from loop seal fluidization. Argon is present in air and its concentration is slightly increased because the total amount of gas decreases in the air reactor.

Also, the concentrations of CO and CO_2 in the air reactor off gas have been monitored. No carbon species could be detected at all during the sets of experiments shown here. This is remarkable because no steam was added to the natural gas feed except for the part of the lower loop seal fluidization that might have been directed back into the fuel reactor. However, this would only amount to a global steam to organic carbon ratio of less than 0.4 in the worst case that all the lower loop seal steam would enter the fuel reactor, which is highly unlikely considering the direction of particle movement and the local pressure situation [13].

A somewhat clearer picture is obtained at an operating temperature of 1176 K, reported in Figure 4. CH_4 conversion is high even for high air/fuel ratios. The gas concentrations follow the equilibrium predictions within measurement accuracy. On the air reactor side it can be observed that O_2 is taken up quantitatively by the particles for any air/fuel ratio below 0.95. This means that the air reactor exhaust consists only of N_2, Ar and some steam from the loop seals.

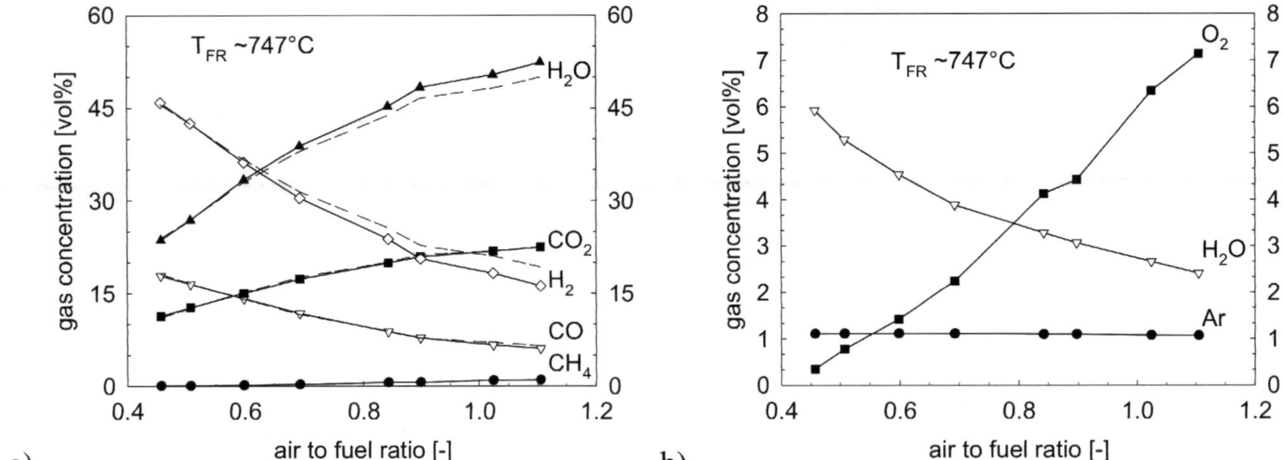

Figure 3. Gas compositions from fuel reactor (a) and air reactor (b) at a fuel power of 140 kW and a fuel reactor operating temperature of 1020 K. The missing amount to 100% in (b) is N_2.

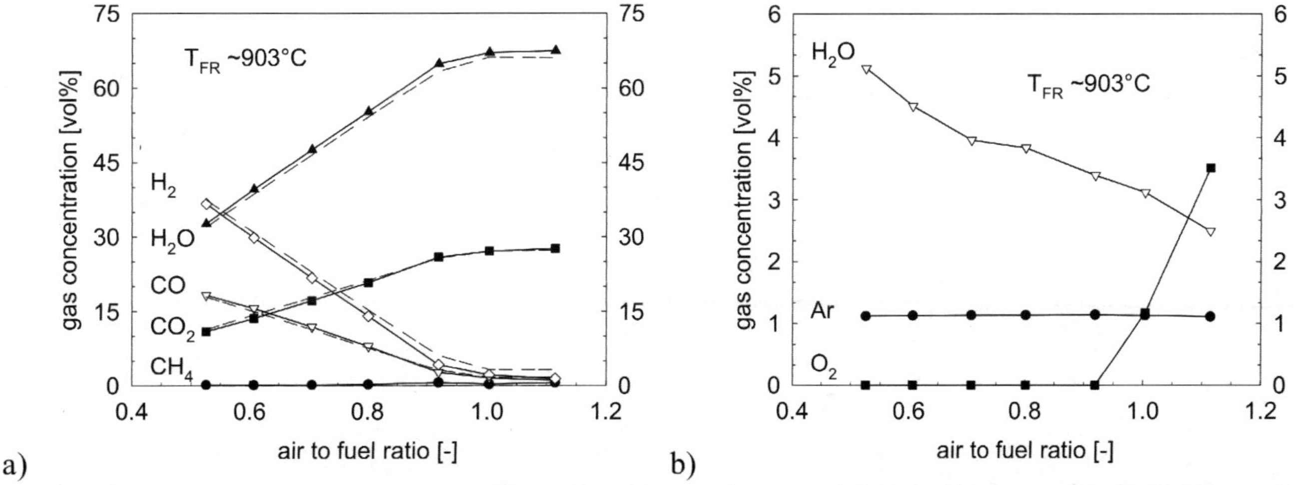

Figure 4. Gas compositions from fuel reactor (a) and air reactor (b) at a fuel power of 140 kW and a fuel reactor operating temperature of 1176 K. The missing amount to 100% in (b) is N_2.

The minimum air/fuel ratios shown in Figures 3a and 4a (0.46 and 0.52 respectively) are the steady state points where the reactor cooling was finally switched off completely. Neither air nor fuel could be preheated during the experiments. Below this minimum air/fuel ratio shown, only non-steady state operation with the system continuously loosing temperature could be run. Such points are not shown here, but it can be reported that with the same parameters operated as for the points shown, i.e. no extra steam to the fuel, carbon formation started at global air/fuel ratios of 0.4. This was detected in a reproducible way by CO_2 measurements in the air reactor off gas.

Discussion

First results with CLR at 140 kW

Based on the experimental findings, the following assumptions are acceptable for the oxygen carrier used at least for air/fuel ratios between 0.4 and 0.7 and steam/carbon ratios larger than 0.4:

- the fuel reactor exhaust gas is in thermodynamic equilibrium
- the air reactor exhaust gas is practically free of oxygen
- no carbon is lost to the air reactor (no carbon deposition on the particles)

Within the validity range of the above assumptions, prediction of the streams out of a CLR unit is possible.

Comparison of CLR to methane steam reforming and autothermal reforming

According to the experimental findings, CLR works today at atmospheric pressure. In the following, some pros and cons of CLR compared to technologies that are currently used are summarized. The main advantages of CLR compared to methane steam reforming are:

- no external combustion is required to provide heat to the reforming process
- all carbon involved is available in the synthesis gas

- less steam is required
- less concern with respect to sulfur contaminants
- heat transfer is very high in the fluidized bed, no limitation of the reaction rate by heat transfer
- smaller reactor volumes and less catalyst required per unit fuel feed
- the temperature can be increased because the reactors can be refractory lined
- higher CH_4 conversion possible even at higher pressure because temperature can be increased
- a by-product consisting of N_2 and Ar is obtained from the air reactor

The main advantages of CLR compared to autothermal reforming are:

- no air separation unit is required, no energy costs for air separation occur
- less concern with respect to sulfur contaminants
- Ar from air goes with the N_2 to the air reactor exhaust and can be separated as a by-product

The main challenges connected to CLR are:

- effective dust removal
- fluidized bed system operation at increased pressure

Both challenges mentioned have much better chances to be solved today than in the 1950-ies, where CLR was initially looked at. Resistant oxygen carrier particles are available today. New filter materials are available that allow operation at increased temperatures. Fluidized bed technology as a whole has witnessed a great progress since then. However, pressurized CFB is not yet a standard technology and some research and development effort will be necessary to make such systems work.

Coupled production of H_2, CO_2 and N_2+Ar

For production of pure hydrogen along with 100% CO_2 capture, a process configuration according to Figure 5 can be advantageous. Both gases, CO_2 and H_2, are obtained at high purity by using adequate separation technologies. If the CLR system is operated at atmospheric pressure, these would likely be a gas wash with chemical solvents (amines) followed by a pressure swing adsorption (PSA) plant for H_2 separation. Arrangement of two consecutive separation steps and recycling the off gas allows operation at reduced separation efficiencies. The same is valid for the conversion efficiency of the CO-shift unit. Though, with the exception of a small bleed stream, all hydrogen provided with the fuel will finally end up in H_2 and all carbon provided in CO_2. The necessary amount of bleed is largely dependent on the concentrations of non-hydrocarbon impurities in the fuel feed. Dilution of the fuel reactor gas stream with N_2 from the air reactor side in the CLR system has been shown to be very low in the pilot unit [13].

Preliminary process calculations based on the assumption of full conversion to equilibrium have been done. It turns out that much lower global air/fuel ratios in the range of 0.2 – 0.3 are thermally indicated if the feed streams are preheated and heat losses in the CLR unit are low. It is expected to be crucial for process efficiency how much steam needs to be added to the fuel reactor feed in order to avoid carbon formation on the particles at these air/fuel ratios.

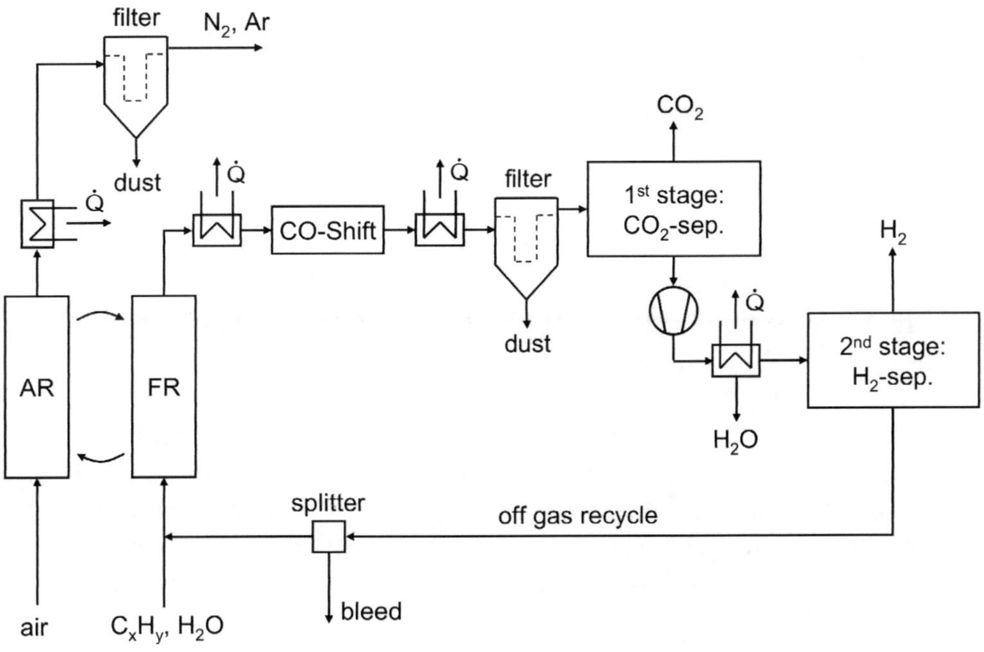

Figure 5. Production of pure gas streams from CLR.

Conclusions

Chemical looping reforming was tested at 140 kW fuel power in a dual circulating fluidized bed installation. A designed nickel-based oxygen carrier was used in the experiments. The results show that thermodynamic equilibrium is reached in the reformer at low global air/fuel ratios or sufficiently high temperature, respectively. No carbon formation occurred for air/fuel ratios larger than 0.4 even though the steam to organic carbon ratio was less than 0.4. On the air reactor side, oxygen can be quantitatively absorbed by the oxygen carrier particles and a potentially valuable side stream consisting of N_2 and Ar is obtained.

A process configuration is proposed for production of pure H_2 and CO_2 from chemical looping reforming. Two separation steps where the pure gases can be obtained are arranged consecutively and the final off gas is largely recycled into the CLR fuel reactor. The performance of the system will depend on the necessary amount of steam to prevent carbon deposition on the particles at low global air/fuel ratios. This remains to be experimentally determined in the future. Such an investigation is best arranged along with optimization of oxygen carrier particles for CLR. The particles used at present have been optimized towards CO_2 yield in CLC operation. For CLR, however, particles with high reforming reactivity but with a low tendency towards carbon formation at low oxygen availability are to be preferred. Another interesting topic for the future is to assess the possibilities for pressurization of fluidized bed systems suitable for performing CLR.

Acknowledgements

This work was part of the EU financed project CACHET (FP6 Contract No. 019972), coordinated by BP. The project is also part of Phase II of the CO2 Capture Project (CCP). The oxygen carrier has been produced by the Flamish Institute for Research and Technology (VITO), Belgium under the guidance of Chalmers University of Technology, Sweden in the context of the EU financed project CLC GAS POWER (FP6 Contract No. 019800).

References

1. Dybkjaer, I., "Tubular reforming and autothermal reforming of natural gas – an overview of available processes," Fuel Processing Technology, 42 (1995), 85-107.
2. Lyngfelt, A., Johansson, M., Mattisson, T., "Chemical looping combustion – status of development," in *Circulating fluidized bed technology IX*, Werther, J., Nowak, W., Wirth, K.E., Hartge E.U., eds., TuTech, Hamburg, Germany. 2008.
3. Hossain M.M., de Lasa H.I., "Chemical-looping combustion (CLC) for inherent CO2 separations – a review", Chemical Engineering Science, 63 (2008), 4433–4451.
4. Johansson, M., Mattisson, T., Rydén, M., Lyngfelt, A., "Carbon capture via chemical-looping combustion and reforming," at *International seminar on carbon sequestration and climate change*, Rio de Janeiro, Oct. 24–27, 2006.
5. Linderholm, C., Mattisson, T., Lyngfelt, A., "Long-term integrity testing of spray-dried particles in a 10 kW chemical-looping combustor using natural gas as fuel," Fuel, 2009, in Press, doi:10.1016/j.fuel.2008.12.018.
6. Kolbitsch, P., Pröll, T., Bolhàr-Nordenkampf, J., Hofbauer, H., "Operating experience with chemical looping combustion in a 120kW dual circulating fluidized bed (DCFB) unit," at: *9th International Conference on Greenhouse Gas Control Technologies (GHGT-9)*, Nov. 16-20, 2008, Washington DC, available online: http://www.co2captureproject.com/ghgt-9/GHGT9_Kolbitsch.pdf.
7. Welty, A.B.Jr., "Apparatus for conversion of hydrocarbons," United States Patent No. 2550741, 1951.
8. Mattisson T., Lyngfelt, A., "Applications of chemical looping combustion with capture of CO2", in *Proceedings of the 2nd Nordic Minisymposium on Carbon Dioxide Capture and Storage*, Göteborg, Sweden, 2001.
9. Rydén, M., "Hydrogen production from fossil fuels with carbon dioxide capture, using chemical-looping technologies," PhD thesis, Chalmers University of Technology, Göteborg, Sweden, 2008.
10. de Diego, L.F., Ortiz, M., García-Labiano, F., Adánez, J., Abad, A., Gayán, P., "Hydrogen production by chemical-looping reforming in a circulating fluidized bed reactor using Ni-based oxygen carriers," Journal of Power Sources, in Press, doi:10.1016/j.jpowsour.2008.11.038.
11. Kolbitsch, P., Pröll, T., Bolhàr-Nordenkampf, J., Hofbauer, H., "Design of a chemical looping combustor using a dual circulating fluidized bed (DCFB) reactor system," Chemical Engineering and Technology, 32 (2009), 398-403.
12. Bolhàr-Nordenkampf, J., Pröll, T., Kolbitsch, P., Hofbauer, H., "Comprehensive Modeling Tool for Chemical Looping Based Processes," Chemical Engineering and Technology, 32 (2009), 410-417.
13. Pröll, T., Kolbitsch, P., Bolhàr-Nordenkampf, J., Hofbauer, H., "A novel dual circulating fluidized bed (DCFB) system for chemical looping processes," submitted to AIChE Journal, 2009.

Sorbent-Enhanced Water Gas Shift Reaction Studies

Robert W. Stevens, Jr.[1,2], Abolghasem Shamsi[1], Stephen Carpenter[1,3], and Ranjani Siriwardane[1]

[1] *U.S. Department of Energy, National Energy Technology Laboratory, 3610 Collins Ferry Road, P.O.Box 880, Morgantown, WV 26507-0880*

[2] *Parsons, P.O. Box 618, South Park, PA 15129*

[3] *REM Engineering Services, 3537 Collins Ferry Rd., Morgantown, WV 26505-2374*

Introduction

The water gas shift reaction reacts water with CO and is used to change the CO/H_2 ratio for chemical production and for producing pure hydrogen. The water gas shift (WGS) reaction in combination with CO_2 absorption will play a significant role in reducing CO_2 from fossil fuel power generating systems such as IGCC by increasing system efficiencies and reducing fuel consumption. The exothermic and reversible WGS reaction,

$$CO(g) + H_2O(g) \leftrightarrow CO_2(g) + H_2(g) \quad (\Delta H_{298} = -41 \text{ kJ/mol}) \qquad (1)$$

is not thermodynamically favorable at higher temperatures therefore excess steam is added and several reactors are used in series to obtain higher conversions. Addition of excess steam and the required number of reactors to drive the reaction is expensive and it increases the cost of operation.

The water gas shift reaction can be enhanced by actively removing the reaction product(s); thus enabling conversion of the reactants to proceed to a greater degree. One of the techniques for removing the products from the reaction medium is absorption and separation of CO_2. As long as the partial pressure of CO_2 is below its equilibrium pressure, the reaction will continue forward consuming CO and H_2O. The removal of CO_2 uses less steam by eliminating the need to increase the partial pressure of the reactant to force the forward reaction.

Enhancement of the water gas shift (WGS) reaction by solid sorbent has received a lot of attention recently in literature.[1-5] NETL researchers have developed novel CO_2 capture sorbents for the moderate temperature application of the WGS reaction. The sorbents were evaluated for their ability to enhance the WGS reaction at temperatures of 300 – 600 °C, both in the presence and absence of catalyst. Testing at 500 – 600 °C in the absence of catalyst resulted in CO conversions of nearly 100% prior to sorbent saturation, surpassing the thermodynamic equilibrium CO conversion of the gas-phase WGS reaction under identical conditions. The sorbents were also tested during the WGS reaction at 400°C and 150 psig in the presence of a Pt/Ce_2O_3 catalyst.

Experimental setup

A schematic of the laboratory-scale fixed-bed reactor system is shown in Figure 1. The reactor system features three zones: (i) feed delivery, (ii) the reactor zone, and (iii) analysis. The feed delivery zone of the system consists of mass flow controllers (Brooks model 5850i) for gas feed as well as a Teledyne-Isco syringe pump (model 260D). The reactor zone consists of a packed bed reactor with a 3.5 ml heated volume and an additional furnace for preheating the water feed. The analysis section consists of a condenser to trap unreacted H_2O, an online mass spectrometer (Pfeiffer Quadrupole Prisma), and a gas chromatograph (HP model 5820). The catalyst and sorbent solids were ground with a small mortar and pestle, (30-40 mesh) to promote uniformity within the fixed bed, and subsequently charged to the reactor to allow for a gas hourly space velocity of 1000 h^{-1} relative to the sorbent and 20,000 h^{-1} relative to the catalyst. Inert silica (SiO_2) was used to ensure a constant bed volume for catalyst-only and sorbent-only baseline runs. In each test, simulated syngas (36% CO, 27% H_2, bal. He) and water (H_2O/CO = 2:1) was directed into the the reactor at 400 °C and 150 psig. Reactivity was monitored through product gas analysis via mass spectrometry (MS) and/or gas chromatography (GC).

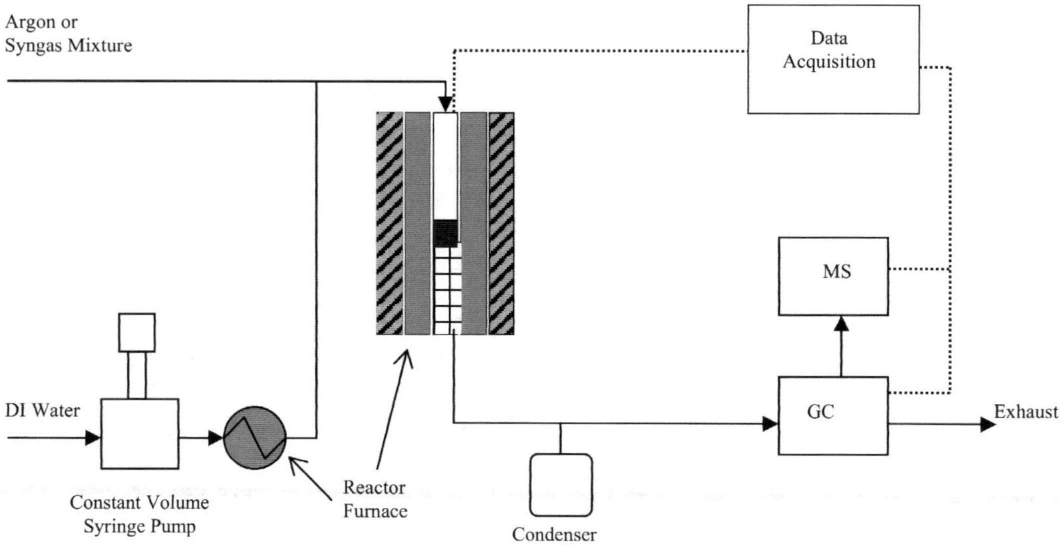

Figure 1. Schematic of laboratory scale fixed-bed reactor system.

The 2 wt% Pt/ZDC-50 (zirconia-doped ceria) catalyst sample was obtained from NexTech Materials. Three sorbents were synthesized by NETL researchers, containing calcium oxide (CaO) and sodium hydroxide (NaOH) as a promoter, at ratios of 4:1, 10:1, and 20:1, respectively, by weight. A pure calcium oxide sorbent was also employed in the study as a baseline.

Regeneration studies were also conducted via heating the catalyst/sorbent mixture from 150 – 1100 °C at 5 °C/min at 1 atm. Mass spectrometry was again utilized to measure the desorption profile of CO_2 as a function of temperature.

Results and discussion

Figure 2 illustrates the equilibrium conversion of CO during the WGS reaction as a function of temperature at different levels of CO_2 capture with a feed ratio of 2:1 H_2O:CO. The exothermic and reversible WGS reaction is not thermodynamically favorable at higher temperatures. Thermodynamic calculations show that removing carbon dioxide from reaction products enhances the reaction of water with carbon monoxide, which results in higher production of hydrogen. While equilibrium conversion may be maximized at lower temperatures, the kinetics of the reaction are much slower at lesser temperatures. Figure 2 shows that successful employment of a CO_2 capture sorbent may allow for the same equilibrium conversion at a much faster rate. For example, a sorbent that is capable of removing 75% of the CO_2 produced can offer a 90% CO conversion at ~675 °C while in the absence of sorbent, the reactor would need to be maintained at ~450 °C.

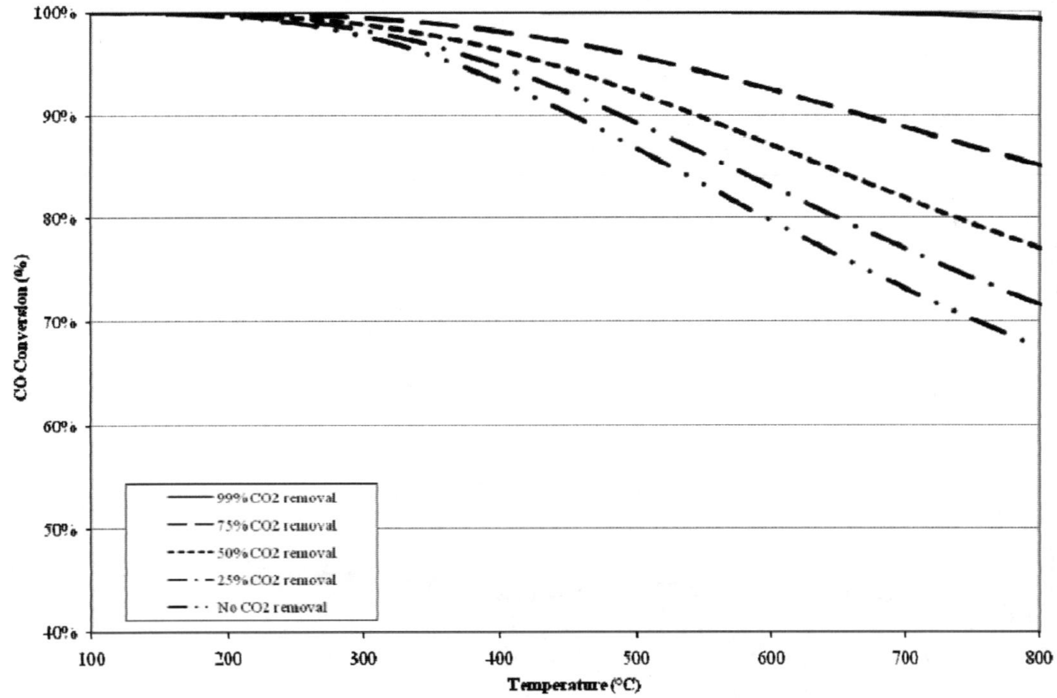

Figure 2. WGS equilibrium CO conversion as a function of temperature with various levels of CO_2 removal. H_2O:CO feed ratio = 2:1. Valid at any pressure.

Figures 3-5 illustrate the results of sorbent-enhanced WGS testing over the catalyst/sorbent mixtures. In each figure, the product profiles for H_2, CO, and CO_2 were monitored. Each figure can be divided into three zones: (i) sorbent-enhanced, (ii) sorbent breakthrough, and (iii) steady state WGS reaction. During the sorbent-enhanced portion of the study, all three sorbents led to 100% CO conversion, elevated H_2 production, and no CO_2 was observed in the reactor effluent. The 4:1 sorbent test (Fig. 3) appeared to yield the lowest H_2 concentration while the 10:1 sorbent (Fig. 4) afforded the longest duration prior to breakthrough. The breakthrough profiles were very similar among all catalyst/sorbent mixes.

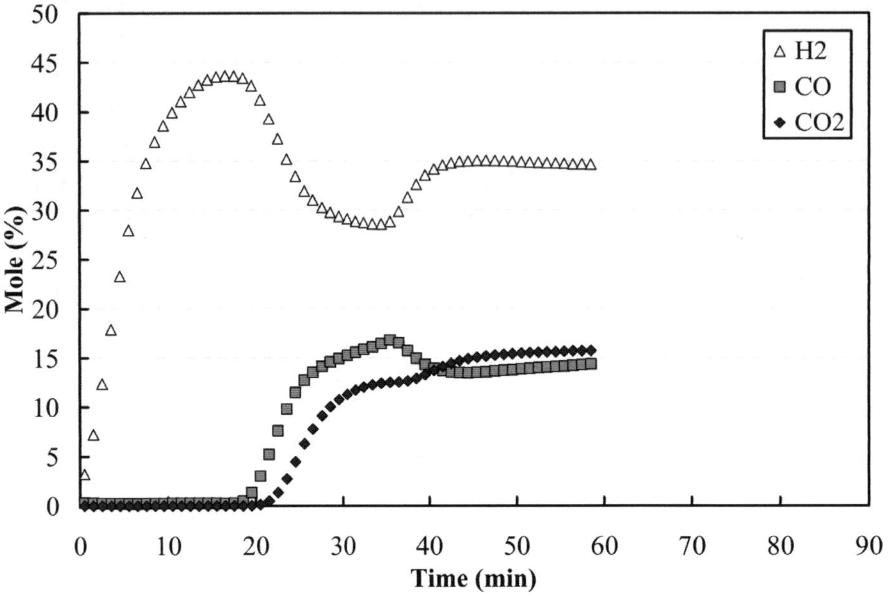

Figure 3: Concentrations of H_2, CO, and CO_2 taken from the MS data of the SE-WGS reaction for 4:1 CaO/NaOH sorbent and 2%Pt/ZDC-50 catalyst.

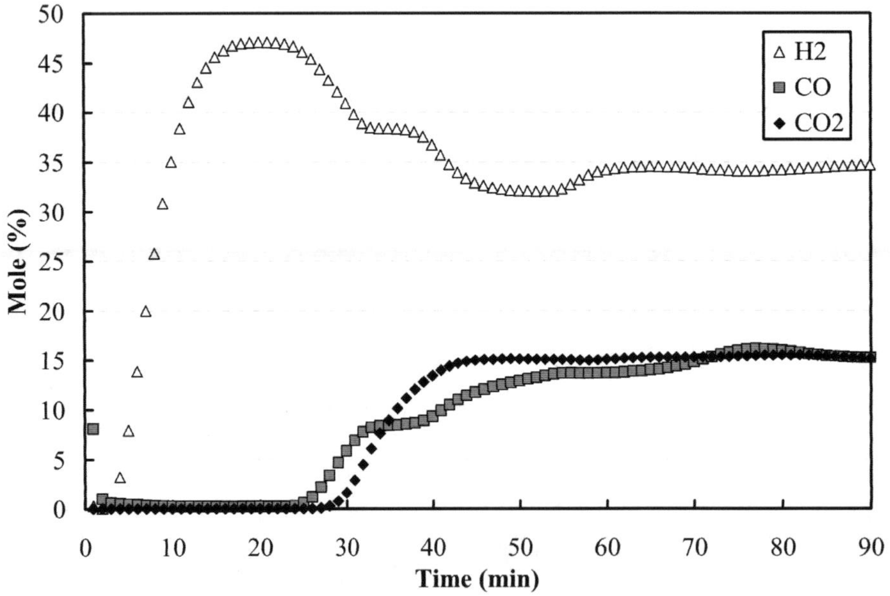

Figure 4: Concentrations of H_2, CO, and CO_2 taken from the MS data of the SE-WGS reaction for 10:1 CaO/NaOH sorbent and 2%Pt/ZDC-50 catalyst.

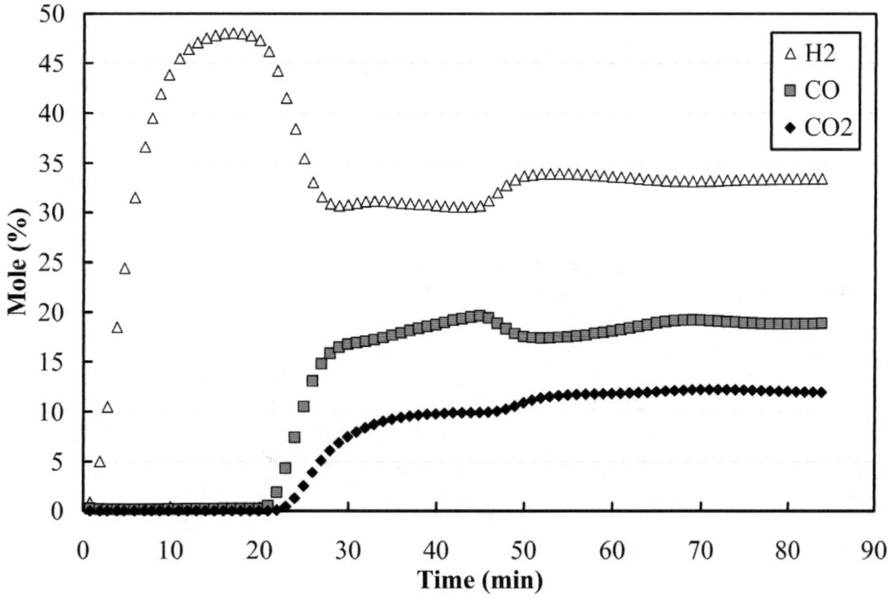

Figure 5: Concentrations of H_2, CO, and CO_2 taken from the MS data of the SE-WGS reaction for 20:1 CaO/NaOH sorbent and 2%Pt/ZDC-50 catalyst.

The apparent CO_2 capture capacity was calculated over the course of each test; these results are shown in Figure 6. The CaO baseline was also tested, although the product profiles are not shown for brevity. The 10:1 CaO:NaOH sorbent possessed the highest CO_2 capture capacity of those studied, with a capacity of 10.5 mol CO_2/kg sorbent, which approaches the theoretical limit for CaO of 17.9 mol CO_2/kg.

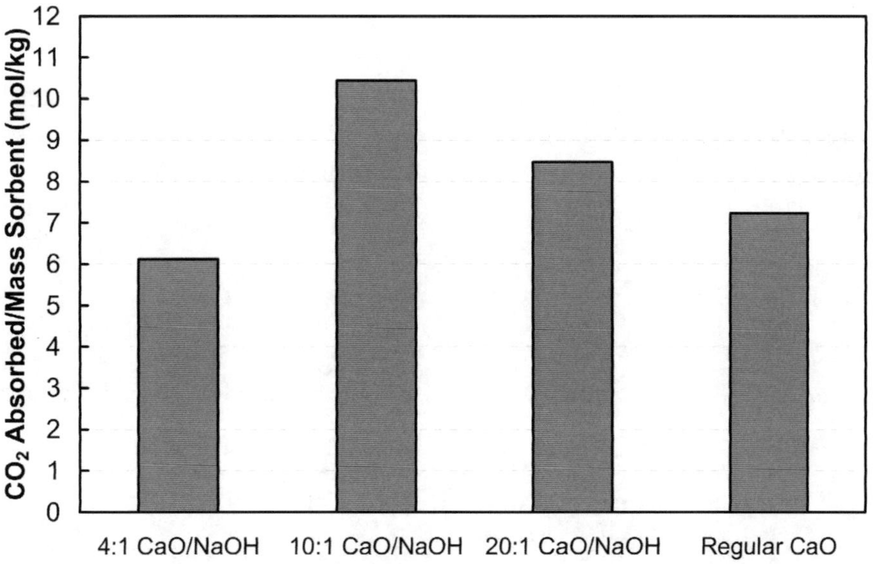

Figure 6: Moles of CO_2 absorbed per mass of sorbent for each study at 400°C, 150psig.

Figure 7 depicts the normalized decomposition profiles during sorbent regeneration of the CaO (i.e., baseline) and 20:1 CaO:NaOH sorbents. It can be clearly seen that the sodium-promoted sorbent regenerates at a lower temperature. This benefit may lead to a significant reduction in operating costs of the SE-WGS reactor.

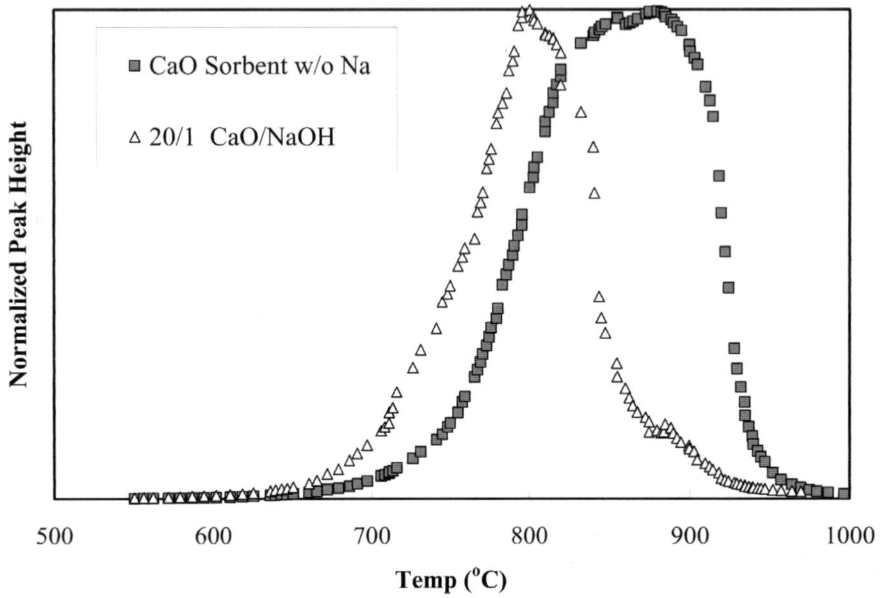

Figure 7: Temperature profiles showing the decomposition of CO_2 released for 20:1 CaO/NaOH sorbent and a similarly processed sorbent without Na.

Conclusions/Summary

- NETL sorbents were successful in enhancing the Pt-catalyzed WGS reaction at 400 °C and 150 psig
- Sodium appears to reduce the sorbent regeneration temperature

References

(1) Johnsen, K, H. Ryu, J. Grace, C. Lim, Chemical Engineering Science, **2006**, 61, 1195-1202.
(2) Xiu, Li, A.E, Rodrigues, Chemical Engineering Science, **2002**, 57, 3893 – 3908.
(3) Ding, Y., E. Alpay, Chemical Engineering Science, **2000**, 55, 3929-3940.
(4) Gauer, C., W. Heschel, Principles and Materials, VGB Powertech, **2005**, 81 – 85.
(5) Hufton, J., S. Weigel, W. Waldron, S. Nataraj M. Rao, S. Sircar, Proceedings of the 1999 USDOE Hydrogen Program Review, **1999**, 244 – 258.

76b High-Pressure Hydrogen Purification USING Electrical Swing Adsorption

John P. Hurley, Gregory T. Dvorak, Dustin P. McNally, and Nikhil Patel. Energy & Environmental Research Center, University of North Dakota, 15 North 23rd Street, Stop 9018, Grand Forks, ND 58202-9018

The University of North Dakota Energy & Environmental Research Center (EERC) is developing technologies for hydrogen production and use at U.S. military installations. Hydrogen will be produced from JP-8 and renewable feedstocks. The work is being funded by the U.S. Army Corps of Engineers Engineering Research and Development Center (ERDC) at the Construction Engineering Research Laboratory (CERL) in Champaign, Illinois.

In initial work, a high-pressure water-reforming (HPWR) concept for on-demand production of high-pressure proton exchange membrane fuel cell-quality hydrogen from JP-8 has been developed at the bench scale. Tests on aromatics- and sulfur-free Syntroleum-produced "S-8" fuel as feedstock produced hydrogen concentrations of up to 56%. The HPWR system operates at up to 600 standard cubic feet/hour (1.5 kilograms/hour) hydrogen production capacity and is currently being upgraded to operate at 12,000 psi. By reforming liquid hydrogen carriers and purifying the gas stream at pressure, 10,000 psig hydrogen for use in fuel cell vehicles can be created with up to a 15% energy savings over low-pressure production followed by compression of the hydrogen. To purify the hydrogen, a bench-scale system is being built and tested to also operate at 12,000 psi. The purification method being tested is electrical swing adsorption (ESA), which was developed at the Oak Ridge National Laboratory. The process is similar to pressure swing adsorption in that the gas is passed through a material that adsorbs contaminants, allowing purified hydrogen to flow through. However, instead of regenerating the adsorber by reducing the pressure in the system, pressure is maintained, and an electric current is passed through the adsorber, causing desorption of the contaminant gases. To permit the flow of electric current through the adsorber, it must be monolithic and electrically conductive. To reduce the large standard volume of hydrogen that would be used to purge the system at such high pressures, the monolith must also have a much higher density than standard granular adsorbers which may contain as much as 80% porosity on a bulk basis.

The EERC has developed a method for making electrically conductive high-surface-area activated carbon monoliths that are approximately twice as dense as granular adsorbers. Initial tests of the ESA process are being performed in a small laboratory-scale rig with a 1"-diameter by 4"-long monolith operating at up to 1000 psi. Concurrently, the 12,000-psi system is being built that can operate with a 1.5"-diameter by 24"-long monolith or a 2.5"-diameter by 30"-long monolith. In this paper, we will describe the overall research program, provide results of testing with the laboratory rig, and give a general system description of the larger bench-scale rig, which is undergoing shakedown testing.

Optimization of oxygen feed membranes in autothermal steam-reformers

J. Hüppmeier[1], S. Barg[2], M. Baune[1], J. Thöming[1]

(1) Centre for Environmental Research and Sustainable Technology – UFT, University of Bremen, Bremen, Germany

(2) Ceramic Materials and Components – CERAMICS, University of Bremen, Bremen, Germany

Introduction and previous work

In conventional concepts for autothermal steam-reformers the spatial temperature distribution is disadvantageous. The different rates of the exothermic oxidation and endothermic reforming reactions lead to the problem of temperature hot-spots, which can damage or even destroy the catalyst or other elements of the reformer. This is especially critical in the case, when the required oxygen is premixed in the feed gas stream resulting into a high temperature peak in the entrance region (fig. 1).

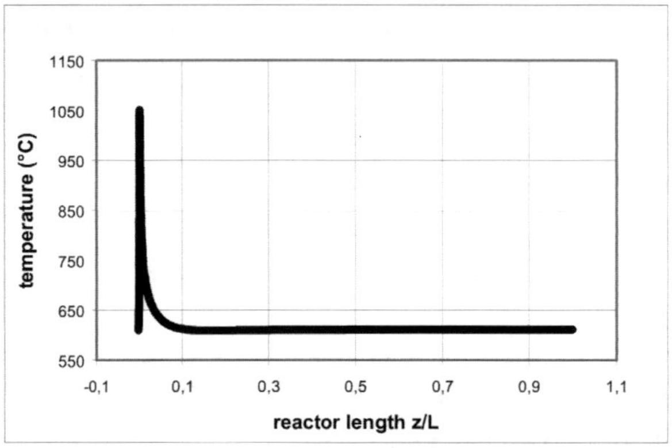

Figure 1: Temperature peak in entrance region of a conventional autothermal steam-reformer. Worst-case scenario calculated from simulations [Hüppmeier2008].

In the past, some first attempts have been made to solve this hot-spot problem. One concept is that the oxygen is not premixed to the feed gas but injected at different positions distributed over the length of the reformer. This leads to more but much smaller temperature peaks [Tiemersma2006, Barrio2006]. Extending this idea, one can use also a continuous oxygen injection over the length of the reformer. This can for example be realized by an open porous membrane. By varying the permeability of such a membrane we can adjust the spatial oxygen flux distribution (fig. 2).

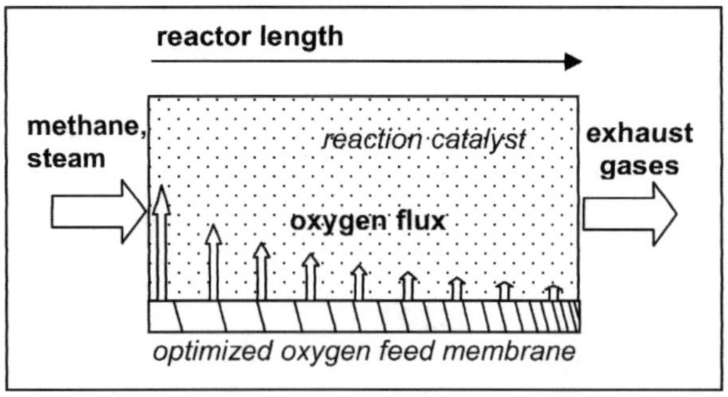

Figure 2: Oxygen feed strategy for autothermal steam-reformers. Key element is an oxygen feed membrane with varying permeability over reactor length.

One of our previous investigations [Hüppmeier2008] focused on the optimization procedure of the membrane structure to achieve an isothermal behaviour of the adiabatic autothermal steam-reformer. In the present work mainly the experiments resulting from this optimization procedure are in focus. For a better understanding, we give a short summary of this procedure here. Because the temperatures inside the reformer depend strongly on the local amount of oxygen, key element in the simulations is a system for controlling the air supply to the reformer. In contrast to normal control systems we use not the time as independent variable but the location z along the length of the reformer (Fig. 3).

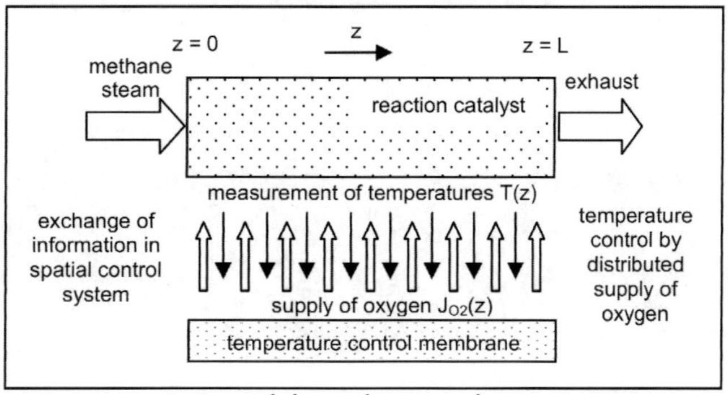

Figure 3: Spatial control system consisting of the reformer, the temperature measurement on location z and a locally supply of the oxygen controlled by a membrane.

At each location z we measure the temperature and compare it to the desired isothermal value. If the temperature is too low, we supply more air, and if it's too high, we supply less. Of course, this optimization procedure can be done more exactly and faster by a digital controller. We found a controller with proportional and differential characteristic (PD-controller) giving the best results for our system. In Figure 4 the whole procedure is depicted. First, we build our "spatial control system", containing of a one-dimensional steady-state model of the autothermal steam-reformer and a digital PD-controller (1). Then we tried to reach an isothermal distribution in the reformer (2) and extracted the used oxygen flux distribution supplied by the controller (3). The last step was to transform the flow distribution to a membrane permeability distribution via Darcy's law (4).

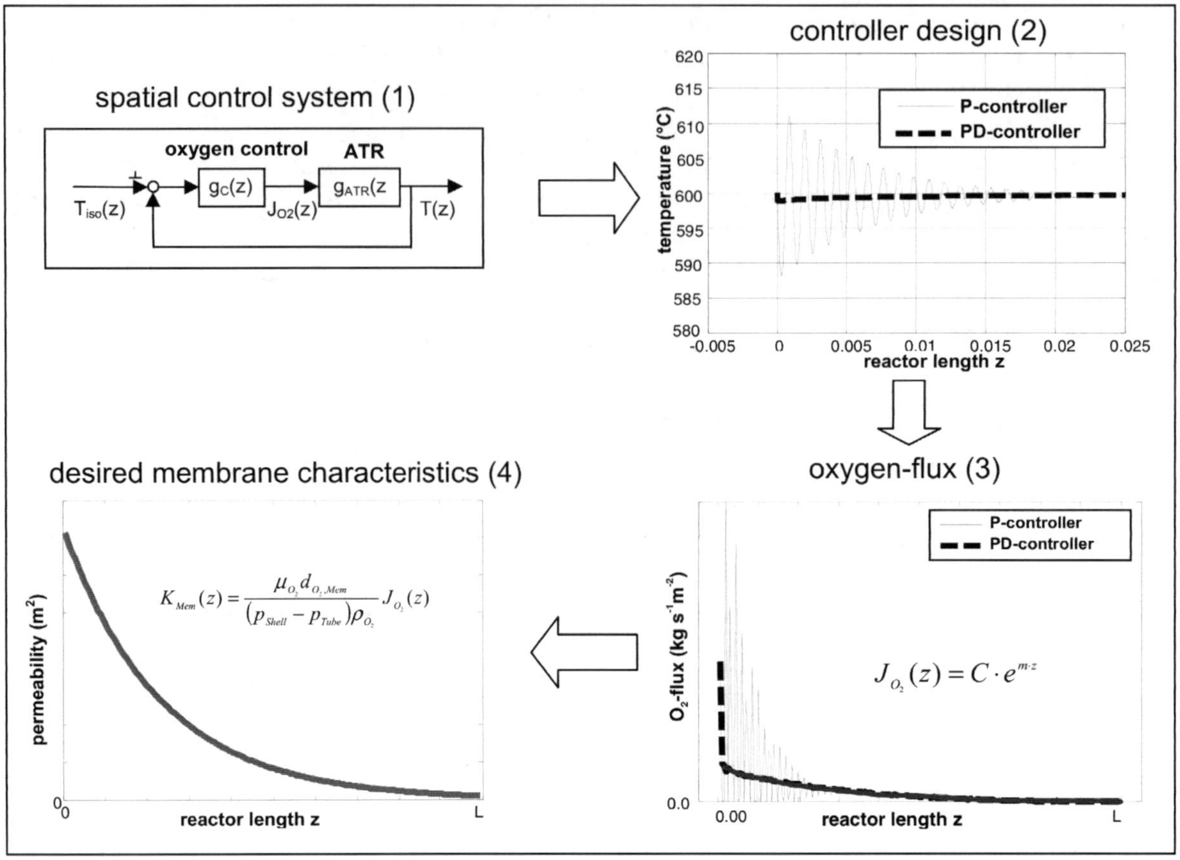

Figure 4: optimization procedure for an oxygen supply membrane to achieve an isothermal behaviour of the autothermal reformer. From building a model of the spatial control system (1) to designing the controller by simulations (2), to extracting the desired oxygen flux leading to the desired permeability profile of the membrane (4) [Hüppmeier2008].

The outcome of this investigation was that a membrane with an exponential decreasing permeability is needed and that we have to premix some amount (about 5 to 10 %) of the oxygen to the feed gases.

Experimental Setup

Autothermal Steam-Reformer

All the experiments were done with the setup described in Figure 5. The three reactants for the autothermal steam-reforming, namely methane, air and liquid water were supplied to the system via two mass flow controllers and a flow controlled syringe pump, respectively. The gases were preheated and the water vaporized and preheated by an external electric heater. Before entering this heater, the air mass flow was split via the adjustable valves V1 and V2. The stream through V1 is dedicated to flow through the membrane and the stream through V2 is premixed with the other reactants methane and steam. After entering the reaction chamber, doing their reactions and leaving it, the composition of the exhaust gases was measured by gas chromatography. The valve V3 is a security valve and closed for all operations.
With controlling the mass fluxes and the valves V1 and V2 we can achieve all necessary operations for our experiments.

The reaction chamber and therefore the catalyst was of the size 66x20x18 mm. For measurement of the temperature profile inside the reformer four thermocouples were located on the centerline of the catalyst. The distances of each thermocouple from the entrance were 8, 21.5, 35 and 62 mm.

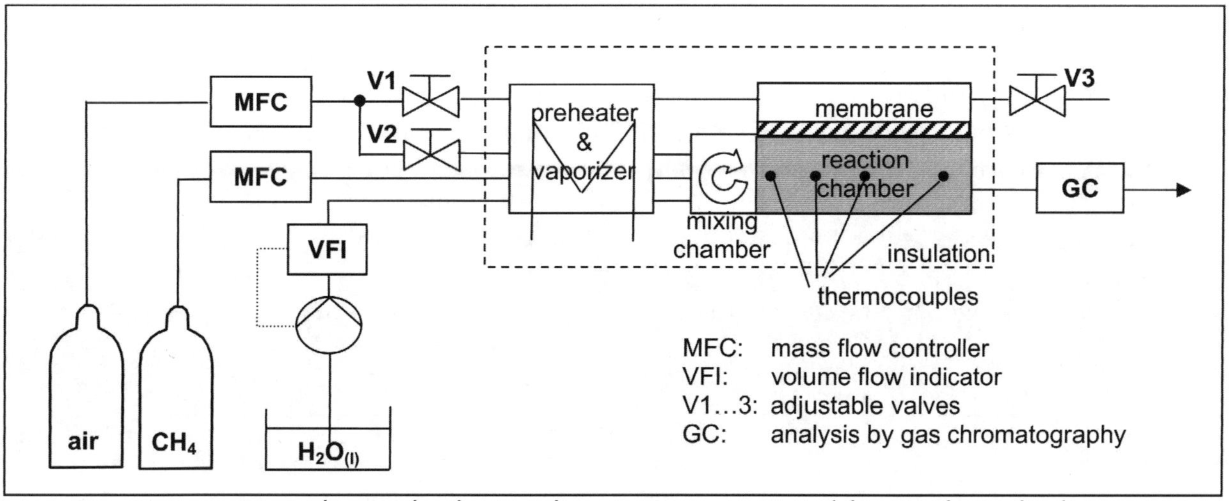

Figure 5: experimental setup for the membrane tests consisting of the autothermal reformer, a preheater and vaporizer and the controlling and measurement equipment.

Membranes

All membranes used were Al_2O_3 – foams built by the HAPES foaming process [Barg2009]. Three different membranes with different mean pore sizes (21, 9 and 5 mm) and the corresponding permeabilities (fig. 6) were used. We tested pure membranes with constant permeability and a combination of these membranes. With this combination, we tried to achieve the optimal permeability distribution from our simulation results like it is shown in figure 6.

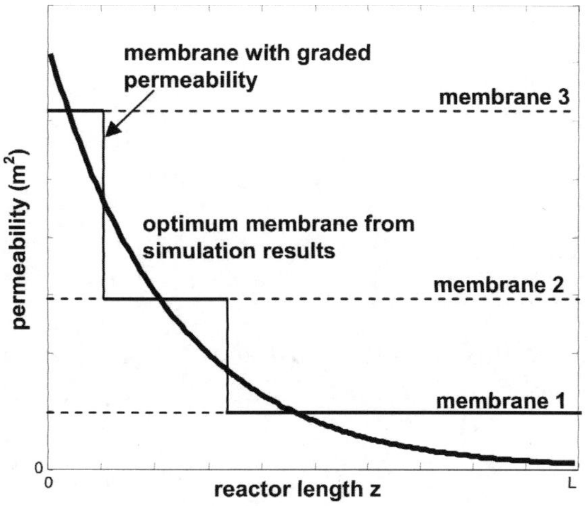

Figure 6: desired permeability profile over the length of the reformer compared to real membranes and a combination of these membranes

Experimental results

Our first experiments investigate the occurrence of the temperature peak in our experimental setup under several operation conditions. This was necessary, because from temperature peak, its location and its maximum value depend strongly on the individual setup and especially on the amount of catalyst used.
In figure 7 the temperature profiles in our autothermal steam-reformer are shown for different O_2 to CH_4 ratios and different H_2O to CH_4 ratios. It can be seen that the H_2O to CH_4 ratio has only small effects on the temperature profiles. The temperature in the entrance region is about 710 to 730°C in contrast to about 625 to 640 °C in the outlet region.
Of course, the amount of oxygen has a much greater influence on the temperature profile. For an O2 to CH4 ratio of 0.6 the temperature in the entrance region raises up to over 760°C. We also see a great difference in the temperatures shortly after the entrance region. This is an indicator of location changing of the hot-spot towards the interior of the reformer.

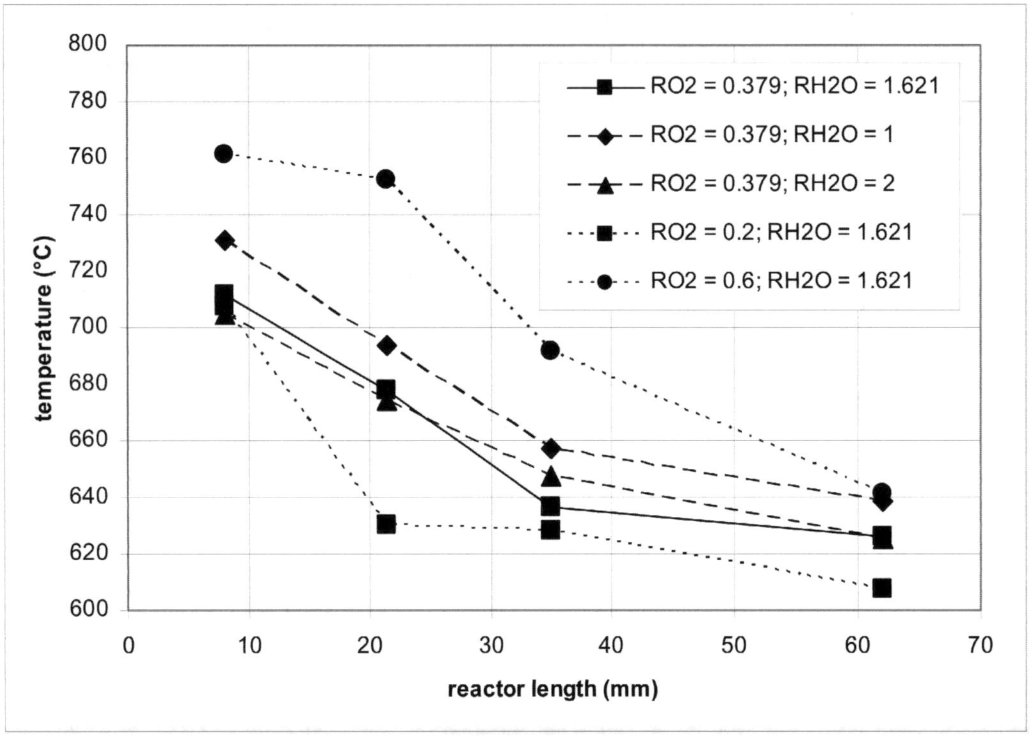

Figure 7: temperature profiles along the length of the reformer for different O_2 to CH_4 ratios R_{O2} and different H_2O to CH_4 ratios R_{H2O} in contrast to figures 8 and 9 without a temperature control membrane

In Figure 8 the temperature profiles are shown using a temperature control membrane compared to the operation without a membrane. The operating conditions were all the same for these experiments. Due to our simulation results, we added a small amount of the used oxygen to the premixed to the feed gases, so that we have an optimal temperature in the entrance region compared to the temperatures in the interior of the reformer.
For the membrane with the constant permeability, we can avoid the hot-spot in the entrance region but because the ongoing feed of oxygen over the length of the reformer, there appears another hot-spot at the end. This temperature peak is even higher than the original one for the case without a membrane.
The best result we get with our graded membrane. All temperatures are in the range of 645 and 655 °C.

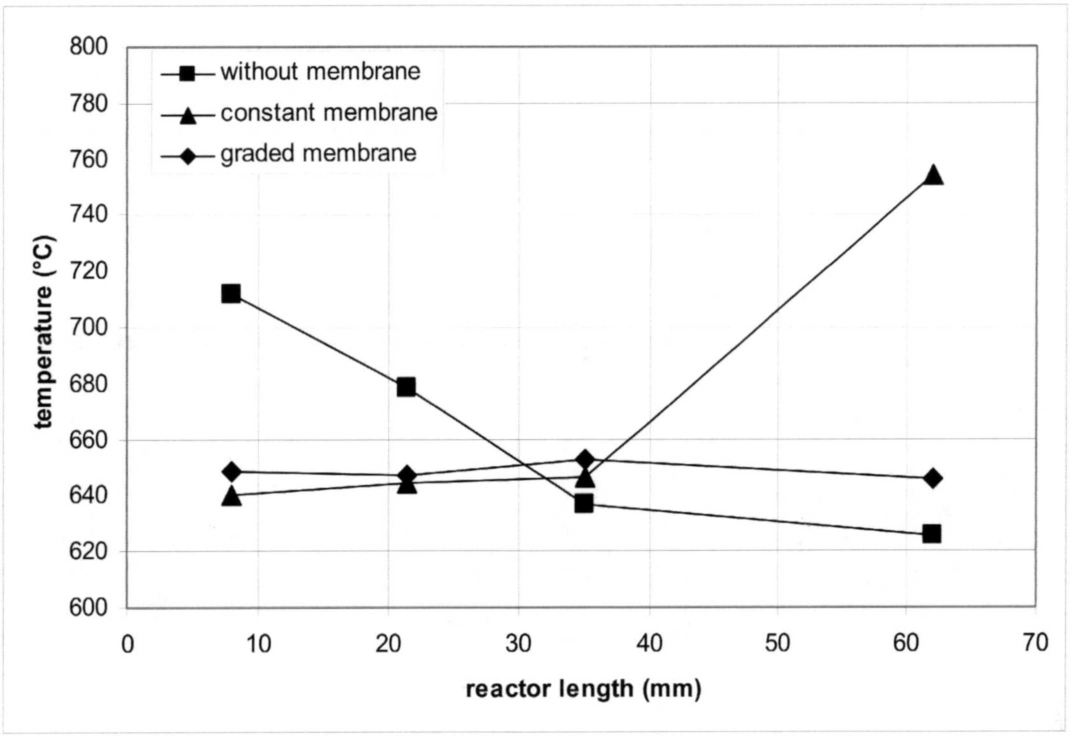

Figure 8: temperature profiles along the length of the reformer using different temperature control membranes at their optimal operation points, i.e. the flux through the membrane was added by a small amount of oxygen premixed to the feed gases (5.2 % for the constant membrane and 6.5 % for the graded membrane)

For our best membrane, the graded one, we also investigated the effect of premixing different amounts of oxygen to the feed gas. In the entrance region a high supply of oxygen is needed to start the partial oxidation reaction. This can not be achieved only by the membrane itself, because its permeability is too low. By adjusting the valves V1 and V2 (fig. 5) we can vary the amount of premixed oxygen.

What can be seen from figure 9 is, that the amount of premixed oxygen has a strong influence on the temperature in the entrance region. The influence on the total temperatures in the interior of the reformer is also detectable but not that significant.

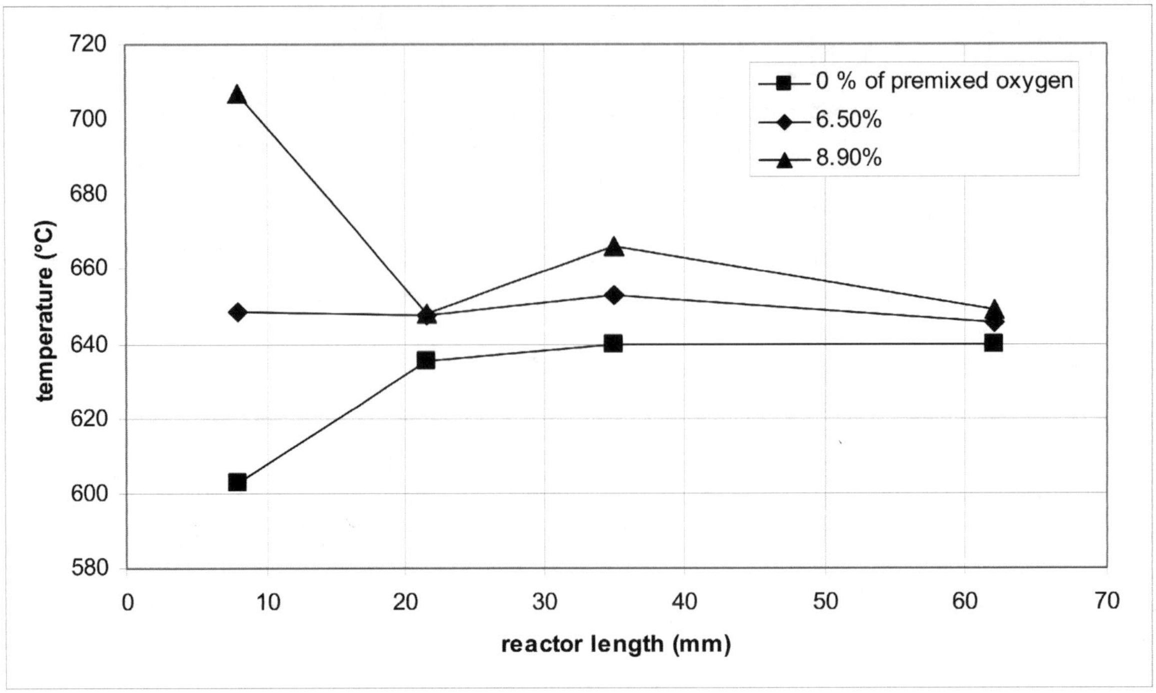

Figure 9: temperature profiles along the length of the reformer using the graded temperature control membrane with different amounts of oxygen added to the feed gases

As a consequence of our experimental results, it can be stated that the graded membrane designed by the mentioned optimization procedure leads to quite good results concerning the temperature peak. Also, the result from our simulations, that we need a small amount of oxygen premixed to the feed gases, could be validated in our experiments.

Acknowledgements

The funding of the research project by the Deutsche Forschungsgemeinschaft through the Research Training Group "PoreNet" is gratefully acknowledged.

References

T.P. Tiemersma, C.S. Patil, M. van Sint Annaland, J.A.M. Kuipers Modelling of packed bed membrane reactors for autothermal production of ultrapure hydrogen, Chemical Engineering Science 61 (2006) 1602–1616

V.L. Barrio, G. Schaub, M. Rohde, S. Rabe, F. Vogel, J.F. Cambra, P.L. Arias, M.B. Güemez, Reactor modeling to simulate catalytic partial oxidation and steam reforming of methane. Comparison of temperature profiles and strategies for hot spot minimization, International Journal of Hydrogen Energy (2006), doi: 10.1016/j.ijhydene.2006.10.022

J. Hüppmeier, M. Baune, J. Thöming, Interactions between reaction kinetics in ATR-reactors and transport mechanisms in functional ceramic membranes: A simulation approach, Chemical Engineering Journal 142 (2008) 225–238, doi:10.1016/j.cej.2008.03.004

S. Barg , E. G. de Moraes, D. Koch, G. Grathwohl, New cellular ceramics from high alkane phase emulsified suspensions (HAPES), *J. Eur. Ceram. Soc.* (2009), doi:10.1016/j.jeurceramsoc.2009.02.003

A MEMBRANE REACTOR CONCEPT FOR SIMULTANEOUS PRODUCTION OF PURE HYDROGEN AND SYNTHESIS GAS OVER CERAMIC MATERIALS

Antigoni P. Evdou, Dr Lori T. Nalbandian, Dr Vassilis T. Zaspalis
Chemical Process Engineering Research Institute, Centre for Research and Technology-Hellas, Thessaloniki, Greece

Abstract

The thermochemical water splitting, in particular when combined to solar energy supply, is one of the most promising routes for long-term sustainable hydrogen production. The technology however has major drawbacks such as the batchwise non-isothermal character of the process and the high temperatures required for material regeneration. In this paper experimental demonstration is provided of a novel concept for hydrogen production by a two-step thermochemical water splitting process simultaneously with synthesis gas production by a stepwise redox methane oxidation reaction. It is based on a dense membrane reactor which ensures isothermal and continuous production of high purity hydrogen. Perovskite materials having the formula $La_{1-x}Sr_xFeO_3$ are investigated as potential redox materials for the membrane preparation since they exhibit mixed type conductivity for the necessary transfer of anions, vacancies and electrons. Their ability to accommodate large concentrations of vacancies in their structure and to reversibly pick up and deliver oxygen at high temperatures was studied by thermogravimetric oxidation/reduction experiments which indicated that the materials are able to loose and uptake reversibly oxygen from their lattice and the weight loss is proportional to Sr content. Conventional reactor experiments indicated that the materials can be used as catalysts in a redox process where water is dissociated giving rise to the production of pure hydrogen. Dense, disc shaped membranes of the materials were synthesized and placed in a membrane reactor. Experiments at 1000°C revealed the possibility of performing the reduction and oxidation steps simultaneously and isothermally on each side of the membrane reactor. A steady-state situation was thereby achieved where hydrogen was continuously produced on one side while the material was simultaneously regenerated by a carbon containing species, on the other side. The created oxygen vacancy gradient formed the driving force for a continuous flux of vacancies from the membrane reduction surface to the membrane oxidation surface. The system is also able to operate on partial pressure based desorption without the need of a carbon containing reductant, so that a process towards hydrogen production, based only on renewable hydrogen source such as water, can be established. The hydrogen production rate under enhanced oxygen ions removal by a reductant is estimated to be up to 145 cm^3 H_2 (STP)m^{-2} min^{-1} while under unforced reduction conditions a hydrogen production rate of 47.5 cm^3 H_2 (STP)m^{-2} min^{-1} was measured.

Introduction

The thermochemical water splitting, in particular when combined to solar energy supply, is one of the most promising routes for long-term sustainable hydrogen production [1]. The technology

however has major drawbacks such as the batchwise character of the process, the high temperatures required for material reduction (regeneration) and the low hydrogen yields.
In this paper experimental demonstration is provided of a novel concept for hydrogen production by a two-step thermochemical water splitting process simultaneously with synthesis gas production by a stepwise redox methane oxidation reaction. It is based on a dense membrane reactor which ensures isothermal and continuous production of high purity hydrogen [2].
A schematic diagram of the membrane reactor concept is shown in Figure 1. Water vapour enters compartment 1 of the reactor where splitting takes place producing hydrogen, while the released oxygen atoms fill the anion vacancies of the perovskite structure, according to the reaction shown in Equation (1).

$$H_2O + V_o^{\cdot\cdot} + 2e' \rightarrow O_o + H_2 \uparrow \quad (1)$$

Due to the chemical potential difference oxygen ions diffuse to the opposite membrane side where either they desorb because of the low oxygen partial pressure difference as shown in Eq. (2), or are removed by e.g. a carbon containing reductant such as CO or CH_4, as shown in Eq. (3).

$$O_o \rightarrow \frac{1}{2}O_2 + V_o^{\cdot\cdot} + 2e' \quad (2)$$

$$O_o + CX \rightarrow CXO + V_o^{\cdot\cdot} + 2e' \quad (3)$$

In the first case i.e. Eq.(1) and (2) the overall process reaction is the water splitting while in the second case i.e. Eq.(1) and (3) the overall process reaction is either the water gas shift (if CX=CO) or the steam reforming (if CX=CH_4).

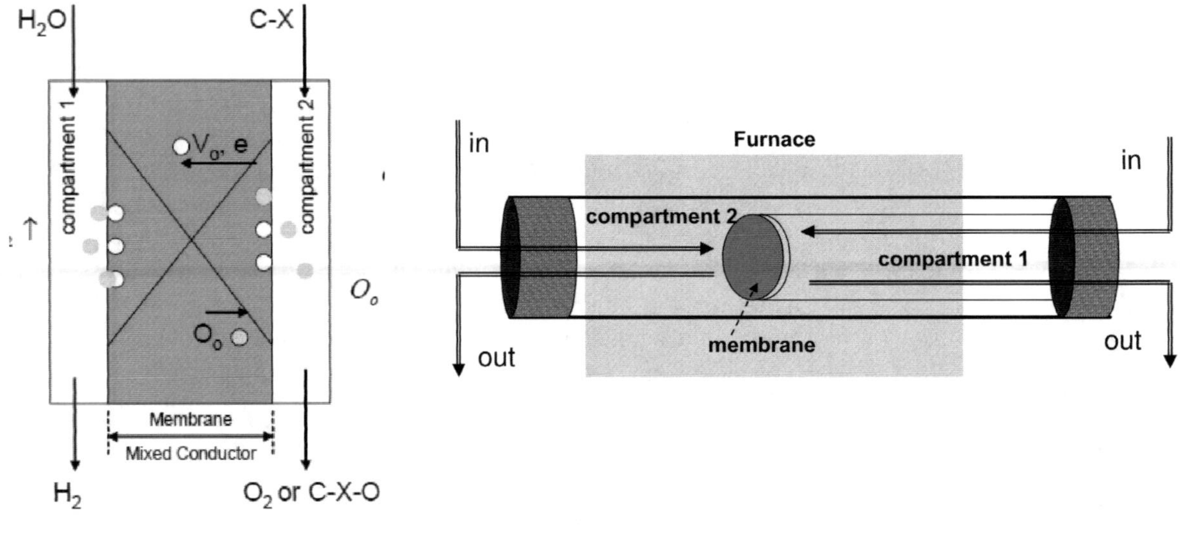

Figure 1. Schematic diagram of the (a) membrane reactor concept and (b) the laboratory constructed membrane reactor

Perovskite materials of the type $La_xSr_{1-x}MO_{3-\delta}$ (M=Mn, Fe) are selected since they are known to be able to accommodate large concentrations of vacancies in their structure, to reversibly pick up and deliver oxygen at high temperatures and to exhibit mixed type conductivity for the necessary transfer of anions, vacancies and electrons [3].

Experimental

Powdered material synthesis, membrane preparation

Materials with the general formula $La_{1-x}Sr_xFeO_3$, x=0, 0.3, 0.7, 1 are synthesized by precipitation of the metal precursor salts with the additions of hydrogen peroxide (H_2O_2) and aqueous ammonia [4]. The solid product is then dried, homogenized and finally fired in air atmosphere to temperatures up to 1300°C. The preparation of the membranes involves roll-granulation, uniaxial compaction and sintering. The final membranes are disc shaped pellets with a diameter of 10mm and thickness 1-5 mm.

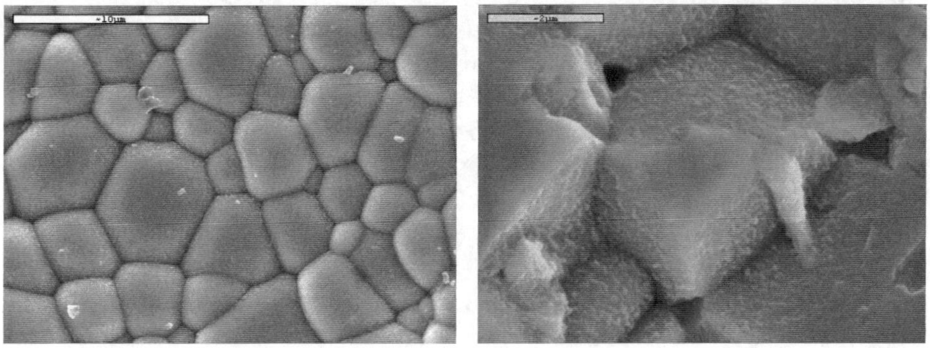

Figure 2: Scanning electron microscopy image of the surface of a dense $La_{0.7}Sr_{0.3}FeO_3$ membrane, after calcination at 1300°C

Material Characterization

The basic physicochemical characterization of the prepared samples includes crystal structure (XRD), surface area determination and Scanning Electron Microscopy observation of the sintered membrane surface. The redox properties of the materials are studied at a SDT2956 (TA instruments) thermobalance and at a multicomponent mixture thermogravimetric analyzer with simultaneous qualitative chemical analysis of the exit stream by mass spectrometry (Hiden-Isochema, IGA). Reaction experiments with the materials in powder form are performed in a reaction unit (Altamira AMI-I) using a U-type quartz reactor. The reactor products are quantitatively analyzed with a quadrupole mass-spectrometer (Balzers-Omnistar) that has been earlier calibrated for all reactants and products of the process.

Membrane Reactor

The membrane reactor consists of two co-axial tubes as shown in Figure 3. The membrane specimen is gas tightly sealed on top of the inner α-Al_2O_3 tube and divides the membrane reactor into two compartments. An inlet and an outlet stream exist in each compartment, all connected individually to a mass spectrometer (Balzers-Omnistar) for chemical analysis. To control leakage free operation during the entire duration of the experiment, different inert carrier gases are used; helium is used in compartment 1 and argon in compartment 2.

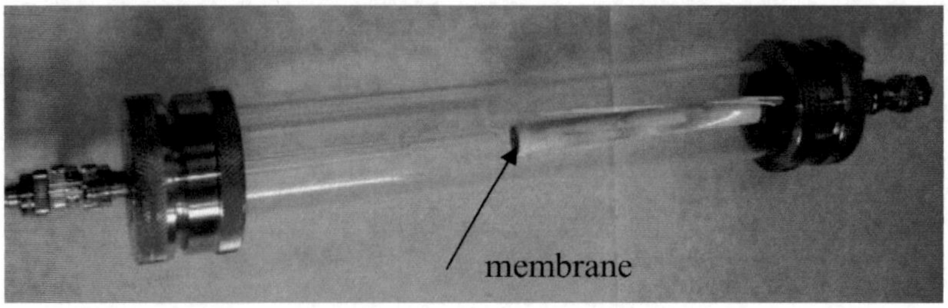

Figure 3: Image of the laboratory constructed membrane reactor

Results and Discussion

Redox behavior of materials

The redox properties of the materials used for membrane preparation, $La_{1-x}Sr_xFeO_3$, are investigated as a function of the composition (different Sr content, x), the temperature and the O_2 partial pressure.

The mass variation with temperature, which indicates the oxygen loss, was studied in air, during 4 heating-cooling cycles. The mass loss for the sample $La_{0.3}Sr_{0.7}FeO_3$, reaches to 98.57% of its original mass (corresponding to 0.187 oxygen atoms per "mole") at 1000°C. Oxygen can be completely reabsorded into the sample when temperature descends to 400°C. The mass changes show a very good repetition in 4 cycles. Figure 4a shows the isothermal oxygen desorption and reabsorption, for the same sample, at 1000°C, when gas flow is changed from argon to air and from air to argon. The amount of oxygen desorbed and reabsorbed (0.5% weight, 0.065 oxygen atoms per "mole") shows a remarkable stability over the 4 cycles of the experiment.

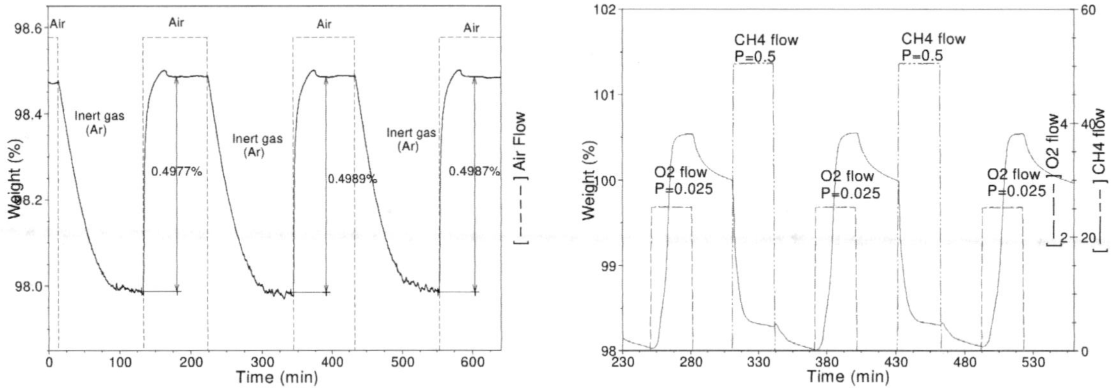

Figure 4. Redox behavior of $La_{0.3}Sr_{0.7}FeO_3$ (a) effect of O_2 partial pressure (b) reduction/oxidation cycles in the presence of CH_4

Analogous stability is also shown by the materials at more severe reduction conditions, in the presence of a reductant such as methane or CO. The weight variations of a $La_{0.3}Sr_{0.7}FeO_3$ sample during 3 successive reduction/oxidation cycles are shown in Figure 4b. Upon reduction with methane the

sample looses weight delivering lattice oxygen that is being almost entirely taken up when oxygen is introduced at the same temperature.
Similar results have been obtained from all the prepared samples. Conclusively, the results shown in Figure 4 indicate that lanthanum strontium ironates have the potential to be used as redox materials. It is therefore of interest to investigate their ability in dissociating water vapor at their oxygen deficient state, taking up the oxygen and giving rise to the production of pure hydrogen.

Hydrogen production capability

Prior to testing in the membrane reactor, the performance of the prepared powdered materials is evaluated in a batch reactor unit. In this case the two different steps of the water splitting reaction, activation of the membrane material by reduction and deactivation by water splitting are performed sequentially. Since the materials are initially inactive, the first step includes activation with CH_4 or CO, down to the maximum oxygen loss point. Subsequently water vapour is injected to the reactor through a constant volume pulse system, until no conversion of water is measured and the catalyst is deactivated.
During the activation step the material oxidises the CH_4 or CO, at very high conversion, delivering its lattice oxygen:

$$La_{1-x}Sr_xFeO_{3-\delta} + a_1CH_4 \rightarrow La_{1-x}Sr_xFeO_{3-\delta-4a_1} + a_1CO_2 + 2a_1H_2O \quad (4)$$

or $$La_{1-x}Sr_xFeO_{3-\delta} + \beta CO \rightarrow La_{1-x}Sr_xFeO_{3-\delta-\beta} + \beta CO_2 \quad (5)$$

or written according to the Kröger-Vinck notation:

$$4a_1O_o + a_1CH_4 \rightarrow a_1CO_2 + 2a_1H_2O + 4a_1V_o^{\cdot\cdot} + 8a_1e' \quad (4a)$$

or $$\beta O_o + \beta CO \rightarrow \beta CO_2 + \beta V_o^{\cdot\cdot} + 2\beta e' \quad (5a)$$

where α_1, β are stoichiometric coefficients. Both the above equations (4) and (5) are forms of the more general equation (3) written in the introduction.
Subsequently and when some big portion of the available lattice oxygen has been consumed and a considerable oxygen deficiency has been reached the conversion of CH_4 or CO is reduced. When CH_4 is used, CO and H_2 appear in the products, at this state, probably originating from the reaction:

$$La_{1-x}Sr_xFeO_{3-\delta} + a_2CH_4 \rightarrow La_{1-x}Sr_xFeO_{3-\delta-a_2} + a_2CO + 2a_2H_2 \quad (6)$$

or alternatively:

$$a_2O_o + a_2CH_4 \rightarrow a_2CO + 2a_2H_2 + a_2V_o^{\cdot\cdot} + 2a_2e' \quad (6a)$$

where α_2 is a stoichiometric coefficient.
Finally a situation is reached where the solid reaches its maximum possible oxygen deficiency. Known volumetric portions of water are injected over the solid and the water is converted to hydrogen according to equation (7);

$$La_{1-x}Sr_xFeO_{3-\delta-a} + \beta H_2O \rightarrow La_{1-x}Sr_xFeO_{3-\delta-a+\beta} + \beta H_2 \quad (7)$$

or

$$\beta H_2O + aV_o^{\cdot\cdot} + 2ae' \rightarrow \beta O_o + (a-\beta)V_o^{\cdot\cdot} + 2(a-\beta)e' + \beta H_2 \quad (7a)$$

where α,β are stoichiometric coefficients. The first water pulse has the highest conversion since the solid is at its highest activation state. The oxygen atoms fill the anionic vacancies of the lattice and progressively the conversion of water is decreased up to the point where it becomes zero, when the material is completely deactivated. In Figure 5a, the results of a typical experiment at 1000°C are presented which includes activation of the sample $La_{0.7}Sr_{0.3}FeO_{3\pm\delta}$ with methane and water splitting.

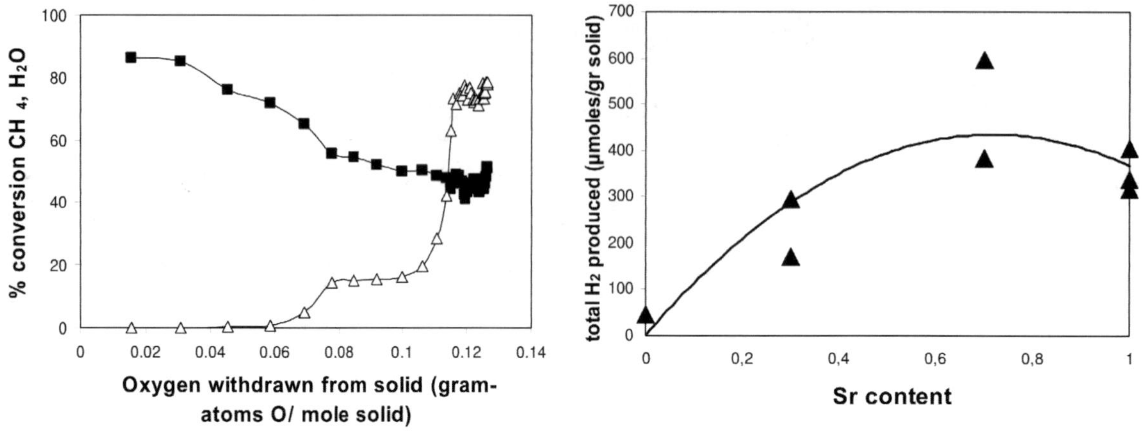

Figure 5. (a)Conversion of CH_4 (■) and water (△) as a function of oxygen withdrawn from the material $La_{0.7}Sr_{0.3}FeO_{3\pm\delta}$ at 1000°C. (b) total quantity of produced hydrogen at 1000°C vs. Sr content

The initial water conversion, as well as the total quantity of produced hydrogen, are functions of the stoichiometry of the material and the temperature of the reaction. In Figure 5b the total quantity of produced hydrogen at reaction temperature 1000°C is presented, as a function of Sr content for the materials with general formula $La_{1-x}Sr_xFeO_{3-\delta}$. At low Sr content, increasing x results in increased hydrogen production. A maximum is observed at x=0.7, thus the $La_{0.3}Sr_{0.7}FeO_{3\pm\delta}$ sample is considered thus far the optimum material for the water splitting reaction.

Membrane Reactor experiments

In the membrane reactor the two steps of the water splitting reaction, lattice oxygen removal-activation and hydrogen production–deactivation, are performed simultaneously at the different compartments of the reactor.

Because of technical reasons related with the complexity of mass spectrometry analysis, the experiments towards demonstrating the membrane reactor principle are performed with carbon monoxide as the reductant at the activation step, instead of methane which was used during the batch reactor experiments.

During a typical experiment, the membrane reactor is heated to 860°C, initially with inert gas flow in both compartments, while the signals of water and hydrogen in compartment 1 are continuously monitored. The injection of water in compartment 1 does not involve any changes in the signal of hydrogen in compartment 1, which maintains its background value, since the membrane is initially inactive.

The injection of CO in compartment 2, results in the reduction of the membrane surface in side 2, creating oxygen vacancies. Due to the ionic conductivity of the perovskitic material the oxygen vacancies are transported through the crystal lattice to the surface of the membrane in compartment 1 and activate it. The water that is injected in compartment 1 is thus split, hydrogen is produced, while simultaneously oxygen is delivered to the solid which fills its anion vacancies,. The lattice oxygen is transported once again via the material to the surface of membrane in compartment 2, where it is continuously consumed oxidising CO. Globally a clean flow of oxygen ions, originating from water, is created, from side 2 to side 1 of the membrane, while simultaneously oxygen vacancies flow from side 1 to side 2 of the membrane. Thus a steady state is reached where the surface of the membrane in compartment 2 is rich in oxygen, continuously oxidising CO, while the membrane surface in compartment 1 is rich in oxygen vacancies, continuously splitting water at the highest initial conversion. By optimising the process parameters e.g. gas flows, temperature, membrane thickness, it is possible to keep the membrane material at its highest activation state, during the steady state operation of the reactor.

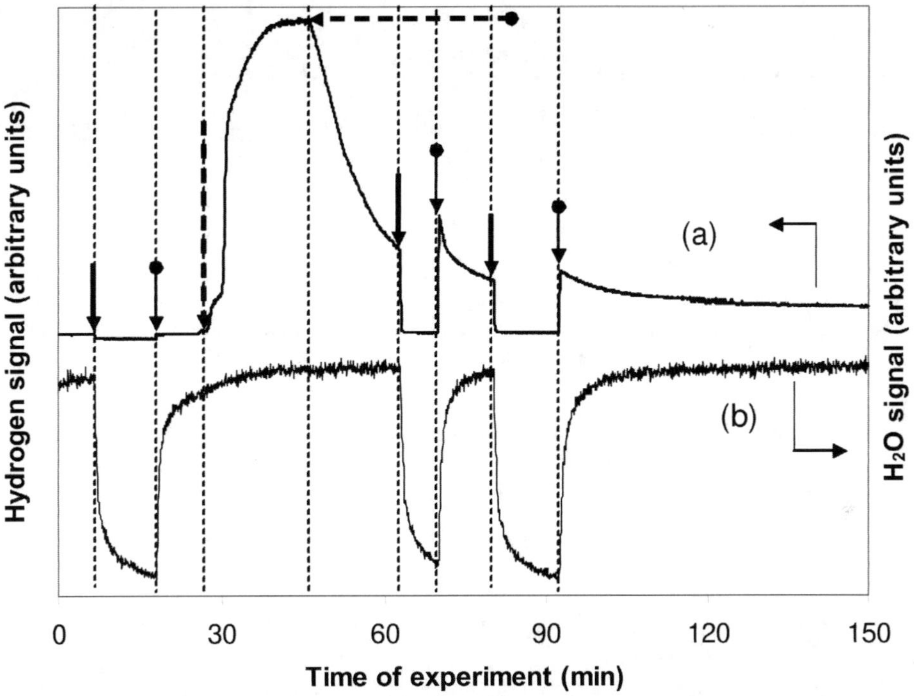

Figure 6: Hydrogen and water MS signals in compartment 1 during a typical experiment with $La_{0.3}Sr_{0.7}FeO_{3\pm\delta}$ (a) produced H_2 (b) injected H_2O,
⟶: water off, ●⟶ : water on, – –►: CO on, ●– –►: CO off

The interruption of CO injection in compartment 2 results in decreased hydrogen production rate in compartment 1. However, hydrogen yield does not become zero, instead it reaches a new steady value, which is of course lower than before (Figure 6). In this case the oxygen ions are desorbed from the membrane surface in side 2 as molecular O_2, only under the effect of oxygen partial pressure difference between the two sides of the membrane. The flow of oxygen ions is in this case smaller, however it is not insignificant. The water continues be split, producing hydrogen which originates only from renewable water.

Periodical water feed shut downs in compartment 1, were performed during this steady state, in order to excluded the possibility of base-line shifting and quantify the obtained results. As shown in Figure 6, interruption of H_2O injection results in a significant drop of the H_2 signal, down to its background value, which proves that the observed hydrogen signal is due to real hydrogen produced from the decomposition of the injected water.

Similar experiments with inactive membrane materials (a-Al_2O_3 and fused silica) did not show any significant change in the hydrogen signal during water feed shut downs, either before or after CO injection in compartment 2. The hydrogen production rate under unforced reduction is estimated from the loop pulsing calibration experiments in the membrane reactor to be 47.5cm^3 H_2 (STP)m^{-2} min^{-1} The rate reaches 145 cm^3 H_2 (STP)m^{-2} min^{-1} if metal oxide reduction in compartment 2 is enhanced by carbon monoxide oxidation.

Conclusions

The operation of a membrane reactor for continuous and isothermal hydrogen production according to the concept of Figure 1a is feasible. The material oxidation by water dissociation, and the material reduction by CH_4 or CO oxidation, processes are combined on either side of a dense perovskite membrane. At 860°C a continuous hydrogen production of 145 cm^3 H_2 (STP)m^{-2} min^{-1} has been obtained with a $La_{0.7}Sr_{0.3}FeO_3$ membrane.

The membrane reactor process can also run, though at lower hydrogen production rates, without the need of a carbon containing reductant, giving thus rise to a process that produces hydrogen entirely from water. The hydrogen production in this case approaches 47.5 cm^3 H_2 (STP)m^{-2} min^{-1}.

Further research is nessecary towards optimization of the membrane materials as well as the reactor design and process parameters in order to attain hydrogen production rates that are commercially acceptable.

References

1 Steinfeld Aldo (2005), "Solar thermochemical production of hydrogen—a review", *Solar Energy*, Vol. 78, pp. 603-615

2 Evdou A., Zaspalis V. and L. Nalbandian (2008), "Perovskite membrane reactor for continuous and isothermal redox hydrogen production from the dissociation of water", *Journal of Membrane Science,* Vol. 325 pp. 704–711

3 Evdou A., Zaspalis V. and L. Nalbandian (2008), "$La_{1-x}Sr_xMnO_{3-\delta}$ perovskites as redox materials for the production of high purity hydrogen", *International Journal of Hydrogen Energy,* Vol. 33, pp. 5554–5562

4 Chen C.H., Kruidhof H., Bouwmeester H.J.M. and A.J. Burggraaf (1996), "Preparation of gas-tight strontium-doped lanthanum cobaltate by an aqueous sol-gel process", *Material Science and Engineering B*, Vol. 39, pp. 129-132.

Hydrogen Separation with Palladium-Copper Membranes on Porous Stainless Steel with Sol-Gel Derived Ceria as Diffusion Barrier

Ailing Qiao[a], Ke Zhang[a], Ye Tian[a], Huajiang Luo[a], Y. S. Lin[a,b], Yongdan Li *[a]

[a] *Tianjin Key Laboratory of Applied Catalysis Science and Technology and State Key Laboratory for Chemical Engineering (Tianjin University), School of Chemical Engineering, Tianjin University, Tianjin 300072, China*

[b] *Department of Chemical Engineering, Arizona State University, Tempe, AZ 85287-6006, USA*

Abstract

Pd-Cu alloy membranes were prepared by electroless plating on porous stainless steel (PSS) support. Sol-gel derived ceria was introduced as the intermediate layer by a sol-dip-coating method to prevent intermetallic diffusion and to enhance the affinity between the support and membrane. Ceria layer moderates the pore size and porosity of the support effectively to meet the needs of the alloy layer deposition. High temperature permeation was carried out from 573 K to 773 K with a pressure difference of 1 atm. The results showed that properly prepared ceria layer was effective as the diffusion barrier in the temperature range examined, which is suitable for water-gas shift reaction. The performance of the membranes in this work is also compared with that in our previous work in which pure Pd membrane and yttrium stabilized zirconia intermediate layer were examined.

Keywords: Palladium-copper alloy membrane; intermediate layer; hydrogen permeation; electroless plating; ceria, sol-gel.

1. Introduction

Dense palladium-based membranes have attracted extensive attention because of their high hydrogen permeability and perfect hydrogen selectivity, as well as their thermal stability and catalytic activity [1-2]. Compared with pure palladium membrane, Pd-Cu alloy membrane demonstrates better resistance to hydrogen embrittlement and poisoning by gaseous impurities (H_2S, CO and H_2O, etc) and lower cost [3-4]. Overall, an intermediate layer is usually introduced on PSS surface to decrease average pore

* Corresponding author: ydli@tju.edu.cn

diameter, make PSS surface more smooth and avoid intermetallic diffusion.

Porous ceramic materials are frequently used as intermediate layer materials, such as γ-alumina, titania, zirconia and ceria [5-7]. Among these materials, ceria has a stable structure of fluorite in temperature range from room temperature to its melting point (2750 K), and its linear thermal expansion coefficient is the closest to palladium. However, information on sol-gel derived ceria as intermediate layer for Pd-based membranes are rather limited, which is mainly due to the difficulty in obtaining stable ceria sol [8]. Up to now, few reports on stable ceria sol for mesoporous membranes are published and the concentration of sol is relatively low.

In this work, stable ceria sol was synthesized with higher concentration than the reported value in literature and then dense Pd-Cu composite membrane was prepared on ceria-coated PSS substrate by sequential electroless plating method. Hydrogen permeation through the as-prepared Pd-Cu membranes was also examined.

2. Experimental

PSS 316L disks of 25.4 mm diameter and 1.2 mm thickness were cleaned ultrasonically with Na_2CO_3 solution, acetone and de-ionized water at 333 K sequentially at first. The material was then treated further with 1.0 mol·L^{-1} nitric acid solution for 15 min, and then with 1.0 mol·L^{-1} sodium hydroxide solution for 15 min at 353 K, in order to improve the surface roughness.

Stable ceria sol was prepared by dissolving 20 g 99.5% $Ce(NO_3)_3 \cdot 6H_2O$ in 40 ml de-ionized water at 343 K. 1.0 mol·L^{-1} NH_4OH solution was added slowly to this solution. The precipitate was filtrated, washed with de-ionized water and peptized with dropwise addition of 1.0 mol·L^{-1} HNO_3 solution under stirring and heating at 343 K until a milky sol was obtained at a pH of 1.5-2.5 with concentration of 11.4 g CeO_2/L. Then PSS was dip-coated with one side being covered with an adhesive tape at room temperature. The coated support was dried at 313 K with controlling the humidity in an oven, to avoid crack development. After drying, the material was calcined in a furnace at 723 K for 3 h with a heating rate and a cooling rate at 0.6 K/min. The coating-drying-calcining process was repeated for three times.

Prior to electroless plating, the surface of the ceria side was sensitized and activated by dipping in a bath containing a commercial reducer (OPC-150) and palladium chloride [9]. The composition of the Pd/Cu electroless plating bath are listed in Table 1 and Table 2, respectively. The as-deposited Pd-Cu/CeO_2-PSS composite membrane was rinsed with de-ionized water to remove salts adsorbed or entrapped, and then dried overnight in an oven at 333 K. Finally, the composite membrane was heated in nitrogen to 753 K with a rate of 0.6 K/min, and then annealed in hydrogen at the same temperature for 6 h.

Table 1 Composition of the electroless plating bath for palladium

Component	Amount
$PdCl_2$ (g/L)	16.7
Ammonia (ml/L)	175
$Na_2EDTA.2H_2O$ (g/L)	37.2
Sodium Dodecyl Sulfate (0.01%, ml/L)	3
$N_2H_4 \cdot 2H_2O$ (0.206M) (ml/L)	24

Table 2 Composition of the electroless plating bath for copper

Component solution	Amount
$Cu(NO_3)_2 \cdot 3H_2O$ (g/L)	9.67
$Na_2EDTA \cdot 2H_2O$ (g/L)	29. 8
NaOH (g/L)	8.65
Formaldehyde (ml/L)	50

The gas permeation experiment was carried out in a home-made steady-state permeation set-up. A permeation cell with an effective area of 2.01 cm^2 was employed. A high temperature membrane cell was designed and used to measure the permeation properties at high temperatures. Pure hydrogen or nitrogen (99.999% purity) was introduced to the upstream side of the membrane, and the pressure was measured with an accurate pressure gauge. The downstream flow rate was measured with an electronic soap-bubble flow meter, and the downstream side pressure was atmospheric (101 kPa) without purging gas. The equipment and membrane cell used in this work has been shown in a previous paper [10].

3. Results and discussion

Fig.1 shows the surface morphology of PSS coated by ceria by one or three times. After three-time modification, the big pores were covered by ceria particles and thus the smoothness of support surface improves. Fig 2 shows the XRD patterns of ceria surfaces before (a) and after (b) calcinations at 723 K for 3h. It is shown that ceria particles demonstrate the pattern of the fluorite (FCC) structure. Calcined samples show higher crystalline intensity than that of the samples before calcinations.

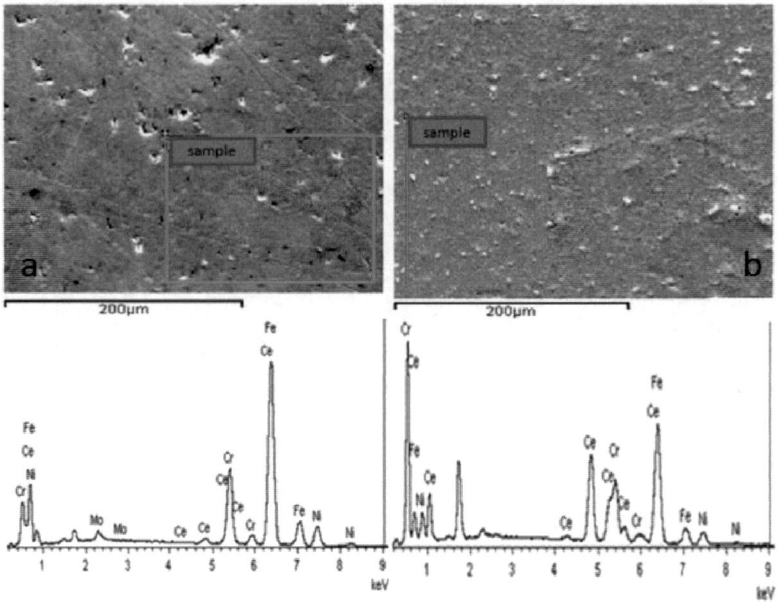

Fig.1 Surface morphology and EDS analysis of one-time coated CeO_2 (a) and three-times coated CeO_2 (b) on PSS

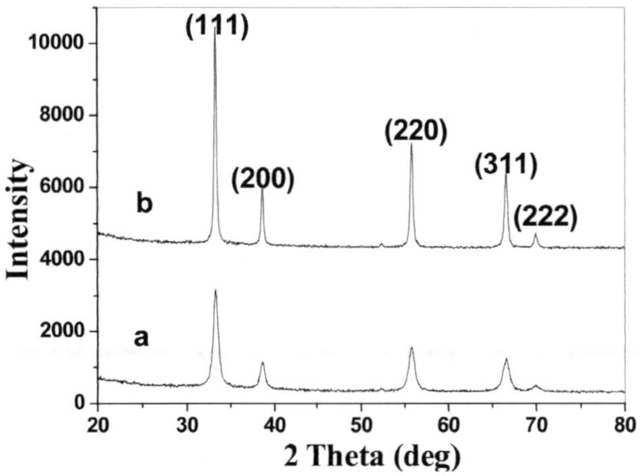

Fig.2 XRD patterns of CeO_2 layer (a) before calcination (b) after calcination at 723K for 3h

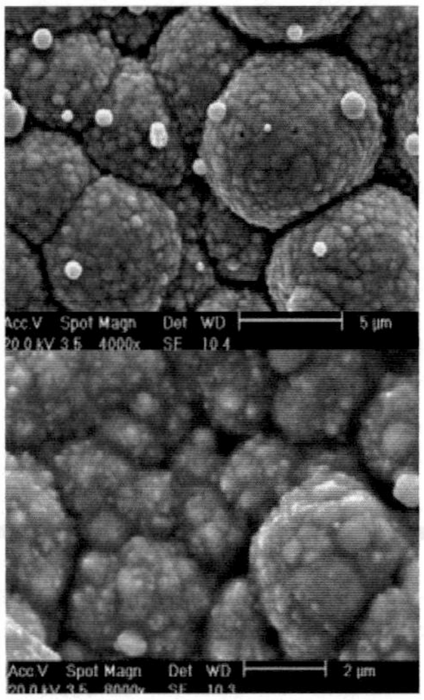

Fig.3 Surface morphology of palladium layer

Fig.3 and Fig.4 show the surface morphology of Pd and Cu layer after being electroless plated, respectively. Pd membrane demonstrates typical "cauliflower" structure that sphere agglomerates formed by small Pd particles' interconnection, while Cu surface exhibits much smaller particles directly that also closely connected. In Fig.5, the surface and the cross-section of Pd-Cu membrane are illustrated, respectively. The cross-sectional morphology (Fig.5a) proves the formation of homogeneous alloyed Pd-Cu layer and excellent adhesion of the alloy film to the modified PSS support. A strong interaction between palladium and copper in which bonds between "sd" levels of Pd and resonant "dsp" levels of Cu can form. The formation of metallic bonds would lead to a redistribution of the electrons between d and sp levels of both Pd and Cu. The originally empty conduction band levels of Cu becomes partially occupied upon alloying whereas Pd losses more or less a part of its "sd" metallic character due to the interaction with Cu [10,11]. Therefore, a homogeneous palladium-copper alloy film can be formed by thermal diffusion.

Fig.4 Surface morphology of copper layer

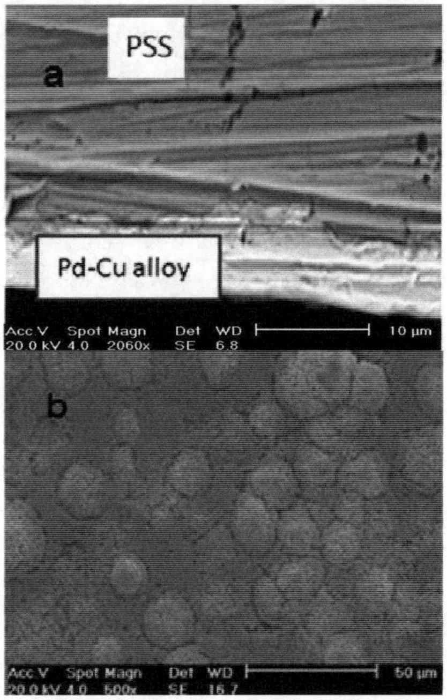

Fig.5 The cross-section (a) and surface (b) morphology of Pd-Cu alloy membrane

The gas-tightness of the as-prepared Pd membranes was confirmed by zero measurable nitrogen permeation flux at the pressure difference of 0.1 MPa at room temperature, which gives a nitrogen permeance smaller than 10-10 mol/(m^2·s·Pa) based on the permeation equipment limit. Fig.6 shows hydrogen flux through membrane versus trans-membrane hydrogen pressures at different temperatures. As shown, hydrogen flux is linearly proportional to the difference of hydrogen pressure, which indicates that surface reaction is the rate-limiting step. At the condition of 773 K and 0.1 MPa, the hydrogen flux of this membrane is 9.9 cm^3/cm^2·min, compared to the flux of 14.6 cm^3/cm^2·min for Pd/YSZ/PSS in our previous work [5] and 12.0 cm^3/cm^2·min for Pd membrane prepared by Anwu Li et al. [12] at the same condition. It is mainly because that the Pd-H bond is weaker in Pd-Cu alloy than in pure Pd [11].

For Pd-Cu alloy membrane, the adsorptive properties of Pd are affected by the alloying effect. For instance, for hydrogen adsorption, it is clear that the Pd-H bond is weaker in Pd-Cu alloy than in pure Pd. This is due to the effects of Pd electronic structure and a change in the adsorption sites. As mentioned before, a strong interaction between palladium and copper leads to a redistribution of electrons between d and sp levels of both Pd and Cu. This in turn results in a weak overlaps of the “sd” levels of Pd and the “s” level of H. Furthermore, hydrogen can be assumed to adsorb and dissociate mainly on palladium atoms in Pd-Cu system [11,13]. The Cu atoms in the composite membrane are non-effective for hydrogen dissociative adsorption on the surface of membrane. This is the reason why hydrogen permeation flux on Pd-Cu composite membrane is lower than the pure Pd membrane under a similar condition.

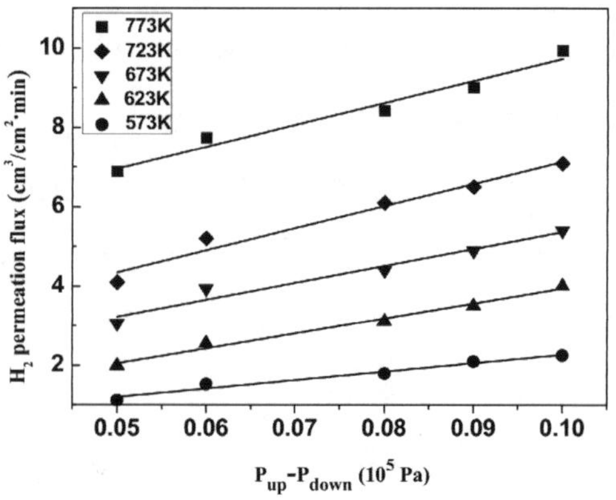

Fig.6 Dependency of H_2 flux as a function of pressure difference

4. Conclusions

Dense electroless-plated Pd-Cu alloy membranes were prepared on porous stainless steel (PSS) supports with sol-gel derived ceria as the intermediate layer. Ceria is suitable for PSS surface modification that can make its surface smoother and its average pore size smaller. High temperature permeation was performed from 573 K to 773 K with the pressure difference of 1 atm. The results showed that the hydrogen permeation flux through Pd-Cu/CeO2/PSS was lower than that of pure palladium membrane obtained in our previous study under the same conditions. It is mainly because that the adsorptive properties of Pd in Pd-Cu system are affected by the alloying effect where H_2 can be assumed to adsorb and dissociate mainly on palladium atoms. The Cu atoms in the composite membrane are non-effective for hydrogen dissociative adsorption on the surface of membrane.

Acknowledgements

This work has been supported by the Natural Science Foundation of China under contract numbers 20425619. The work has been also supported by the Program of Introducing Talents to the University Disciplines under file number B06006, and the Program for Changjiang Scholars and Innovative Research Teams in Universities under file number IRT 0641.

References

1. S. Uemiya, Sep. Purif. Methods 1999, 28, 51-85.
2. S. N.Paglieri, J. D.Way, Sep. Purif. Methods 2002, 31, 1-169.
3. B.D. Morreale, M.V. Ciocco, B.H. Howard, R.P. Killmeyer, A.V.Cugini, R.M. Enick, J. Membr. Sci. 241 (2004) 219-224.
4. B.H. Howard, R.P. Killmeyer, K.S. Rothenberger, A.V. Cugini, B.D. Morreale, R.M. Enick, F. Bustamante, J. Membr. Sci. 241 (2004) 207-218.
5. K. Zhang, H.Y. Gao, Z.B. Rui, P. Liu, Y.D. Li and Y.S. Lin, Ind. Eng. Chem. Res., 2009, 48 (4), 1880-1886.
6. K. Zhang, X.T. Wei, Z.B. Rui, Y.D. Li, Y. S. Lin, AIChE. J., 2009, 55, 630-639.
7. J. Kim, Y. S. Lin, J. Membr. Sci. 1998, 139, 75-83.
8. N. Rane, H. Zou, G. Buelna, Y. S. Lin, J. Membr. Sci., 2005, 256, 89-97.
9. K. Zhang, H.Y. Gao, Z.B. Rui, Y. S. Lin, Y. D. Li, Chin. J. Chem. Eng., 2007, 15(5), 643-647.
10. H.Y. Gao, Y. S. Lin, Y.D. Li, B. Q. Zhang, J. Membr. Sci., 2005, 263, 142-152.
11. A. Rochefort, M. Abon, P. Delichre, J. C. Bertolini, Surf. Sci., 1993, 294, 43-52.
12. A.W. Li, W.Q. Liang, R. Hughes, J. Membr. Sci., 1998, 149, 259-268.
13. A. Noordermeer, G.A. Kok, B. E. Nieuwenhuys, Surf. Sci. 1986, 172, 349-362.

HYDROGEN PRODUCTION BY METHANE DECOMPOSITION USING ORDERED MESOPOROUS CARBONS AS CATALYSTS WITH REMARKABLE ACTIVITY AND STABILITY

D.P. Serrano[1,2], *J.A. Botas*[1,2], *J.L.G. Fierro*[3], *R. Guil-López*[1], *P. Pizarro*[1] *and G. Gómez* [1]

[1]*Department of Chemical and Environmental Technology, ESCET, Universidad Rey Juan Carlos.*
c/ Tulipán s/n, 28933 Móstoles, Madrid, Spain
email: david.serrano@urjc.es
[2] *IMDEA Energía. c/ Tulipan s/n, 28933 Móstoles, Madrid, Spain*
e-mail: david.serrano@imdea.org
[3]*Instituto de Catálisis y Petroleoquímica, CSIC, c/ Marie Curie 2*
28048 Madrid, Spain

Abstract

The properties of ordered mesoporous carbons for hydrogen production by methane decomposition have been compared to those corresponding to a wide variety of carbon materials (carbon blacks, activated carbon, carbon nanotubes and graphite). The major aim of this study is to identify the carbon properties which control its initial catalytic activity. The latter was determined and compared using temperature programmed experiments in a thermobalance. The initial activity was followed through the threshold temperature, defined as the temperature at which hydrogen production starts being detected.

The highest initial catalytic activity was observed with the CMK materials, followed by activated carbon and carbon blacks. On the other hand, at longer reaction times activated carbon was quickly deactivated. Deactivation process seems to be greatly linked to the presence of micropores whereas long-term activity is retained in those materials formed by mesopores (CMKs) or nanoparticles (carbon blacks), which make them more resistant to deactivation by the formation of carbonaceous deposits.

Whereas no clear dependence is observed between the initial catalytic activity and the surface area neither with the graphite order or the presence of polar groups in the carbon catalysts, characterization of these materials by XPS shows that a direct relationship exists in regard to the amount of defects present on the graphene layers. This fact strongly supports that these defects are the main active sites for methane decomposition over carbon catalysts.

Keywords: hydrogen production, methane decomposition, carbon black, activated carbon, carbon nanotubes, ordered mesoporous.

1. INTRODUCTION

Hydrogen is a clean energy carrier as it just leads to the formation of water during its controlled oxidation, either by direct combustion or when fed to a fuel cell [1]. However, hydrogen is not a primary energy source, since it cannot be found free in our planet, hence, it must be produced from other precursors. Although water is the ideal H_2 source, in short-mid term, its production must be considered from whatever available sources, including fossil fuels [2].

Methane, main component of natural gas, is the hydrocarbon with the highest H/C ratio and, nowadays, the most used hydrogen source. Conventional H_2 production is carried out by steam reforming of methane (SRM) [3], although this process emits large quantities of CO_2 (at least 1 mol of CO_2/mol of converted CH_4). Thereby, additional stages to separate and capture the CO_2 so produced must be included, increasing the price of the SMR process whereas its overall efficiency is reduced.

A different approach for H_2 production from methane is the one based on the methane decomposition reaction ($DeCH_4$) [4, 5]:

$$CH_4\ (g) \rightarrow 2{\cdot}H_2\ (g) + C(s) \qquad \Delta H_{1073} = 90.1\ kJ/mol\ (CH_4) \qquad \text{(Equation 1)}$$

The main advantage of $DeCH_4$ is the production of pure H_2, free of CO or CO_2, since the carbon contained in the methane molecule is transformed into a carbonaceous solid [6,7]. Accordingly, the hydrogen so produced is optimal for being fed to a fuel cell [8].

Traditionally, methane decomposition is catalyzed by nickel, copper, iron and a number of noble metals [9,10]. This process is often aimed to the formation of carbon nanotubes, which requires the Ni-catalyst to be separated and eliminated at the end of the process [10]. However, when the goal of the process is H_2 production, the metal catalyst must be recovered and reused to improve the economy of the process. Metal catalysts deactivate quickly by the carbon produced, which settles on the metal surface covering the active sites, then, regeneration stages are needed. Catalyst regeneration could be accomplished by combustion of the carbon deposits, although it would produce CO_2 emissions similar to those of the steam reforming [9, 11].

A more appealing option may be the use of carbonaceous materials as catalysts as they present a lower cost than the metallic catalysts [12]. Moreover, since in this case the catalyst and one of the reaction products possess quite similar properties, this alternative opens the possibility of preparing the catalyst by activation of the carbon produced during the reaction. Moreover, the reaction could become an autocatalytic system provided that the properties of the deposited carbon may be tailored for the latter to be catalytically active in methane decomposition. Nevertheless, since large amounts of solid carbon are obtained by $DeCH_4$, the possible commercial application of this material is also an essential factor to consolidate the technology, which in turn will depend on the properties of the produced carbon [13].

A variety of carbon materials have been recently investigated in the literature as $DeCH_4$ catalysts such as activated carbons [14], carbon blacks [15] and coal chars [16]. Different proposals have been stated trying to assign the differences in catalytic activity observed with those materials to a variety of parameters (surface area, polar groups, defects, etc), but the origin of the activity is still not well established. More recently, we have reported preliminary results about the use of ordered mesoporous carbons as catalysts in this reaction. In particular, two carbon materials belonging to the CMK family [17] have shown to be highly active catalysts for hydrogen production via methane decomposition reaction [18]. The best results were obtained over the CMK-5 carbon, characterized by having high surface area and a double mesopore system. This material showed to be a promising catalyst due to its high activity and stability.

In this context, the present work has been outlined to study the main factors governing the catalytic activity of carbon catalyst in $DeCH_4$. Thereby, a wide variety of carbon materials, including the aforementioned CMK-type carbons, have been selected. A detailed characterization of the carbon catalysts has been carried out in order to check and correlate their main properties with their activity.

2. EXPERIMENTAL

2.1. Carbon catalysts

Ordered mesoporous carbons, CMK-3 and CMK-5, were synthesized using a nanocasting pathway with pure silica SBA-15 [19] as hard-template according to a synthesis procedure reported in literature [20]. Furfuryl alcohol (FA, Acros Organics) was used as carbon precursor, trimethylbenzene (TMB, Aldrich) as solvent and oxalic acid (OA, Acros Organics) as polymerization catalyst. The synthesis procedure comprised the following steps: a) filling the pores of SBA-15 with a solution containing 0.012 g OA / 3 mL FA in TMB, by the wetness impregnation technique; b) polymerization of the carbon precursor within the SBA-15 pores by heating treatment at 50 °C for 1 day and, then, at 90 °C for 2 days; c) carbonization under inert atmosphere by first heating at 150 °C for 3 h, secondly heating to 300 °C at 1 °C/min, then heating to 850 °C at 5 °C/min, and finally keeping the last temperature for 3 h; d) removal of the silica template by dissolution with an ethanol-water mixture containing 1 M NaOH; e) filtration and washing of the black powders and f) drying at 90 °C overnight.

CMK-3 sample was obtained by using pure FA in the impregnation solution, while for CMK-5 preparation a solution containing 50 % FA in TMB was used. As a consequence, in the last case the pores in SBA-15 are not completely filled by the carbon precursor.

In addition, a number of industrially manufactured carbons were evaluated as catalysts for methane decomposition. These materials were selected as representatives of a variety of carbon types exhibiting a high range of properties: two carbon blacks (Vulcan XC72 and Black Pearls 2000, both from *Cabot Corp.*), one activated carbon (Merck), one sample of multiwalled nanotubes (Sun Nanotech.) and one graphite sample (Fluka).

2.2. Carbon catalyst characterization

The possible presence of metals in the carbon catalysts was checked by ICP spectrophotometry using a Varian VISTA AX CCD instrument. Textural properties and BET surface area of the catalysts were measured by N_2 adsorption/desorption at 77 K, by means of a Micromeritics TRISTAR 3000 instrument. The surface area was calculated by applying the BET equation, whereas the contribution of micropores and external surface to both volume and surface area was evaluated by means of the t-plot method. XRD patterns of commercial samples were recorded on a Philips PW 3040/00 X'Pert MPD/MRD diffractometer using Cu-K_α radiation at a scan rate of 0.2 $^{o}\cdot s^{-1}$, in the 2θ range of 5°–60°. The degree of arrangement of graphene planes for each carbon-catalyst was studied using the characteristics graphitic C (002) and C (101) peaks. XPS measurements of carbons were acquired with a VG Escalab 200R spectrometer equipped with a hemispherical electron analyzer and a MgK_α (hν=1253.6 eV) X-ray source. Survey and multi-region spectra were recorded at C_{1s} and O_{1s} photoelectron peaks. The peaks were fitted by a non-linear least square fitting program using a properly weighted sum of Lorentzian and Gaussian component curves after background subtraction according to Shirley and Sherwood. The constant charging of the samples was corrected by referencing all energies to the C_{1s} peak contribution at 284.8 eV.

2.3. CH_4 decomposition experiments

Catalytic tests were carried out in a thermobalance (TA Instruments TG/DTA SDT 2960 thermal analyser), by following the sample weight increase during the reaction caused by the carbon deposition on the catalyst (Eq. 1). The catalytic tests were carried out at atmospheric pressure, by switching a flow of 100 mL·min^{-1} of methane (10 % CH_4 in Ar), using about 30 µl of each catalyst (i.e. different weights for each carbon) placed in a ceramic pan (90 µl of capacity). Carbon was produced as the reaction progressed, filling the ceramic pan. Prior to the reaction, the catalysts were dried under N_2 flow until 250 °C. After drying, the samples were purged under the flow of the reaction mixture at 50 °C, and the temperature was increased in the range 50-1100 °C with a constant heating rate of 10°C·min^{-1}, being kept constant at 1100 °C for 40 min. The initial catalytic activity was evaluated for each catalyst by using the threshold temperature (T_{th}, defined as the temperature at which the hydrogen production reached 10^{-4} mol·g^{-1}). The hydrogen production was calculated from Eq. 1.

3. RESULTS AND DISCUSSION

3.1. Carbon catalyst properties

The carbonaceous solids used as catalysts for $DeCH_4$ were selected in order to include in the study materials having a wide range of physicochemical properties, as it is summarized in Table 1.

TABLE 1. Properties of the carbon catalysts.

Catalysts	Metal content (wt%)	Textural properties S_{BET} / $m^2 \cdot g^{-1}$	S_{ext} / $m^2 \cdot g^{-1}$	$S_{\mu pore}$ / $m^2 \cdot g^{-1}$	XRD ratio C(101)/C(111)
Activated carbon (AC)	0.21	1152	81	1071	0.84
Carbon black (CB-bp)	< 0.005	1285	297	988	0.94
Carbon black (CB-v)	< 0.005	196	168	28	0.65
Multiwall nanotubes (MWNT)	14.8	65	45	20	0.51
Graphite (GRAPH)	< 0.005	8	8	---	0.03
CMK-3	< 0.005	1323	1323	---	0.53
CMK-5	< 0.005	1940	1940	---	0.52

The textural properties of the catalysts have been obtained from the N_2 adsorption-desorption isotherms (see Figure 1). AC, CB-bp and CMKs are high surface area materials. In the case of the AC, the surface area is originated mainly by the presence of microporosity, as denoted by the type I isotherm corresponding to this material. For the CB-bp sample a combination of microporosity and external surface area is present as this material is formed by nanoparticles. On the other hand, the CB-v sample exhibits little microporosity, which is consistent with being formed mainly by non-porous nanoparticles. The multiwall nanotubes present a small surface area, whereas this parameter is even lower for the graphite sample.

Both CMK samples exhibit type IV isotherms, denoting clearly the presence of mesoporosity. CMK-3 shows a single adsorption step featured by an inflection point at a relative pressure about 0.30, which indicates that this material is formed by just one type of mesopores. In contrast, two steps can be observed for the CMK-5 sample. Thus, in addition to the main step with inflection point at a relative pressure about 0.30, a second adsorption step occurs at higher relative pressures (c.a. 0.55). As earlier reported, both CMK samples have in common a pore system of 3 nm diameter size. This porosity corresponds to the cavities generated by the removal of the silica template. Accordingly, these mesopores are a replica of the silica walls in SBA-15, which implies that their size is determined by the width of the latter. A second type of mesopores, located at 5.8 nm, appears for CMK-5. This additional mesoporosity is due to the partial filling of the SBA-15 pores with the carbon precursor during the CMK-5 preparation. In this case, the size of these mesopores is determined by the diameters of the mesopores in SBA-15, although having a smaller size due to the partial filling, polymerization and carbonization of the carbon precursor.

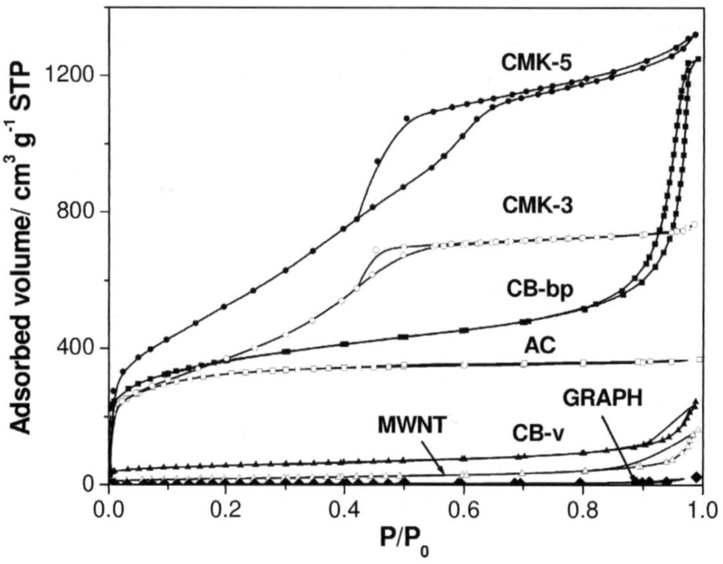

Figure 1. N_2 adsorption-desorption isotherms of the carbon catalysts.

The graphenic crystallinity of the carbon catalysts can be derived from their XRD patterns (Figure 2). Taken into account the enlargement factor used in this figure, a marked difference of crystallinity can be established between the samples. These differences can be quantified using the ratio between two of the peaks typical of graphenic materials, C(002) at $2\theta = 26°$ and C(101) at $2\theta = 44°$ (Table 1). This ratio is usually taken as indicative of the order of the graphene sheets for each carbon sample [21]. Thus, the highly ordered planes of the graphite (GRAPH), which are piled up, show a low value of this XRD ratio. On the contrary, strongly disordered materials having a high surface area, as it is the case of AC and CB-bp, present a XRD ratio close to 1. Likewise, the broadness of these reflections indicates that CMK carbons must be considered as essentially amorphous materials, with just small domains of graphitic crystallites. The (002) reflection shifts towards lower angles for the carbon black and the CMK carbons, which indicates a higher curvature of the graphene layers, as typical of turbostratic carbons [22,23].

In order to further quantify the degree of arrangement of the graphene layers, as well as to characterize the chemical nature of the catalyst surface, XPS-spectra were recorded. Figure 3 displays C_{1s} core-level spectra of a representative CB-bp sample. The C_{1s} profiles were fitted to four main components: (i) asymmetrical so-called *graphite peak* at 284.8 eV linked to carbon polyaromatic structures, (ii) carbon atoms single bonded to oxygen at 286.3 eV, e.g. alcohols and phenols (C-OH), (iii) double C=O bonds from ketones and aldehydes at 287.8 eV, and (iv) carbon atoms with three C-O bonds (-COO-) at 289.3 eV from carboxylic groups [24, 25]. All samples present the same five peaks, although the difference between them relays on the population of each component. The exception is graphite in which the C_{1s} spectrum only presents four components (no C_{1s} carboxylic component was detected).

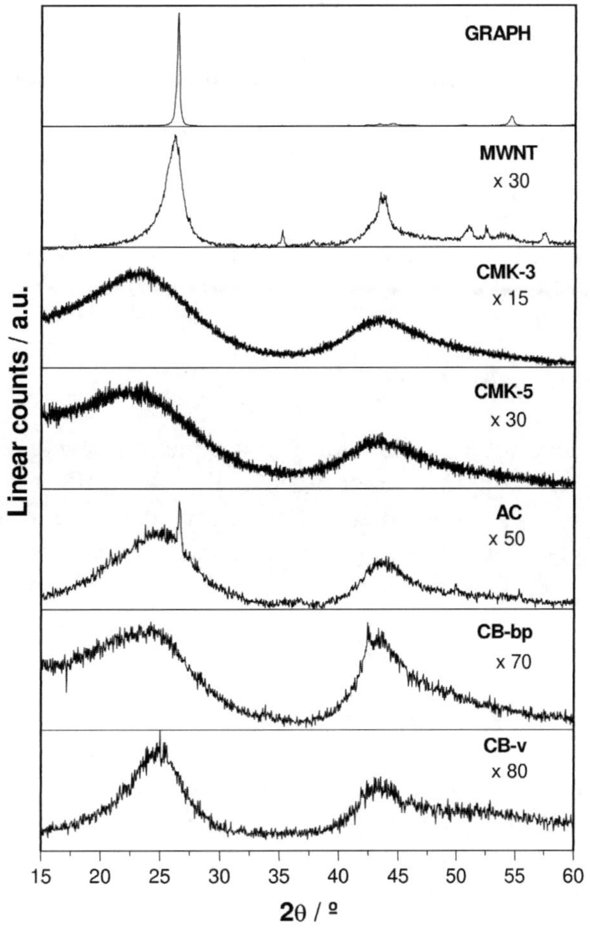

Figure 2. XRD patterns of the carbon catalysts.

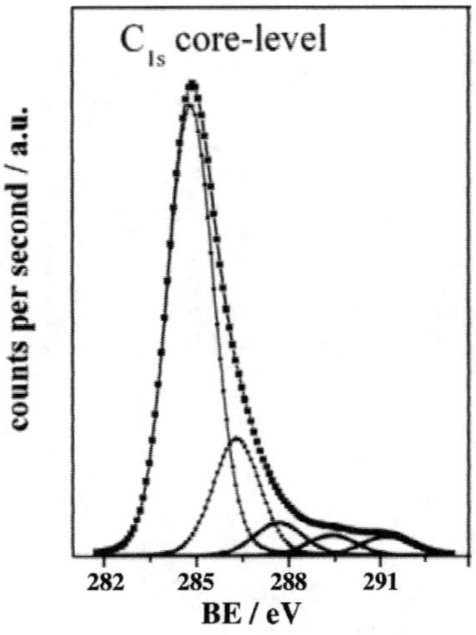

Figure 3. XPS measurements for the fresh CB-bp carbon black catalyst: fitted curves for C_{1s} core-level spectrum.

Besides the chemical nature of the surface, XPS analyses also permit to study the nature of the basal planes of graphene layers. In an ideal graphene layer, the 2p orbitals of the carbon atoms overlap and form a large system of delocalized π electrons. Defects in the graphene layers avoid overlapping of the 2p orbitals [26], which are linked to their degree of arrangement (graphitization). The full width at half maximum (FWHM) of the *graphite peak* at 284.8 eV is directly related to the number of defects in the graphene structure [26,27]. A high concentration of defects in the graphene layers leads to a wide *graphite peak* and to a high value of the FWHM parameter. The trend in the degree of arrangement of the graphene sheets obtained by XPS agrees well with that derived from other characterization techniques (XRD and TEM), being as follows from more to less order:

GRAPH > MWNT > CB-v > CB-bp ≈ AC > CMK-3 > CMK-5

3.2. CH_4 decomposition over carbon catalysts

Figure 4 shows the evolution of the hydrogen production during the methane decomposition obtained in temperature programmed experiments for the different carbonaceous materials employed as catalysts. Table 2 summarizes the $DeCH_4$ catalytic results obtained with these materials. The initial activity for each sample was determined using the threshold temperature (T_{th}, temperature at which the hydrogen production reached 10^{-4} mol·g^{-1}). In this way, a low T_{th} value means high catalytic activity and, on the contrary, a high value means low catalytic activity. The following catalytic activity order is observed from the data shown in Table 2:

CMK-3 > CMK-5 > AC ≈ CB-bp > CB-v > MWNT > GRAPH

As shown in a previous work [18], CMK materials are very active catalysts in methane decomposition. Thus, both CMKs exhibit the lowest threshold temperature with values even quite lower than those corresponding to the AC and CB-bp samples. It must be noted that the latter have been earlier reported in the literature as the carbonaceous materials having the highest initial catalytic activity in methane decomposition.

Table 2. Catalytic activity in the $DeCH_4$ reaction: threshold temperature and amount of hydrogen produced at different temperatures.

Catalyst	w°/mg	T_{th} /°C[a]	Hydrogen produced / mol·g_{catal}^{-1}		
			950 °C	1050 °C	1100 °C
AC	11.9	779	0.023	0.054	0.061
CB-bp	4.41	778	0.020	0.073	0.118
CB-v	3.14	795	0.010	0.044	0.093
MWNT	3.41	865	0.006	0.035	0.082
GRAPH	10.0	920	-	0.004	0.023
CMK-3	2.99	744	0.059	0.186	0.219
CMK-5	1.20	753	0.081	0.317	0.428

[a] T_{th} = Threshold temperature, defined as temperature for H_2 produced = 10^{-4} mol·g^{-1}

The activity order undergoes some changes when the catalyst behaviour is compared at long reaction times. Under those conditions, the activated carbon becomes quickly deactivated. The deactivation process seems to be linked to the presence of micropores whereas long-term activity is retained in those materials formed by mesopores (CMKs) or nanoparticles (black carbons), which make them more resistant to deactivation by pore blockage caused by the formation of carbonaceous deposits. After about 25 minutes of reaction, the activity of both CMK samples slows down, showing a sharp change in the curve slope. Nevertheless, after that change, both CMKs still exhibit an important and almost constant hydrogen production rate. Thus, the CMK-5 material remains by far as the most active catalyst at long reaction times.

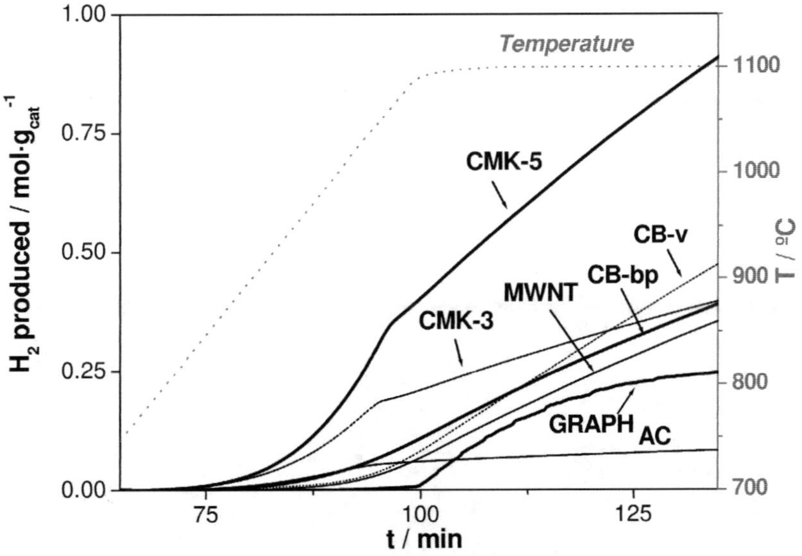

Figure 4. TG methane decomposition over the different carbon catalysts.

In order to evaluate the sustainability of carbon catalysts, long term reactions were also performed under isothermal conditions (Figure not shown). While both commercial activated and black carbons were almost completely deactivated after 24 hours of reaction, a continuous and significant hydrogen production rate was still observed after 72 hours for CMK-5 catalyst. It is important to note that after this time more than 25 g of carbon co-product per gram of raw catalyst were deposited in this last experiment. Although the reaction mechanism is not yet completely understood, the remaining of an important fraction of accessible surface active sites must be presumed to justify this result.

A number of factors have been proposed as the origin of the carbonaceous material activity in $DeCH_4$, one of them being the surface area of the catalyst [7, 14, 28]. The results here obtained indicate that a relationship between surface area and carbon activity could exist (Table 2). Thus, the lowest T_{th} values correspond with carbon materials having the highest BET surface areas (CMKs, AC and CB-bp samples). In contrast, materials with little surface area (graphite and cokes) present high T_{th} values. However, the relationship between surface area and catalytic activity is not so evident when comparing graphically the behaviour of the different carbon catalysts. Figure 5 depicts the catalytic activity, measured through the T_{th} parameter, versus the BET surface area. In any case, a clear and proportional relationship cannot be found, showing that the surface area is not the only one factor determining the catalytic activity of the carbon materials. Thus, the CB-v sample presents relatively low surface area, while its threshold temperature is quite low, denoting the high activity of this material for catalysing the methane decomposition reaction.

Another parameter proposed in the literature as closely related to the origin of the carbonaceous material activity in $DeCH_4$ is the presence of oxygenated groups on the carbon surface [14, 28]. Accordingly, the threshold temperature obtained in the $DeCH_4$ experiments has been represented versus the populations of the different oxygenated groups in the carbon catalysts determined by XPS (Figure not shown). However, no clear relationship can be established between the concentration of the different oxygenated groups and the initial catalytic activity, but on the contrary the points show a high degree of dispersion. Therefore, the concentration of oxygenated groups does not seem to be directly linked to the initial activity for methane decomposition.

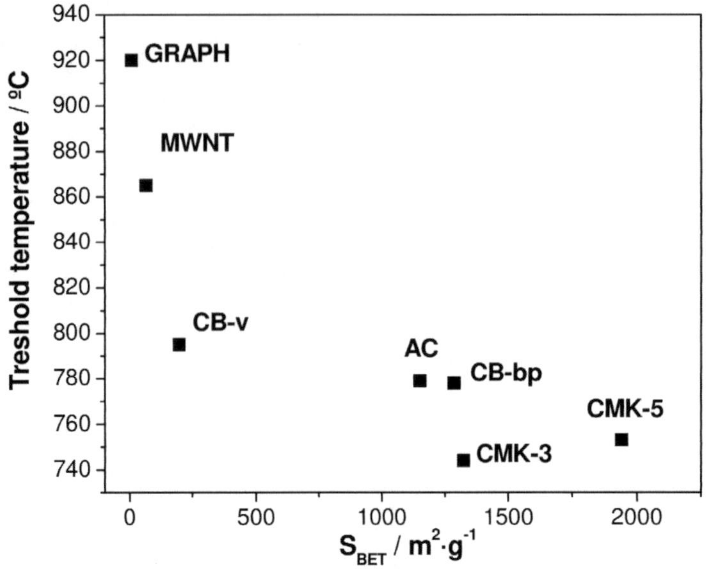

Figure 5. Correlation between the threshold temperature (T_{th}) in the $DeCH_4$ reaction and the BET surface area of the carbon catalysts (from N_2-adsorption).

The more or less pronounced graphitic nature of the carbons can be considered as another property that may influence the catalytic behaviour. To check this alternative, the threshold temperature has been represented (Figure not shown) against the relative intensity of the XRD peaks corresponding to C(002) and C(101) diffractions. As above indicated, this ratio is usually taken as indicative of the order of graphene sheets. It can be concluded that higher values of this ratio, i.e. lower graphenic order, lead to a reduction in the threshold temperature, suggesting that this parameter is related to the catalytic activity in methane decomposition. However, the correlation is far from being perfect. Thus, the values corresponding to the CMK materials deviate strongly from the general trend. Both CMK-3 and CMK-5 samples exhibit threshold temperatures lower than those expected taking into account their degree of graphitic order.

The last carbon property proposed to correlate the initial $DeCH_4$ catalytic activity is the amount of the defects on carbon surface [12, 16, 29], although no quantitative studies on their concentration in the carbon catalysts employed in this reaction have been carried out in the literature. In our case, and as indicated above, the XPS measurement has allowed the defect population in the graphene layers to be quantified through the value of the FWHM parameter in the C_{1s} peak at 284.8 eV. These defects decrease the electronic stability in the vicinity of the graphenic structure, increasing the carbon reactivity. Therefore, it can be envisaged that the origin of the carbon catalyst activity is directly related to the defect population on its surface. In order to check this proposal, the degree of arrangement of the graphene layers (FWHM parameter) has been depicted in Figure 6 *versus* the starting carbon activity (determined as T_{th}). A good correlation can be clearly observed for both parameters, showing that T_{th} is almost inversely proportional to the FWHM values. Nevertheless, some deviation is observed for the CMK-5 sample, as it shows a threshold temperature somewhat higher than that expected according to its defect concentration. Even so, this result supports strongly that the major property governing the carbon activity is the defect concentration present in the graphene layers rather than the surface area. Nevertheless, it must be taken into account that the surface area and the defect concentration are not completely independent variables, as a higher carbon surface area should corresponds with a higher number of defects. This fact explains why both properties show a similar variation among the carbon catalysts, although in quantitative terms the catalytic activity is best correlated with the defect concentration.

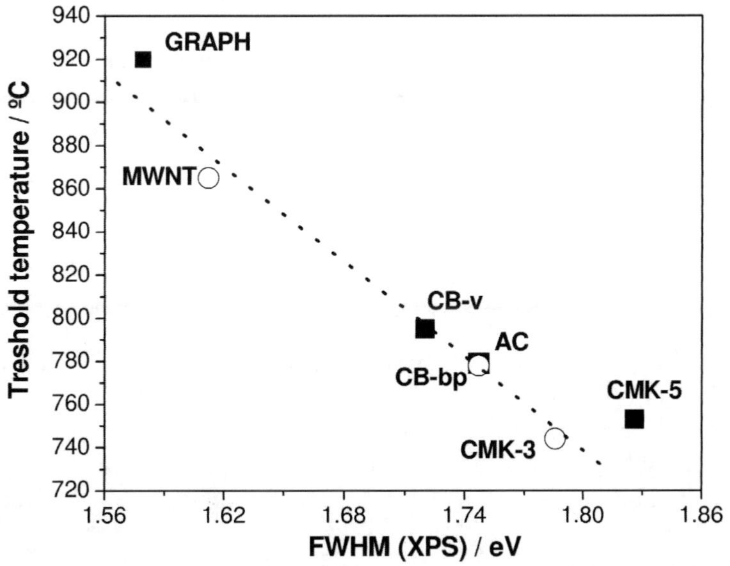

Figure 6. Correlation between the threshold temperature (T_{th}) in the $DeCH_4$ reaction and the full width at half maximum (FWHM) of the XPS graphite peak at 284.8 eV.

4. CONCLUSIONS

The activity of a variety of carbon materials (ordered mesoporous carbons, activated carbon, carbon blacks, multiwall nanotubes and graphites) in hydrogen production by methane decomposition mainly depends on the proportion of defects in the graphene layers of those catalysts. Thus, in temperature-programmed experiments, the highest initial activity (measured as threshold temperature) is exhibited by those carbon materials having a high defect concentration (determined by XPS), such as CMKs, activated carbon or carbon blacks. On the other hand, highly ordered carbon catalysts present little activity. XPS measurements have shown to be a very useful tool for quantifying the amount of defects of the graphene layers present in the different carbon materials. Moreover, a decreasing linear relationship has been found to exist between the threshold temperature and the proportion of defects in the graphene layers quantified by the value of the FWHM parameter in the C_{1s} XPS peak at 284.8 eV.

The activity order undergoes some changes when the catalyst behaviour is compared at longer reaction times. Under those conditions, activated carbon becomes quickly deactivated, which is assigned to the microporous nature of this material. Likewise, CMKs exhibit along the time a sharp change in the slope of the hydrogen production rate curve. Nevertheless, even so CMK-5 remains as the most active catalysts also at long reaction times. The deactivation process is largely determined by the presence of micropores whereas long-term activity is exhibited by those materials having mesopores (CMKs) or formed by nanoparticles (CBs), which make them more resistant to deactivation by pore blockage caused by the formation of carbonaceous deposits.

In order to evaluate the sustainability of carbon catalysts, long term reactions were performed under isothermal conditions. Thus, while commercial carbon was almost completely deactivated after 24 hours of reaction, a continuous and significant hydrogen production was still observed after 72 hours for CMK-5 catalyst. It is important to note that after this time more than 25 g of carbon co-product per gram of raw catalyst were deposited in this last experiment. Although the reaction mechanism is not yet completely understood, the remaining of an important fraction of accessible surface active sites must be presumed to justify this result.

ACKNOWLEDGEMENTS

The authors wish to thank "Comunidad de Madrid" for its financial support to the PHISICO2 Project through the Programme of Activities between Research Groups (S-0505/ENE-404). RGL acknowledges the Spanish 'Ministerio de Educación y Ciencia' and the European Social Fund for a "*Ramon-y-Cajal*" contract.

REFERENCES

[1] S.S. Penner, Energy, 31, (2005) 33-43.
[2] P.V. Zedtwitz, J. Petrasch, D. Trommer, A. Steinfeld, Sol. Energy, 80, (2006) 1333-1337.
[3] W.H. Scholz, Gas. Sep. Purif., 7, (1993) 131-139.
[4] N.Z. Muradov, Int. J. Hydrog. Energy, 18, (1993) 211-215.
[5] M. Steinberg M, Int. J. Hydrog. Energy, 23, (1998) 419–425.
[6] T.V. Choudhary, C. Sivadinarayana, C.C. Chusuei, A. Klinghoffer, D.W. Goodman, J. Catal., 199, (2001) 9-18.
[7] N. Muradov, Int. J. Hydrog. Energy, 26, (2001) 1165-1175.
[8] B.C.H. Steele, A. Heinzel, Nature, 44, (2001) 345-342.
[9] (a) J.I.Villacampa, C. Royo, E. Romeo, J.A. Montoya, P. del Angel, A. Monzón; Appl. Catal. A: Gen., 252, (2003) 363-383; (b) J. Wei, E. Iglesia; J. Phys. Chem. B, 108, (2004) 4094-4103.
[10] N.Q. Zhao, C.N. He, J. Ding, T.C. Zou, Z.J. Qiao, C.S. Shi, X.W. Du, J.J. Li, Y.D. Li, J. Alloys and Comps., 428, (2007) 79-83.
[11] R. Guil-Lopez, V. La Parola, M.A. Pena, J.L.G. Fierro, Catal.Today, 116, (2006) 289-297.
[12] (a) N. Muradov, Energy & Fuels, 12, (1992) 41-48. (b) N. Muradov, F. Smith, A. T-Raissi, Catal. Today, 102-103, (2005) 225-233.
[13] (a) N. Muradov, J. Power Sources, 118, (2003) 320-324. (b) L. P. Biró, C.A. Bernardo, G.G. Tibbetts, Ph. Lambin, editors. Carbon filaments and nanotubes: common origins, differing applications?, Kluwer Academic Publishers, Dordrecht, The Netherlands, 2001.
[14] R. Moliner, I. Suelves, M.J. Lazaro, O. Moreno, Int. J. Hydrog. Energy, 30, (2005) 293-300.
[15] E.K. Lee, S.Y. Lee, G.Y. Han, B.K. Lee, T-J. Lee, J.H. Jun, K.J. Yoon, Carbon, 42, (2004) 2641-2648.
[16] A. Dufour, A. Celzard, V. Fierro, E. Martin, F. Broust, A. Zoulalian, Appl Catal. A: Gen. 346 (2008) 164-173.
[17] R. Ryoo, S.H. Joo, S. Jun, T. Tsubakiyama, O. Terasaki. Stud. Surf. Sci. Catal. 135 (2001) 150.
[18] D.P. Serrano, J.A. Botas, P. Pizarro, R. Guil-López, G. Gómez. Chem. Commun. (2008) 6585–6587.
[19] D. Zhao, J. Feng, Q. Huo, N. Melosh, G.H. Fredrickson, B.F. Chmelka, G.D. Stucky, Science, 1998, 279, 548-552.
[20] P.A. Bazuła, A.-H. Lu, J.-J. Nitz, F. Schüth. Micropor. Mesopor. Mater. 108 (2008) 266-275.
[21] F.R. Feret, Analyst 123 (1998) 595-600.
[22] L. Li, Z.H. Zhu, G.Q. Lu, Z.F. Yan, R. de Marco. Appl. Catal. A 309 (2006) 201-20.
[23] Z.Q. Li, C.J. Lu, Z.P. Xia, Y. Zhou, Z. Luo. Carbon 45 (2007) 1686-1695.
[24] H.P. Boehm, Carbon 40, (2002) 145-149.
[25] M.T. Martínez, M.A. Callejas, A.M. Benito, M. Cochet, T. Seeger, A. Ansón, J. Schreiber, C. Gordon, C. Marhic, O. Chauvet, J.L.G. Fierro, W.K. Maser, Carbon, 41, (2003) 2247-2256.
[26] H. Darmstadt, C. Roy, Carbon, 41, (2003) 2662-2665.
[27] D. Pantea, H. Darmstadt, S. Kaliaguine, L Summchen, C. Roy, Carbon, 39, (2001) 1147-1158.
[28] I. Suelves, M.J. Lázaro, R. Moliner, J.L. Pinilla, H. Cubero. Int. J. Hydrog. Energy, 32, (2007) 3320-3326.
[29] S.Y. Lee, B.H. Ryu, G.Y. Han, T.J. Lee, K.J. Yoon. Carbon, 46 (2008) 1978-1986.

76g Study On a Compact Reactor for Methane Steam Reforming In Fuel Cell Systems

Liwei Pan, Zhongshan Yuan, Changjun Ni, Cheng Zhang, Deyi Li, and Shudong Wang. Lab.for Environmental Engineering, Dalian Institute of Chemical Physics, Chinese Academy of Sciences, Dalian, 116023, China

A compact integrated reactor, in which reactant vaporization, methane steam reforming and combustion are all integrated, is designed in this paper. The reactor, corrugate plates in both reforming chamber and combustion chamber were used to enhance heat transfer of the reformer, offer high methane conversion ratio and efficiency. It will result in simplification of the fuel processing system design and capital cost reduction. The reactor can be started-up at room temperature by the combustion of methane in the combustion chamber,and it also can keep self-sustained, i.e. without any external heating during the operating process. The effects of the ratio of water/methane on the performance of the reactor were investigated, and the distribution of reactants in different chambers was studied. In this reactor with 4L volume, the hydrogen content attained 76%, and the hydrogen output in reforming gas was about 2.3Nm3H2/hr. It is further demonstrated that the catalyst for methane steam reforming designed in our laboratory could be operated continuously with 100% conversion more than 1500hrs.

76h Coal Liquefaction Using High Energy H2 and O2 From Electrolysis Assisted by a Nuclear Source

Gary Otterstrom, Gemini Engineering, 649 North 75 West, Lindon, UT 84042

Water is super heated via helium gas loop with a nuclear reactor (this is a 100% efficient step), the super heated & super pressurized steam is passed through the Ceramatec Electrolysis membrains at 830 degrees C, the water is split into hydrogen and oxygen (a 100% efficient step), the high pressure and high temperature hydrogen is used to catalytically hydrogenate a ground coal and coal oil slurry (a very high conversion rate of liqufaction is acheived for producing aromatic & alphatic compounds of gasoline and kerosene needed is aviation fuels), the high pressure and temperature oxygen is used to burn coal in a parallel process in a syn gas/water gas shift/FT process (a 60% conversion rate of coal liqufaction is acheived for producing alphatic compounds, the waste 40% CO2 is processed through an alge farm to produce high quality tri-glycerides based fuels, this process is the most thermodynamically efficient process for the liqufaction of coal into useful fuels, acheiving a near 100% coal carbon conversion into useful fuels.

80a Carbon Dioxide Capture by Pressure Swing Adsorption

Amal Mehrotra, Armin D. Ebner, and **James A. Ritter**. Department of Chemical Engineering, University of South Carolina, 301 Main St., Columbia, SC 29208

There are many misconceptions about and misleading reports on the viability of pressure swing adsorption (PSA) cycles for CO2 capture from flue gas. These misconceptions and misleading reports stemmed largely from a 1994 International Energy Agency report [1] that evaluated CO2 separation and capture from flue gas using 13X zeolite in both pressure swing adsorption (PSA) and temperature swing adsorption (TSA) processes. They concluded that PSA and TSA are too energy intensive for use with gas and coal fired power systems. This conclusion has led others to extrapolate their findings and further conclude that adsorption systems, in general, are not applicable for CO2 separation and capture from flue gas. This disturbing misinformation has perpetuated for nearly a decade with essentially the same misleading statements being made in a 2003 IEA report [2]. However, it is strongly contended that those reports might have been factually inaccurate, that some grave errors might have been made in the evaluation of cyclic adsorption technologies for CO2 capture from flue gas, and that PSA might be very applicable to the capture of CO2 from flue gas. These claims are substantiated with evidence from works found in the literature and from some of the author's recent works.

First and foremost, a recent report by Izumi [3] from Mitsubishi Heavy Industry indicated that CO2 was being recovered from flue gas commercially using a layered PSA bed containing X and A type zeolites and activated carbon. Although, not much detail about the PSA cycle was provided, this news was exciting. The Japanese must have discovered something positive about using PSA for CO2 capture from flue gas that has been overlooked by investigators in the USA and apparently everyone else around the world!

Second, in a series of works by the Ritter and co-workers [4-7], novel PSA cycles have been explored systematically for the capture of CO2 from flue gas at high temperature using a K-promoted hydrotalcite like compound (HTlc) as the adsorbent. They revealed that a variety of vacuum swing adsorption (VSA) cycles could be used to concentrate CO2 to over 90 vol% at over 90% recovery from a feed stream containing 15 vol% CO2. The most effective cycles were based on the heavy reflux concept, a concept that has been largely missing from the PSA literature. What is really exciting about these high temperature heavy reflux PSA cycles is that they are completely applicable to ambient temperature adsorbents like zeolites and activated carbons. Moreover, Ritter and co-workers are still working on the problem and have discovered some better heavy reflux PSA cycles for CO2 capture from flue gas that will be discussed during this presentation. They also feel that the lack of information on and therefore the lack of understanding of the heavy reflux PSA concept for concentrating the heavy component from a feed stream might be the principle reason why the IEA reports were mistaken in their assessments of PSA for CO2 capture from flue gas. To this end, the heavy reflux concept in PSA will be introduced and discussed during this presentation

Third, some very recent work by Webley and co-workers [8] substantiated the claims made by the Ritter and co-workers over the past few years on the economic viability of PSA cycles for CO2 capture from flue gas, especially VSA cycles. They proclaimed that the information about the cost of PSA provided in the 2003 IEA report [2] was in error. Zhang et al [8] recalculated the cost of PSA to be around $67/tonne of CO2 produced compared to that in the IEA report of $97/tonne. This new cost for PSA, as verified through the IEA, compared much more favorably with amine scrubbing technology at $60/tonne. What is most exciting about this recent set of events is that the PSA cycles analyzed by the IEA were not designed for heavy component recovery, like CO2; instead, they were designed for light component recovery, like CH4 from a CH4/CO2 mixture.

As alluded to above, there are significant differences between these two kinds of PSA cycles, i.e., between the ones designed for light component enrichment and recovery and those designed for heavy component enrichment and recovery, with understanding of the later lacking, even by some of the best PSA researchers in the field. To make matters worse, the design of a multi-bed PSA process with complex cycle sequencing, like that envisioned for CO2 capture from flue gas, is a non-trivial exercise. PSA beds are always coupled together, usually contain more than one layer of adsorbent, and operate sequentially with each undergoing cycle steps such as pressurization, feed, heavy reflux, equalization, depressurization, light reflux, and repressurization. As an added complication, each cycle step is not necessarily of the same duration. These features make the number of possible cycle configurations very large; and unfortunately, design strategies on how to best configure such a complex PSA cycle is more of an art than a science. To this end, this presentation will also introduce some complex PSA cycle sequences that might be applicable for CO2 capture from flue and stack gases.

References:

1. International Energy Agency, “Carbon Dioxide Capture from Power Stations,” 1994.

2. International Energy Agency, “Carbon Dioxide Capture at Power Stations and Other Major Point Sources,” 2003.

3. J. Izumi, “Adsorption technology applications to the industries in Japan,” Abstract and presentation at the 8th International Conference on Fundamentals of Adsorption, Sedona, AR, May (2004).

4. S. P. Reynolds, A. D. Ebner and J. A. Ritter, New Pressure Swing Adsorption Cycles for Carbon Dioxide Sequestration,” Adsorption, 11, 531-536 (2005).

5. S. P. Reynolds, A. D. Ebner and J. A. Ritter, “Carbon Dioxide Capture from Flue Gas by PSA at High Temperature using a K-Promoted HTlc: Effects of Mass Transfer on the Process Performance,” Environmental Progress, 25, 334-342 (2006).

6. S. P. Reynolds, A. D. Ebner, and J. A. Ritter, “Stripping PSA Cycles for CO2 Recovery from Flue Gas at High Temperature Using a Hydrotalcite-Like Adsorbent,” Ind. Eng. Chem. Res., 45, 4278-4294

(2006).

7. S. P. Reynolds, A. Mehrotra, A. D. Ebner and J. A. Ritter, "Heavy Reflux PSA Cycles for CO2 Recovery from Flue Gas. Part I. Performance Evaluation," Adsorption, 14 399-413 (2008).

8. J. Zhang, P. A. Webley, and P. Xiao, "Effect of Process Parameters on Power Requirements of Vacuum Swing Adsorption Technology for CO2 Capture from Flue Gas," Energy Conversation and Management, in press on line (2007).

80b Preparation of new organic-inorganic hybrid adsorbents for post-combustion CO2 capture

Sunho Choi[1], Jeffrey H. Drese[1], Daniel J. Fauth[2], McMahan L. Gray[2], and Christopher W. Jones[1]. (1) School of Chemical & Biomolecular Engineering, Georgia Institute of Technology, Atlanta, GA 30332, (2) U.S. Department of Energy, National Energy Technology Laboratory, 626 Cochrans Mill Road, P.O. Box 10940, Pittsburgh, PA 15236-0940

Recently, separation and sequestration of anthropogenic CO2 from power plant flue gas has been spotlighted due to the growing consensus on its effect on the global climate change. Among several known strategies to remove CO2 from the flue gase, adsorption processes by solid adsorbents have been widely studied. With a particular interest in this approach, we have focused on the potential of hyperbranched aminosilica (HAS) materials as sorbents. This inorganic-organic hybrid material was prepared by ring-opening polymerization of aziridine on mesoporous SBA-15 silica, thereby directly while incorporating aminopolymers onto the silica support. High amine content in the first generation HAS adsorbent lead to a high adsorption capacity of CO2 up to 3.1 mmol/g at 25 °C in the presence of humidified, simulated flue gas, along with adequate regenerability. Following these initial efforts, these novel material have been more deeply characterized with the aim of understanding and improving its CO2 capture ability. This talk will present assessment of the CO2 adsorption capacity of the HAS material as a function of the operating temperature, organic loading, polymer molecular weight, the ratio of 1°:2°:3° amines, and other factors. Experimental conditions such as the reaction temperature and solvents used during synthesis were modified to evaluate an effect of these variables on the composition and structure of the HAS sorbent. Based on these data, we present optimized synthesis methods that enhance the equilibrium capacity of CO2 using the HAS hybrids.

VIABILITY ASSESSMENT OF SOLID SORBENTS FOR POST-COMBUSTION CO_2 CAPTURE

Sharon Sjostrom, ADA Environmental Solutions, Littleton, CO

Introduction

Carbon capture and sequestration (CCS) can be an important component of reducing worldwide CO_2 emissions from stationary point sources, such as coal-fired power plants. Aqueous amines and ammonia are being demonstrated for CO_2 capture in a temperature swing cyclic process. Solid sorbents can also be used in a similar process, but have the potential to drastically reduce the energy required to release the CO_2 during material regeneration. However, the majority of solid sorbent research is being conducted on lab-scale samples without significant consideration of process and/or equipment design.

ADA Environmental Solutions is examining available sorbents as well as process equipment options to assess the viability and accelerate development of solid sorbents as a CO_2 capture option through a DOE Cooperative Agreement and supplemental funding from EPRI and industry. During the first phase of this program, ADA is evaluating materials from top sorbent developers throughout the world. Materials from eleven developers have been tested to-date using a fixed bed screening device and simulated flue gas over multiple adsorption/regeneration cycles. Additional developers have been contacted and materials are scheduled for characterization.

Materials demonstrating the most promise from the initial tests have been characterized on actual flue gas from a lignite-coal fired electric generating unit (EGU). In a parallel effort, large-scale process options are being reviewed and evaluated. An update on the progress of the project, including results from screening tests and options for process equipment will be discussed.

Experimental Apparatus and Procedure

All testing of the sorbents screened during this program was carried out using the same test fixture. This unit was designed to be used in the laboratory on simulated flue gas as well as in the field on actual flue gas with minimal modifications. The flow rate of the simulated or actual flue gas was approximately 0.3 LPM, and the amount of sorbent tested varied from 0.5 to 2.5 g. The sorbent was held in a fixed bed. Although the team does not expect that a fixed bed represents the final contactor design for a large-scale system, this configuration allowed for fast, efficient lab-scale sorbent screening. **Error! Reference source not found.** is a schematic of the sorbent screening testing unit when setup for laboratory testing (only minor modifications are necessary for field testing).

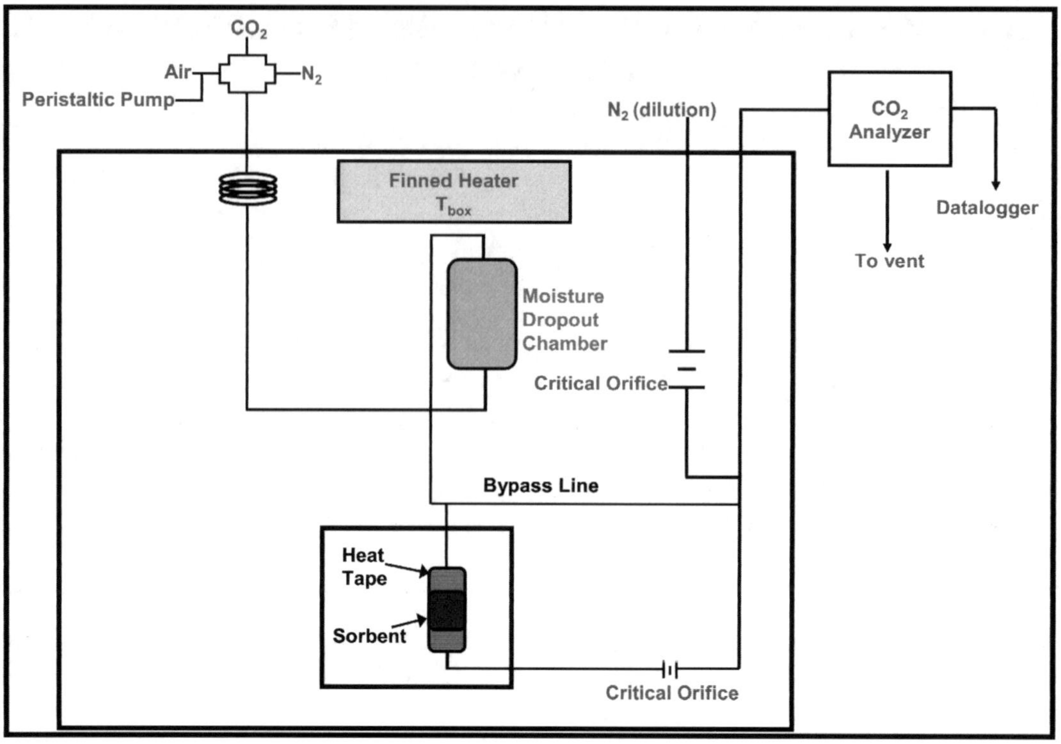

Figure 1: Schematic of Sorbent Screening Test Unit

The steps to carry out the capture/regeneration cycles were the same whether tests were conducted on simulated or actual flue gas. First, the sorbent was placed in the fixed bed. Then, the sorbent was heated to an initial flushing temperature. The initial flushing temperature was set based on the expected regeneration temperature, unless a different temperature was recommended by the sorbent developer. A thermocouple on the outside of the glass fixed bed was used to determine when the bed had reached the desired temperature. When the bed temperature matched the desired capture temperature, the sorbent was flushed with dry N_2 for 10 minutes. Then, the simulated flue gas, composed by mixing compressed gas from tanks, was sent through the bypass line, so that it did not contact the sorbent. The composition of the laboratory sample gas, by volume, was approximately 10 to 14% CO_2, 4% O_2, balance of N_2. The relative humidity, controlled by adding water using a peristaltic pump and then evaporating, was 0%, 45%, or 90%. When the CO_2 monitor reading was stable and represented the correct concentration of CO_2, the datalogger was enabled and the gas flow was directed through the sorbent. After the CO_2 levels returned to their original levels (i.e., the sorbent was saturated with CO_2) the gas was sent through the bypass, which was the end of the capture step.

A temperature swing with N_2 flushing was used to regenerate the sorbents and desorb the CO_2. First, the bed was heated with no gas flow. When the thermocouple on the fixed bed indicated the desired desorption temperature, dry N_2 was sent through the bed to flush out the CO_2. Clearly, this does not represent a final desorption process, but was necessary due to the fixed-bed size restrictions and the amount of gas required by the analyzer.

One key issue that will limit the use of commercial-scale fixed beds for this application is the heat management. Since all the reactions between the sorbents and CO_2 are exothermic, whether physical or chemical, the sorbent will heat up as it reacts with the CO_2. Unfortunately, in most cases the sorbent CO_2 capacity decreases with increasing temperature. Therefore, if the sorbent is allowed to heat during the capture step of the cycle, the test results will show a reduced amount of CO_2 capture. Similarly, if the sorbent is allowed to cool due to the endothermic desorption, all the CO_2 may not be released. To control the temperature of the sorbent during these lab-scale tests, the flow rate of the gas through the fixed bed was kept high proportional to the amount of sorbent in the bed. Therefore, the sample gas during capture maintained lower temperatures, while the hot purge gas maintained higher temperatures during regeneration. However, by using a high ratio of gas to sorbent, the breakthrough curves were affected. Since the gas flow rate is high, the residence time is low (~1 s) and some CO_2 may pass through the fixed bed unreacted, even when adsorption sites are available. **Therefore, the percent removal obtained during laboratory tests does not necessarily represent the total removal that can be realized by the sorbents. In fact, it should be considered a minimum of achievable percent removal.**

When evaluating and comparing sorbents for use in a commercial-scale CO_2 capture and sequestration system, one of the most important evaluation criteria is the energy required for regeneration. For all the sorbents testing in this program, the theoretical energy required for regeneration was calculated. The result of this calculation is used to compare different materials, and assess their respective viability for use in a commercial-scale CO_2 capture system. Note that from previous reports a liquid MEA CO_2 capture system requires approximately 4530 kJ/kg CO_2 (1934 BTU/lb CO_2).[2]

Results

Table 1 lists the twelve sorbents that exhibited CO_2 removal properties during laboratory screening using simulated flue gas. The maximum, minimum, and average CO_2 working capacities for each material are presented in this table. In addition, the theoretical regeneration energy for each of the materials is also given. Note that this theoretical regeneration energy was calculated from the best-case scenario during laboratory screening, the details of which can be found in the individual sorbent section. For those sorbents that were tested on actual flue gas, the theoretical required regeneration energies, based on field results, are presented in Table 2.

Sorbents A, B, C, G, H, and I rely primarily on physical adsorption. Therefore, with the exception of sorbent A, the CO_2 capacities of these materials are lower than the supported amine materials. For physical adsorbents, the energy required during regeneration to counteract the endothermic reaction is low. However, if the CO_2 capacity is low, more sorbent will have to be heated during regeneration. In addition, more material will have to be moved, which is outside the scope of this report, but is certainly an important consideration. Based on the laboratory results, sorbent A is the only physical adsorbent that is currently competitive with the theoretical regeneration energy of aqueous MEA (~4530 kJ/kg CO_2). In fact, sorbent A is significantly lower, and will be extensively tested in the second round of laboratory, and possibly field testing.

Table 1. CO_2 Working Capacities During Laboratory Tests

Laboratory					
Sorbent ID	Sorbent Type	Working CO_2 Capacity			Q_{regen} (kJ/mol CO_2)
		Max.	Min.	Average	
A	Carbon Nanotubes	5.48	5.15*	4.91	812
B	Carbon-based	0.58	0.38	0.48	8820
C	Carbon-based	0.54	0.43	0.49	8020
D	Supported Amine	7.48	3.92	6.10	2190
E	Supported Amine	1.34	0.72	1.04	4140
F	Supported Amine	4.71	3.7*	3.81	2550
G	Carbon-based	0.97	0.28	0.62	11330
H	Carbon-based	1.07	0.66	0.86	6510
I	Carbon-based	1.09	0.57	0.78	7580
N	Zeolite	0.40	0.33	0.37	24510
O	Zeolite	1.61	0.83	1.26	11560
Q	Supported Amine	4.22	1.13	2.79	2990

*Low first cycle neglected (when due to low flushing temp.)

Zeolites rely on their structure to act as a molecular sieve for gases. Thus far, only two zeolites, sorbent N and sorbent O, tested during this program showed any signs of CO_2 removal from simulated flue gas. Sorbent O was included in the field tests and did remove CO_2 during all adsorption cycles.

Sorbents D, E, F, and Q were variations of supported amine materials. Several of these materials were powders, and could only be tested in quantities of 0.5 g. These materials exhibited high CO_2 capacities. It should be noted that the CO_2 capacities of all these materials varied significantly from cycle to cycle. Based on the theoretical energy for regeneration under the laboratory testing conditions, all the supported amines were competitive with aqueous MEA, although sorbent E did show sensitivity to high temperature and possibly amine loss. The other three supported amines were included in the first round of field testing. Figure 2 shows the maximum, minimum, and average laboratory working CO_2 capacities of all the sorbents. For testing conditions and other details, refer to the individual material sections.

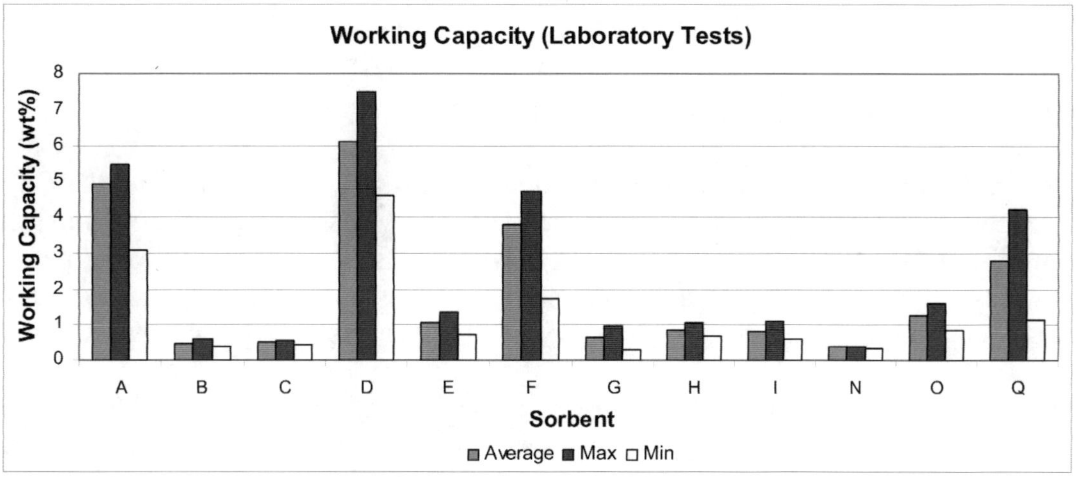

Figure 2. CO_2 Working Capacity During Laboratory Tests

Out of all of the sorbents tested in the laboratory, five were selected for the first round of field-testing. One carbon-based material, three supported amines, and one zeolite were tested for multiple CO_2 adsorption/regeneration cycles using actual flue gas. Table 2 lists the sorbents that were included, the mass tested, regeneration temperatures, the number of field cycles run and key results from these experiments.

Table 2. Field Testing Results

Sorbent ID	$Mass_{sorbent}$ (g)	T_{regen} (°C)	No. Cycles	Max. Working Capacity	Min. Working Capacity	Average Capacity	Q_{regen} (kJ/kg CO_2)	$t_{capt-total}$ (s)	$t_{regen-total}$ (s)	$t_{capt-90\%}$ (s)	$t_{regen-90\%}$ (s)
D	2.5008	120	11	8.20	4.81	6.19	2930	702	1071	427	719
F	0.5010	120	34	5.36	2.99	4.08	3530	171	170	94	94
H	2.5011	120	18	1.10	0.70	0.95	6610	86	180	57	88
O	2.5033	200	18	0.84	0.34	0.54	37080	70	97	50	64
Q	0.5014	120	17	3.55	1.24	2.43	6940	173	287	127	209

The minimum, maximum, and average working capacities recorded during the field test are shown in Figure 3. The three supported amines, D, F, and Q exhibited the highest working capacities, with averages ranging from >6 to 2.5 wt%. The carbon-based material had an average working capacity of nearly 1 wt%, while the zeolite, sorbent O, had an average capacity of approximately 0.5 wt%.

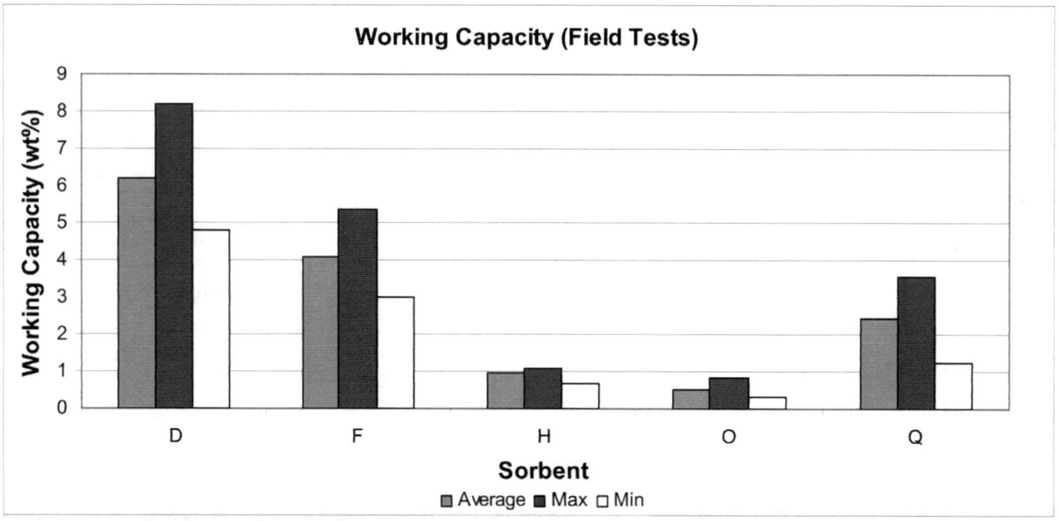

Figure 3. CO_2 Working Capacity During Field Tests

Although the CO_2 capacities are an important component of sorbent performance, the ability of these materials to retain this capacity for a high number of cycles is also a key consideration. In commercial-scale CO_2 capture system, sorbents must retain their capacity for long periods of time, likely thousands of cycles and up to multiple years of operation. Therefore, the field tests included testing the sorbents for many repeated cycles to identify any consistent loss in capacity.

A fresh bed of Sorbent D was used for eleven cycles during field tests. Although this material continued to exhibit high CO_2 capacity, the field tests showed significantly reduced CO_2 removal when compared to the laboratory results. In addition, a slow decrease in the CO_2 capacity of this material was observed, with the last three adsorption cycles resulting in an average of ~5 wt%. Out of the supported amines used during field testing, sorbent D performance decreased the most during field testing. Future laboratory tests will be aimed at determining whether this can be attributed to loss of amine over time, or sensitivity to flue gas constituents and/or conditions.

Sorbent F was also included in field tests due to its high CO_2 capacity and complete regeneration during laboratory testing. This material was tested in 34 adsorption/regeneration cycles on actual flue gas. Between the first cycles and the last cycles, the CO_2 capacity had decreased by approximately 20%. Similar to sorbent D, further testing and discussions with the sorbent developer will be conducted to determine the cause of this loss.

Sorbent Q was the third supported amine material included in the field testing. The first adsorption cycle yielded a CO_2 capacity greater than 6 wt%, but all others were less than 4 wt%. Between cycles 2 and 17, the working capacity of sorbent Q decreased by over 50%, which is certainly unsustainable for any commercial-scale system

Sorbent H is a carbon-based material that was included in field tests. With the exception of a few outlier data points towards the end of testing, this material did not exhibit any loss in capacity over the eighteen cycles that were completed in the field.

Sorbent O is a zeolite that was included in the field testing. For the ten cycles following the initial flushing, the working capacity of decreased dramatically. In fact, between the first cycle and cycle 10, sorbent O lost nearly half its CO_2 working capacity. This is attributed to the preferential selectivity of sorbent O towards H_2O versus CO_2. Towards the end of the tests, the working capacity began to plateau, which likely signaled that equilibrium was achieved under the field operating conditions.

Another important characteristic for commercial-scale CO_2 capture will be the adsorption and desorption reaction rates, because this will influence the necessary residence time for adsorption and regeneration. Table 2 lists the time required for the sorbent to become completely saturated with CO_2 during capture, the time required for complete regeneration at temperature flushing with N_2, and the corresponding times required for 90% of CO_2 capture and 90% of regeneration. These last two terms were included because it is possible that in a commercial operating system the time required to completely saturate the material under operating conditions may be less economical that using 90% or less of the total available capacity. This type of optimization is outside the scope of this program, but will be an important consideration during sequential scale-up and optimization studies. Note that although the times listed for sorbent D are higher than the other materials, this can not be directly compared because the materials because this material removed all of the CO_2 for the flue gas for a period of time. What is important to identify from Table 2 is whether the regeneration time is significantly slower than the adsorption time. If these times are significantly different, the adsorption/desorption process will be less efficient at full-scale.

Finally, the theoretical regeneration energy was calculated based on the working CO_2 capacity and field operating conditions. Of those listed in Table 2, only sorbents D and F have a lower theoretical regeneration energy than the reported value for aqueous MEA. However, two others are close and may surpass this goal with a slight change in sorbent or operating conditions (i.e. regeneration temperature).

Conclusions

Many different materials have been tested thus far in the ADA CO_2 sorbent screening program. All materials that have been contributed to the program either have been, or will be tested in multiple CO_2 adsorption/regeneration cycles. This report discusses in detail the first twelve materials that showed any signs of CO_2 removal from either simulated or actual flue gas. Based on the theoretical regeneration energy, several sorbents are superior to the current state of the art CO_2 capture technologies. Further testing will be performed to determine whether these materials are stable enough to be used in a commercial-scale CO_2 capture system. In addition, follow up reports will contain test results for several more sorbents that have been submitted to the program, but have yet to be tested.

References

1. Hoffman, J.S., Richards, G.A., Pennline, H.W., Fischer, D., Keller, G, (2008) *Factors in Reactor Design for Carbon Dioxide Capture with Solid, Regenerable Sorbents*, Clearwater Coal Conference, Clearwater, FL.
2. Tarka, T.J., Ciferno, J.P., Gray, M.L., Fauth, D. (2006) *CO_2 Capture Systems Using Amine Enhanced Solid Sorbents*. 5th Annual Conference on Carbon Capture & Sequestration, Pittsburg, PA.

80d The Effect of Hydration on the Adsorption of Carbon Dioxide with Tertiary Amidines on Activated Carbon

W. Richard Alesi[1], John R. Kitchin[1], and McMahan L. Gray[2]. (1) Department of Chemical Engineering, Carnegie Mellon University, Pittsburgh, PA 15213, (2) U.S. Department of Energy, National Energy Technology Laboratory, 626 Cochrans Mill Road, P.O. Box 10940, Pittsburgh, PA 15236-0940

Green house gas emissions have come under increased scrutiny in recent years as political and scientific communities begin to understand the impact that human activity has on the environment. Carbon dioxide (CO2) stands at the forefront of this concern due to the copious amounts of CO2 produced anthropogenically. As more that 10 billion tons CO2 are emitted worldwide each year from fossil fuel combustion for energy creation [1], one option to reduce CO2 emissions is the separation, either before or after combustion, of the various gases in the combustion feed stream to facilitate subsequent storage and sequestration of CO2. Recent findings with carbon capture using gas phase adsorption experiments in a plug flow reactor highlight the benefits of this type of capture over membrane or absorption capture processes [2]. In this work, we evaluate the effectiveness of a bicyclic tertiary amidine impregnated on activated carbon in capturing carbon dioxide under post combustion flue gas conditions. Additionally, the effect of hydration on the sorbents was evaluated to help elucidate the mechanism of the interaction and to ascertain whether the presence of water is necessary for CO2 to associate with the tertiary amidine. Experimental Details A custom plug flow reactor (PFR) connected to a mass spectrometer was used for all experimental results. An activated carbon support was chosen, providing a high surface area for the impregnation of the amidines. The amidine, 1,5-Diaza-bicyclo[4.3.0]non-5-ene (DBN), was evaluated in order to determine its effectiveness in capturing CO2 under a range of operating conditions. Upon adsorption, subsequent changes in the operating temperature and partial pressure of CO2 provided the driving force for desorption of the CO2 from the amidine. Initial conditions were specified to closely emulate the conditions expected at the flue of a pulverized coal power plant (30°C - 40°C, 0.1 pCO2/p°, 5% water), while the final operating conditions consisted of an increase in temperature (80°C) and a reduction in the partial pressure of CO2 by switching to a He purge gas. A heated sparger was used to introduce a controlled amount of water in the system. Adsorption breakthroughs and desorption peak responses from a controlled temperature ramp were established from molar concentration measurements in the exit gas stream provided by the mass spectrometer. We establish both the capture efficiency (mol CO2/mol Amidine) and capture capacity (mol CO2/kg sorbent) as a quantitative indicator of the effectiveness of the sorbent. Results and Discussion Breakthrough and desorption plots for DBN are displayed in Figure 2. The deviation from the 10% CO2 feed stream recorded by the mass spectrometer during the adsorption phase is a result of the adsorption of the CO2 to the surface of the sorbent. We see in (2.b) that DBN has a close to negligible interaction with CO2 in the absence of water, however, when operating the sparger at room temperature (2.a, 40°C, 6.3% mol H2O), we see a much larger concentration change in both the adsorption and desorption phases. This result coincides with that found by Heldebrant et al [3], wherein it is believed that the amidine interacts with CO2 in the formation of a bicarbonate salt, which would require the presence of water for the interaction to commence. Besides the increased capture capacity obtained in our hydrated samples, we find from thermogravimetric measurements over long time scales, result in a continual mass gain in the presence of a hydrated feed stream most likely as a result of the hydrophilic activated carbon support. This might ultimately result in a long term reduction in capture efficiency unless this water can be removed in an additional process. We calculate that DBN has a overall capture efficiency of 37%, slightly less than the 40-56% attributed to the state-of-the-art liquid amine systems [4], however this efficiency is only at the temperatures and pressures mentioned previously and could be manipulated by adjusting the operating conditions of adsorption and desorption. Conclusion This work provides an insight on the interaction of the tertiary amidine, DBN, with CO2. We find that hydration is necessary for these tertiary amidines to interact, however the hydrophilic nature of the activated carbon support signifies that alternate, less hydrophilic supports might be better equipped to operate in the humidified flue streams of a coal power plant. Additionally, by changing the operating conditions of adsorption we can modify the capture efficiency of DBN for use in alternative CO2 separation processes. References 1. B. Metz, IPCC Special Report on Carbon Dioxide Capture and Storage (Cambridge University Press, 2005). 2. M. L. Gray et al., “Capture of carbon dioxide by solid amine sorbents,” International Journal of Environmental Technology and Management 4, no. 1 (2004): 82-88. 3. D. J Heldebrant et al., “The Reaction of 1,8 Diazabicyclo[5.4.0] undec7ene (DBU) with Carbon Dioxide,” The Journal of Organic Chemistry 70, no. 13 (2005): 53355338. 4. R. J. Hook, “An investigation of some sterically hindered amines as potential carbon dioxide scrubbing compounds,” Ind. Eng. Chem. Res 36, no. 5 (1997): 1779. Figure 1. The tertiary amidine 1,5 Diazabicyclo [4.3.0] non-5-ene Figure 2. Adsorption and desorption profiles of 1,5 Diazabicyclo [4.3.0] non-5-ene. (a.) capture at 6.3% water content (b.) capture in dry gas conditions

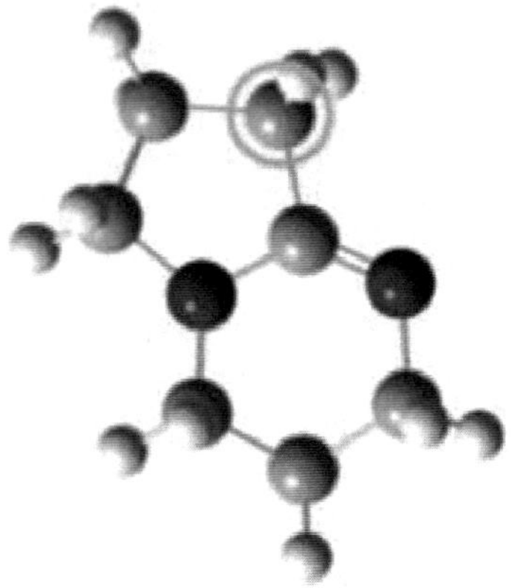

80e A Regenerative Process for CO_2 Removal and Hydrogen Production in Integrated Gasification Combined Cycle (IGCC) Processes

Javad Abbasian and Armin Hassanzadeh-Khayyat. Chemical and Environmental Engineering, Illinois Institute of Technology, 10 W. 33rd Street, Chicago, IL 60616

Advanced power generation technologies, such as Integrated Gasification Combined Cycles (IGCC) processes, are among the leading contenders for power generation conversion because of their significantly higher efficiencies and potential environmental advantages compared to conventional coal combustion processes. Although the increased in efficiency in the IGCC processes will reduce the emissions of carbon dioxide per unit of power generated, further reduction in CO_2 emissions is crucial due to enforcement of green house gases (GHG) regulations. In IGCC processes to avoid efficiency losses, it is desirable to remove CO_2 in the temperature range of 300° to 500°C, which makes regenerable MgO-based sorbents ideal for such operations. In this temperature range, CO_2 removal results in the shifting of the water-gas shift (WGS) reaction towards significant reduction in carbon monoxide (CO), and enhancement in hydrogen production. However, regenerable, reactive, and attrition resistant sorbents are required for such application.

This paper discusses the development and evaluation of a highly reactive and attrition resistant regenerable MgO-based sorbent prepared through dolomite modification, which can simultaneously remove carbon dioxide and enhance hydrogen production in a single reactor. The results of the experimental tests conducted in a high-pressure thermogravimetric analyzer (HP-TGA) and a high-pressure packed-bed reactor indicate that, in the temperature range of 300° to 500°C at 20 atm, more than 95 molar percent of CO_2 can be removed from the simulated coal gas, and the hydrogen concentration can be increased to above 70 percent. Based on the physical and chemical analysis of the sorbent, a two-zone expanding grain model provides an excellent fit to the carbonation reaction rate data at various operating conditions. The modeling results indicate that more than 90 percent purification of hydrogen is achievable. The preliminary economical assessment of the MgO-based process indicates that this process can be economically viable compared to commercially available processes for combined CO_2 removal and enhancement of hydrogen production.

80f A First Principles Evaluation of the Role of Substituent Effects on the Interaction of Carbon Dioxide with Tertiary Amines

W. Richard Alesi and John R. Kitchin. Department of Chemical Engineering, Carnegie Mellon University, Pittsburgh, PA 15213

Introduction To effectively reduce carbon dioxide (CO2) emissions from post-combustion energy sources, there exists a need to develop an optimal sorbent for CO2 capture under flue gas conditions, and alternatively, a means of determining the optimal operating conditions for sorbents to be used under alternative environments. The use of solid amines, in contrast to the traditional liquid amines systems, allows us to analyze the system as a series of adsorption / desorption cycles which can be related to the thermodynamic and kinetic properties of the amine. For these gas phase interactions, the nitrogen acts as a basic site to interact with the acidic CO2, and substituent effects in its vicinity should directly influence both the basicity and nucleophilicity of the capture site. This influence ultimately results in changes to the driving forces involved in the interaction, meaning that each sorbent will require different operating conditions in order to maximize its capture efficiency. Here we investigate the role of substituent effects on the interaction of CO2 with the immobilized tertiary amidine N-Methyltetrahydropyrimidine (MTHP). Through modification of the functional group at the R1 position adjacent to the active nitrogen, we expect changes in the energies of desorption as well as the reaction rates. We use a combination of Density Functional Theory (DFT) calculations and experimental packed bed temperature programmed desorption (TPD) experiments in an attempt to develop a methodology capable of determining the adsorption energies of CO2 with various functionalities of MTHP and determining a range of optimal operating conditions for a particular sorbent. Materials and Methods Recent studies have shown that CO2 – amidine preferentially create bicarbonate complexes over zwitterions complexes when in the presence of water [1]. We apply Density Functional Theory to finite basis set calculations to determine the structure of these complexes in addition to the adsorption energies. Structure optimization calculations were conducted using Dunning's correlation consistent basis sets. Also, the electronic energies obtained through our DFT calculations are adjusted for vibrational energetic and entropic terms derived from the partition function. Finally, a thermodynamic framework is then developed to relate our zero temperature calculations to those experimentally relevant. A plug flow reactor (PFR) was used to obtain packed bed adsorption / desorption experiments in which the exit gas could then be monitored by a mass spectrometer. Adsorption breakthroughs and desorption peak responses from a controlled temperature ramp were established from molar concentration measurements in the exit gas stream provided by the mass spectrometer. From these measurements, we establish both the capture efficiency (mol CO2/mol Amidine) and capture capacity (mol CO2/kg sorbent) in order to form some of the underlying thermodynamic properties of the sorbent for correlation with our DFT results. Results and Discussion Upon comparison of the energies formed in creation of both zwitterion and bicarbonate complexes, we confirm that the bicarbonate complex is more thermodynamically stable than the zwitterion complex. The range of substituent effects on MTHP includes charge transfer, hydrogen bonding, and steric hindrance. We find from our DFT calculations that while both steric hindrance and charge transfer play a role in the interaction, hydrogen bonding proved the strongest of the substituent effects. Additionally, the thermodynamic framework applied to our DFT calculations help to determine the driving forces required to surpass the desorption barrier. From these calculations we conclude that there exists a small range of adsorption energies that result in an optimal interaction for a given set of adsorption and desorption conditions. Conclusions This work provides insight into the role that substituent effects play on the gas phase interaction of CO2 and solid amines. An understanding of these effects, as well as the determination of adsorption energies from the combination of DFT and experimental analysis, provide a basis for determining the optimal conditions for a given sorbent. References 1. D. J Heldebrant et al., "The Reaction of 1,8-Diazabicyclo[5.4.0]undec7ene (DBU) with Carbon Dioxide," The Journal of Organic Chemistry 70, no. 13 (2005): 53355338. Figure 1. N-Methyltetrahydropyrimidine (R1=H), a tertiary amidine where the double bonded nitrogen atom could interact with CO2 in formation of (a) zwitterion complex or (b) bicarbonate complex

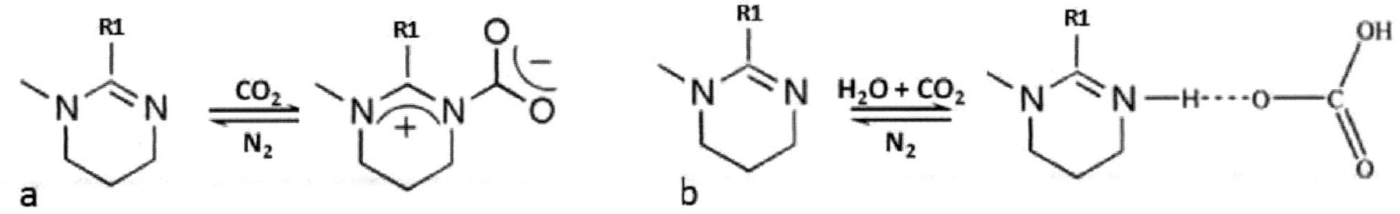

80g Modeling CO_2 Capture Using Ab Initio and Statistical Mechanical Methods

J. Karl Johnson[1], Jin-Chen Liu[2], Rees B. Rankin[3], De-Li Chen[4], Hong-bin Xie[4], and Bo Zhang[1]. (1) Dept.of Chemical and Petroleum Engineering , University of Pittsburgh, 1249 Benedum Hall, Department of Chemical Engineering, Pittsburgh, PA 15261, (2) Chemical Engineering, University of Pittsburgh, 1249 Benedum Hall, Pittsburgh, PA 15261, (3) Department of Chemical and Petroleum Engineering, University of Pittsburgh, Benedum Hall #1123, University of Pittsburgh, Pittsburgh, PA 15261, (4) National Energy Technology Laboratory, Pittsburgh, PA

There is a general consensus that new materials are needed to efficiently and cost effectively capture CO2 from fuel or flue gases. Experimental work is critical to progress in developing and testing new materials, but the number of possible candidate materials makes it impractical to use an Edisonian approach to materials discovery. Molecular modeling is a tool that can be used to help guide experiments and screen materials more cost effectively than experiments alone. We present results from both ab initio quantum mechanical calculations and from semi-empirical statistical mechanical models for evaluating the effectiveness of various materials for CO2 capture. We have studied adsorption separation using nanoporous materials such as metal organic frameworks and zeolitic imidazolate frameworks. We have investigated absorption in liquid oligomers, including primary amines and fluorocarbons. We have also examined the thermodynamics of CO2 capture by solid oxides and hydroxides. We present results from our calculations and comparison with experiments where possible. We discuss the possibility of using molecular modeling to design new materials for CO2 capture.

103a Irreversible Capture of Mercury

David A. Atwood, Department of Chemistry, University of Kentucky, 347 Chemistry-Physics Building, Lexington, KY 40506

There are 1,100 coal-fired electricity generating utilities in the U.S. emitting 48 tons of mercury, a potent neurotoxin, each year. Twenty or more eastern states have either begun, or will begin, regulating Hg emissions from these utilities. A nationwide federal rule is expected by 2013. Thus, there is an immediate need for effective clean-coal technologies that can mitigate mercury release from coal-fired utilities.

For several years now the Atwood group has been working with a set of sulfur-containing compounds designed to bind soft heavy metals through strong, covalent M-S bonds. The best of this group is benzene-1,3-diamidoethanethiol (BDTH2 which is particularly effective for removing the mercuric ion from water. It has been shown to irreversibly bind Hg under a wide range of laboratory conditions, in gold mining effluent, acid mine drainage, for soil-borne mercury, and works for low ppb levels in surface water near former chlor-alkali facilities. Importantly, the resulting BDT-Hg precipitate does not release mercury under either strongly acidic or basic conditions.

This presentation will describe the recent developments in the use of BDTH2 in mercury capture.

103b Update on Global Mercury Emissions and Controls

Leonard Levin, Electric Power Research Institute (EPRI) , 3420 Hillview Ave, Palo Alto, CA 94304

The high vapor pressure of elemental mercury as well as the presence of emissions stacks well above ground level lead to mercury's behavior as a widely distributed global pollutant. Recent inventories by the United Nations Environment Programme demonstrate that China (with the exception of recent economic slowdowns) continues as the largest contributor to the global atmospheric mercury pool, where mercury's mean residence time is on the order of 7 months to one year. The latitude of Chinese sources places them squarely in the prevailing westerlies, leading to wide distribution of mercury throughout the Northern Hemisphere. Indian sources are both lower than previously thought, and located in the trade wind belt closer to the equator; however, Asian monsoon circulation appears to carry Indian emissions into more northerly latitudes annually. South African emissions have also been revised downward, although southern hemisphere emissions from precious metal recovery overall may be much greater than thought. China employs coal as a fuel much more widely than the nearly sole use in the US for power generation, using it for space heating, process heat, steam production, and residence-level enterprise use. The most recent control measures on coal burning facilities in China have included stack controls for particulate matter, which may capture a portion of the mercury as well, and a national effort to reduce coal use in individual homes for more central facilities, of uncertain success so far.

103c Specification Metrics for Activated Carbons Used for Mercury Control

Sid Nelson Jr., Albemarle Sorbent Technologies, 1664 E. Highland Road, Twinsburg, OH 44087

Chemicals typically consumed in air pollution control applications like ammonia for selective catalytic reduction or lime and limestone for scrubbers can be specified fairly simply with small differences in the resulting pollution reduction performance. Heterogeneous activated carbons used for flue gas mercury control, however, are not discrete chemicals, as such, and there can be great differences in their demonstrated performance depending on many material-related factors.

Unfortunately, good correlations between traditional activated carbon specification metrics and mercury performance have proven elusive so far. There has not yet been a "mercury number" developed to gage relative mercury performance, like the molasses number used to gage molasses decolorization efficiency, for example.

This presentation will review the effects that traditional metrics like particle size, surface area, and density have in this application and posit new characteristics such as bromine level, thermal conductivity, electrostatic properties, and concrete-friendliness which can have major impacts on the desirability of particular carbons in particular situations. A number of new metrics will be proposed for consideration in specifying activated carbons for this application.

Screening of Sorbents for Mercury Capture Using Ab initio Computations

A. Jain, S.A. Seyed-Reihani, C. Fischer, D. Couling, G. Ceder, W.H. Green

Forward-looking projections by the EPA indicate that coal will continue as a major source of the world's energy supply. For the time being, coal supplies 25% of the world's energy needs and produces 40% of its electricity[1]. One promising technology to improve the efficiency of energy extraction from coal is the integrated gasification combined cycle (IGCC) plant. While thermochemical analysis suggests the thermal efficiencies of IGCC plants could be around 40%, as compared to 34% for a pulverized coal plant[2], both suffer from pollution footprints, and emission control units can significantly reduce the overall efficiency. The EPA has identified mercury as especially harmful. According to EPA rules from March 2005, mercury emission from coal-fired power plants will ultimately reduce from 48 to 15 tons a year, an approximately 70% reduction[3]. These measures, coupled with the expected rise in the use of gasification in the United States during the next 20 years, have brought the need for a low-cost mercury removal technique– applicable to the IGCC, which does not significantly reduce the plant efficiency[5].

Mercury sorbents currently in use typically operate at or below room temperature, requiring a cooling of the syngas stream to remove pollutants, followed by re-heating. It has been shown[4] that improving sorbent operating temperatures to above the dew point of water in the syngas stream ($\approx$ 165°C for a 20% water syngas stream at typical IGCC pressures) would result in almost a 4% efficiency improvement of a 600 MW IGCC plant. However, removing mercury from the syngas stream has proven challenging due to the high temperature involved and the low concentration of mercury (<100 ppbw in the IGCC syngas stream). The candidate sorbents must draw mercury from a source in which the chemical potential of mercury is low. Current sorbents such as activated carbons, metal oxides and metal sulfides for removing mercury from flue gas operate with a high efficiency at low temperatures[6].

In this work, we demonstrate how first-principle calculations can be applied towards the search for an effective high-temperature mercury sorbent. While previous studies have employed first-principle calculations to understand current sorbents[3,7], this is to our knowledge the first attempt to screen a large number of potential sorbents via quantum chemical calculations. Rather than focus on the detailed mechanism for mercury removal from the gas phase at the screening stage, we restrict our analysis to compound formation of mercury with the tested sorbents. We expect the trends of compound formation to correlate strongly with other mechanisms of mercury capture, i.e. surface adsorption potential as both involve the strength of the metal-mercury bond. This shows how quantum calculations can aid in screening a large chemical space and focusing more rigorous computational and experimental analysis to the most promising candidates.

In this study, we use ab initio calculations to determine formation enthalpies for 87 different known metal-mercury compounds of 35 different elements. These formation enthalpies are used to predict which metals will bind strongly to mercury and thus may be interesting as sorbents. However, based on the presence of the other gas constituents in the syngas stream, we also need to assess the possibility of a competing oxidation reaction which could potentially render the sorbent ineffective for mercury removal. As one step in this stability analysis, we have also calculated binary oxide formation enthalpies of 94 compounds of the form M_yO, where M is an elemental metal. Using these results, we are able to evaluate the metal sorbent candidates most likely capture mercury, while resisting oxidation in the syngas stream.

The results of our analysis demonstrate that the experimentally known successful metal sorbents can be identified from a larger list of all metals by our screening procedure. It has been also noted that the trends in binary oxide formation energy correlate strongly to those for binary mercury compounds. The observed trends indicate that amalgam formation and oxidation are not independent variables that can be freely varied in the space of possible chemistries. Such limitations are quite troubling for stable sorbent design, as it suggests that potential sorbents either have strong amalgam formation energy or are resistant to oxidation, but not both.

With this study, we show that quantum calculations can be applied to systematically screen large material databases for improved mercury sorbents. We also show that our result provides a strong indicator of the behavior in conditions significantly more difficult and computationally expensive to model accurately, such as surface catalysis and surface adsorption in the syngas stream environment. Though the current study has been limited to mercury sorption by metals, we believe that our procedure can be easily extended to guide experiments towards other types of high-temperatures mercury sorbents and to solutions for other emission control challenges.

References

(1) Powell, C. A.; Morreale, B. D., *MRS Bull* **2008**, *33*, 309-315.

(2) Nexant, Inc. **2006**, *EPA-430/R-06/006*.

(3) Podak, B.; Brunetti, M.; Lewis, A.; Wilcox, J., *Environ. Prog.,* **2006**, *25*, 319-326.

(4) Klett, M. G.; Maxwell, R. C.; Rutkowski, M. D., *The Cost of Mercury Removal in an IGCC Plant, Parsons Infrastructure and Technology Group Inc.; DOE-NETL,* **2002**.

(5) Granite, E. J.; Myers, C. R.; King, W. P.; Stanko, D. C.; Pennline, H. W., *Ind Eng Chem Res* **2006**, *45*, 4844-4848.

(6) Granite, E. J.; Pennline, H. W.; Hargis, R. A., *Ind Eng Chem Res,* **2000**, *39*, 1020-1029.

(7) Steckel, J. A., *Phys.Rev.B,* **2008**, *77*, 115412.

103e Feasibility of Removal of Trace Impurities from Coal Derived Raw Syngas to Produce Ultra-Clean Syngas

Ty Nguyen, Queensland Centre for Advanced Technologies, CSIRO Energy Technology , PO Box 883, Kenmore , 4069 QLD, Australia

In advanced low emission power generation systems based on coal gasification, the resulting syngas needs to be cleaned at high temperatures and pressures in order to maximise the operating efficiencies and lifetimes of downstream process equipment, including catalytic shift converters, hydrogen separators and fuel cells turbines. Removal of trace impurities of Na, K, Cl, S, Se, As, Sb, Cd, Pb, etc. becomes essential to produce syngas of ultra pure grade within required tolerances. In the advanced dry gas cleaning process being developed at CSIRO, the majority of impurities from raw syngas are extracted in stages by various sorbents after which the syngas is finally polished in a packed bed reactor called a polisher or guard bed. At CSIRO we have evaluated the thermodynamic feasibilities of a number of chemical reactions to remove trace quantities of H2S, Se, and As in the guard bed. The main objective of this study is to identify through thermodynamic modelling the best materials and chemicals for this process.

103f Effect of Flue Gas Oxygen and Water Vapor Content on High Temperature Capture of Mercury by MinPlus Sorbent

Sung Jun Lee, Chemical Engineering, ICSE, University of Utah, 50 S. Central Campus Dr., MEB Rm. 3290, Salt Lake City, UT 84112 and Jost O.L. Wendt, Department of Chemical Engineering & Institute for Clean and Secure Energy, University of Utah, 3290 MEB, 59 S. Central Campus Dr., Salt Lake City, UT 84112.

An experimental study was undertaken to determine whether MinPlus, an inorganic sorbent that has been shown to capture mercury at high temperatures, might be useful for warm and hot cleanup of gasification off gases. Disperse phase experiments were conducted, using an externally heated quartz tube reactor, with sorbent feeding rates ranging from 1g/h to 6g/h. Experimental techniques, from the feeding of the sorbent to the measurement of ~μ g/m3 levels of Hg++, and Hg in the exhaust, are described in detail. Changes of mercury concentration in the reactor off gas were passed to a Tekran mercury analyzer with a wet bubbler system (modified by thiosulfate addition). The entire Hg sampling, conditioning and analysis system were thoroughly validated in this and other preceding and contemporaneous mercury research programs.

A most important result was that both capture mechanisms (short time and long time) failed to occur in the complete absence of oxygen. It appears that previous results described in the original proposal were clouded by the (probable) presence of a small oxygen leak in the original packed bed apparatus. The current work has discovered that although some oxygen is required to allow the process to occur, the minimum amount of oxygen required might be as low as, and most probably much lower than, 4000 ppm. These results have important implications for the use of this sorbent in gasification processes, and these have been addressed in a companion volume to this Final Report. Experiments were also conducted to explore the effects of added water vapor, and although results here were somewhat ambiguous, the sorbent was still effective in the presence of water.

106a CO2 Capture Using Nanoparticle Ionic Materials and Task-Specific Ionic Liquids

Ah-Hyung Alissa Park and Kunyi Andrew Lin. Earth and Environmental Engineering and Chemical Engineering, Columbia University, 918 S.W. Mudd Hall, MC 4711, 500 W. 120th Street, New York, NY 10027

Since the industrial revolution in the late 18th and early 19th century, the amount of CO2 in the atmosphere has risen from 280 ppm in 1800 to 370 ppm in 2000, mainly due to the consumption of fossil fuels. Unfortunately, CO2 is one of the greenhouse gases that are considered to be responsible for global warming. Potential effects of global warming include sea level rises caused by the melting of glaciers and ice caps, increased frequency and intensity of precipitation and severe weather, poor agricultural yields, and species extinction due to the environmental changes. Moreover, the increased atmospheric CO2 concentration will acidify the ocean and will change the chemistry of the surface ocean. Once the shift in the oceanic chemical balance becomes significant, it will affect the ecosystems.

A number of CO2 capture and storage technologies are currently being developed to meet this challenge. For example, amine-based solvents such as monoethnolamine (MEA) and chilled ammonia are already being implemented for the CO2 separation from the post-combustion flue gas stream. These solvents have very high capacity to capture CO2, however they have some drawbacks including high vapor pressure, which leads to fugitive emissions during regeneration. MEA is also corrosive in nature, and thus, only dilute solution of MEA can be used for CO2 capture.

In order to overcome the shortcomings faced by amine-based solvent technologies, a novel carbon capture technology utilizing an amine-functionalized Task-Specific Ionic Liquid (TSIL) has been developed. Molecular ILs are made by pairing an asymmetric organic cation (e.g., dialkylimida-zolium or alkylpyridinium) with an inorganic anion (e.g. hexafluorophosphate) or an organic anion (e.g alkylsulfates). A high operating temperature is important for the economic feasibility of any CO2 separation technologies developed for industrial processes, and unlike conventional organic solvents that require ambient or lower operating temperatures, ILs are often non-volatile and stable over a very wide temperature range (-40 to > 300 °C).

This study focuses on the development of CO2 capture fluids based on the nanoparticle ionic materials (NIMS) and TSILs. Their absorption isotherms are investigated as a function of CO2 partial pressure and temperature (i.e., combustion and gasification conditions). The effects of the volume fraction of nanoparticles and the different functional groups on the CO2 capture efficiencies of NIMS are also investigated. The experiments are carried out using a high-pressure high-temperature reactor setup and a micro balance system. The NIMS results are then compared with those for the commercially available ILs as well as MEA solvent, in order to optimize the design of NIMS for the maximum CO2 capture.

106b Reversible Ionic Liquids as Double-Action Solvents for Efficient CO2 Capture

Vittoria M. Blasucci[1], Ryan Hart[1], Cerag Dilek[1], Hillary A. Huttenhower[2], Veronica Llopis Mestre[1], Pamela Pollet[3], Eduardo Vyhmeister[1], Charles Eckert[4], and Charles L. Liotta[5]. (1) School of Chemical & Biomolecular engineering, Georgia Institute of Technology, 2277 Peachtree Rd NE Unit 601, Atlanta, GA 30309, (2) School of Chemistry and Biochemistry, Georgia Institute of Technology, 311 Ferst Drive, NW, Atlanta, GA 30332-0400, (3) Chemistry and Biochemistry, Georgia Tech, 311 Ferst Drive NW, Atlanta, GA 30332-0100, (4) Chemical & Biomolecular Engineering, Georgia Tech, 315 Ferst Dr, Atlanta, GA 30332-0363, (5) Chemistry & Biochemistry, Georgia Institute of Technology, 311 Ferst Dr, Atlanta, GA 30332-0100

We have developed a new class of smart solvents with the goal of recovering post-combustion CO2 from fossil fuel-fired power plants. Smart solvents are those which undergo large step changes in properties in response to an external stimulus, such as light, heat, or even pH. Because of this abrupt property change, one can build into the solvent its own separation or regeneration methodology.

We have previously reported the development and application of reversible ionic liquids, as from an amadine or guanidine plus and alcohol. With these simply bubbling CO2 at ambient pressure through the mixture gives almost complete conversion to the ionic liquid, and either inert gas sparging or heat achieve reversal. However, the use of these for separation processes is made more complex by the requirement of two starting materials.

Here we report the application of single-component reversible ionic liquids derived from siloxylated amines and carbon dioxide. Many articles have highlighted the selectivity of conventional ionic liquids in separating CO2 from mixed gas streams. Also, ionic liquids are known to have a large capacity for dissolving gases. Our reversible solvents improve upon conventional ionic liquids by enabling a dual capture mechanism. Specifically, they function not only as physical absorbents but also as chemical adsorbents for CO2 offering high selectivity of CO2/N2 in dilute flue gas streams. Two moles of solvent react with one mole of CO2 to form the ionic liquid, which then acts as a solvent to physically absorb additional CO2.

Further, our one-component systems simplify processing requirements when compared to alcohol amine systems due to the large liquid range of our solvents. These novel solvents produce a clean CO2 stream upon reversal. Further, we show that solvent properties, such as regeneration temperature, can be tuned by simple structural modifications. These changes enable us to design our structures to meet desired goals such as high adsorption and absorption capacity, low energy use for CO2 desorption and precursor regeneration, low viscosity, low corrosion and low cost.

106c Accelerated Carbonation of Steel Slag via Two-Stage Carbon Mineral Capture Technology

Ah-Hyung Alissa Park, Earth and Environmental Engineering and Chemical Engineering, Columbia University, 918 S.W. Mudd Hall, MC 4711, 500 W. 120th Street, New York, NY 10027, Laura Simone, Department of Hydraulics, University of Rome "La Sapienza", Rome, Italy, and Dong Hyun Lee, Department of Chemical Engineering, Sungkyunkwan university, 300 chunchun, jangan, Suwon, South Korea.

The steel manufacturing facilities are one of the major point sources of greenhouse gas emission. Another problem faced by steel industries is the waste disposal (i.e., steel slag) since the amount of steel slag produced during the steel-making process can be as high as 10 – 15 % by weight. Thus, much effort has been focused on improving the environmental sustainability of the steel manufacturing process, for example, by utilizing steel slag as an aggregate and a filler for cement. To achieve this goal, the steel slag has to be conditioned to meet the requirements of the physical and chemical properties. For example, the amounts of free CaO and MgO in steel slags need to be minimized, since free CaO leads to swelling of cement via hydration and carbonation reactions. Steel slag generally contains CaO, SiO2, MgO, Al2O3 and MnO, and is considered to be non-hazardous waste. On the other hand, Alloy steel slags, including stainless steel slag (SSS), often contain high Cr (2 – 5 %) and Ni (0.02 – 0.55 %), which classify SSS as hazardous material upon their release into the environment.

Therefore, the main focus of this study is to accelerate the carbonation of SSS for the production of value-added products, while capturing CO2. The SSS was procured from an Italian steel company and both a batch reaction system and a fluidized bed reactor are used to carry out the two-step carbon capture process: (a) the extraction of Ca from SSS and (b) the capture of CO2 using a high pH Ca-rich solution. The carbonated solid products are then characterized to determine the overall capture capacity as well as the environmental safety of the proposed method. The effects of mixing rate, reaction time, temperature, pH, and chemical additives on the chemical composition, particle size and morphological structure of the solid products are investigated.

Ramgen's Low-Cost, High-Efficiency CO_2 Compressor Technology

> *The CO_2 compressor power required for a pulverized coal power plant is 8-12% of the plant rating.*

CO_2 compressors represent a large fraction of the enormous capital and operating cost penalties of any CCS system. The CO_2 compressor power required for a pulverized coal power plant with an amine-based capture is approximately 8-12% of the plant rating, depending on conditions. A 1,000-MW PC plant would require 120 MW, or 160,000 hp, at an estimated $180 million for today's 3 x 50% configuration. The ammonia-based processes consume about 1/2 of the power at less than ½ the cost, but still represent a considerable expense.

The CO_2 compressor power required for an IGCC power plant is approximately 5% of the plant rating. A 600-MW PC plant would require 30 MW, or 40,000 hp, at an estimated $45 million for the same 3 x 50% configuration.

Both of these values are based on current estimates of the "state-of-the-art" integrally geared turbo compressor at nominal discharge pressure of 2200 psia, and do not include installation costs at 35 to 50 percent increment.

Ramgen's shock compression technology represents a significant advancement in the state of the art for many compressor applications, and specifically for CO_2 compression. The principal advantage of Ramgen's shock compression is that it can achieve exceptionally high compression efficiency at very high single-stage compression ratios, resulting in a product simplicity and size that will lower both manufacturing and operating costs.

The novel Ramgen technology concept addresses the two greatest objectives identified by the Department of Energy for the Capture and Storage of CO_2 – lower costs and better efficiency.

Figure 1 Ramgen's Super-Sonic Shock Wave CO_2 Compressor

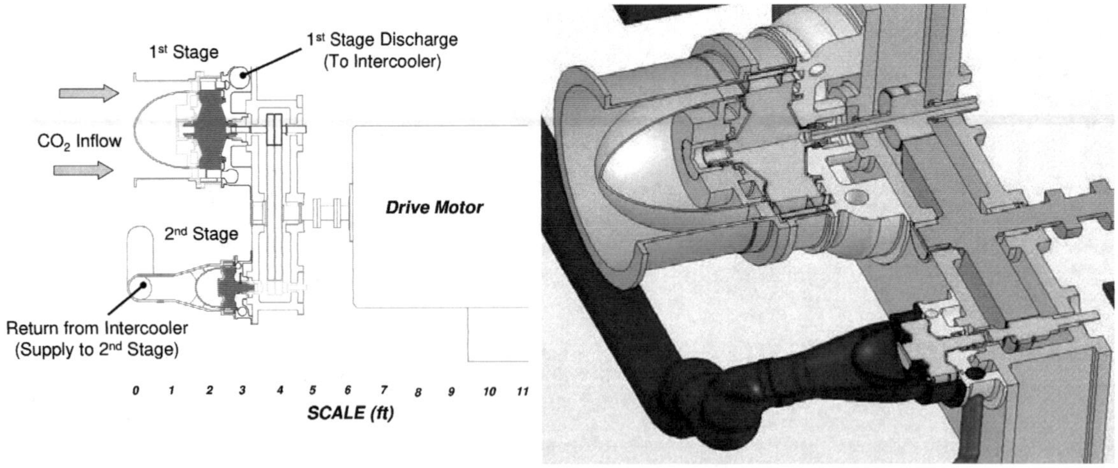

Ramgen Power Systems, Inc. received a 4-year $11.0 million DOE grant to begin the development of a CO_2 compressor capable of developing the required 100:1 pressure ratio in two stages of compression.

Ramgen based its original design concept on a two-stage, integrally geared compressor, with a common driver, following the pattern of the conventional designs, but with two stages instead of eight. Figure 1 shows a drawing of one such candidate configuration, along with a dimensional scale to indicate its approximate size for a 10,000 hp unit. The final specifications of the system will be defined as a deliverable of the program.

The reason that existing CO_2 compressor designs are so expensive is, in part, because the overall pressure ratio is 100:1, and, in part, because CO_2 requires stainless steel construction in the presence of water vapor. But by far, the most significant impact on cost is an aerodynamic design practice that limits the design pressure ratio per stage on heavier gases such as CO_2.

Standard turbomachinery design practice is to limit the inlet flow Mach number to less than 0.90 at the inducer blade tip to avoid generating shock waves in the blade passages and their accompanying losses. This is typically done by adjusting the stage speed. The Mach number itself is a function of molecular weight and therefore the effect is more pronounced on the heavier-than-air CO_2. This inducer blade tip speed limit results in a pressure ratio per stage limits of approximately 1.7 to 2.0:1 on CO_2. At these stage pressure ratios, eight stages of compression are typically required to reach an overall pressure ratio of 100:1.

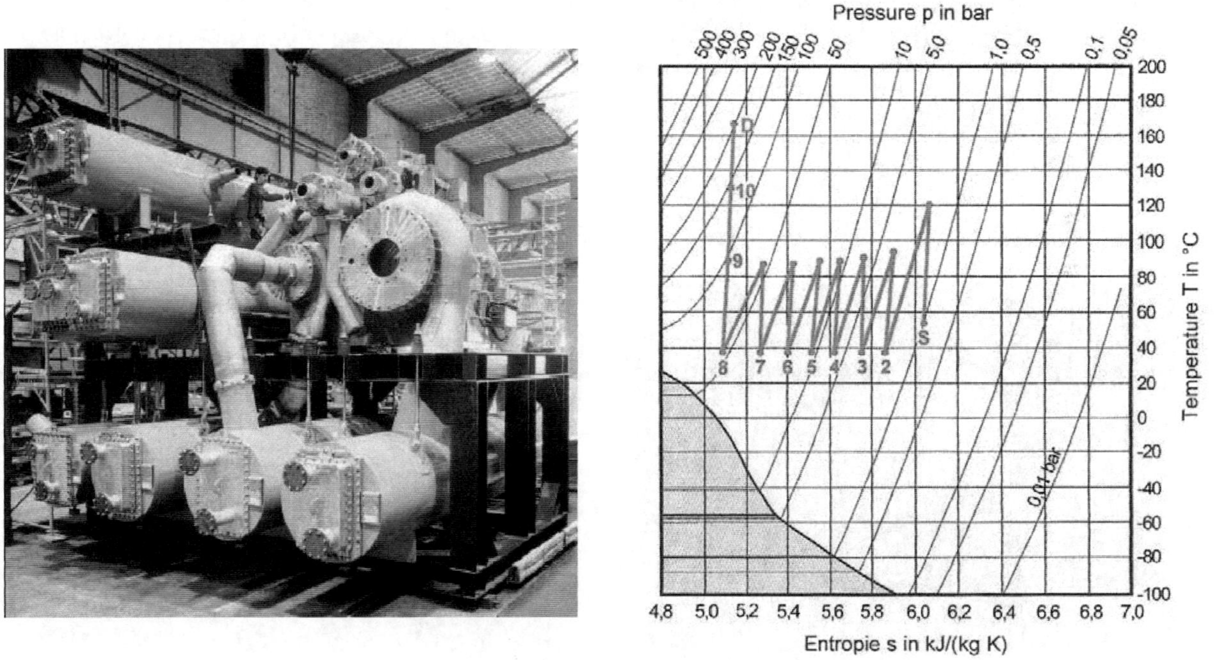

Figure 2 – MAN Turbo 10-stage 200:1 CO_2 Compressor

This issue is further complicated by the need to intercool the CO_2 between each compression stage. The heat of compression discharge temperature associated with these very low stage pressure ratios is approximately 200°F, which, as an inlet to the next stage, is too hot to achieve good efficiency, but lacks the thermal driving force for cost-effective heat exchanger selection. This heat is also of insufficient quality to be of practical use elsewhere in the process. The only option is to reject virtually all the compressor electrical input power through intercooler heat exchangers and supporting cooling towers, themselves a significant capital and installation expense.

Intercooler selection is made even more difficult by the need for low-pressure drop designs and the requirement to use low heat transfer effectiveness 304 corrosion-resistant, stainless steel construction. Air cooled heat exchangers, often required in arid climates, exacerbate the problem with their generally lower approach temperatures and require substantial fan horsepower, often overlooked in the compressor power evaluation.

Ramgen, on the other hand, designs its rotors to create and manage shock structures and can realize the full effect of shock waves to generate substantial pressure ratios, efficiently. The proposed Ramgen CO_2 compressor concept would achieve the required 100:1 pressure ratio in two stages of compression, each rated at 10:1 (10 x 10 = 100). This configuration would feature a conventional intercooler between the first and second stages as well as an aftercooler, if required by the application. In the CCS application, it is likely and preferred that the capture process itself would become the intercooler and aftercooler.

Ramgen intends to offer its compressors at approximately 40-60% of the price of the current integrally geared, state-of-the-art designs.

In addition to the obvious cost advantages and as a direct result of the Rampressor being able to achieve single-stage compression ratios of 10-12:1, stage discharge temperatures are estimated to range between 450-500°F, depending on inlet gas and cooling water temperatures. This offers the potential for significant heat integration, without compromising compressor performance. The combined compressor and heat recovery creates an even more impressive energy efficiency advantage by recovering 70-80% of the electrical input energy in the form of useful heat. Potential uses for the available heat are to regenerate amine solutions or pre-heat boiler feedwater.

Ramgen's Discrete-Drive Single-Stage Configuration
The Ramgen concept is so interesting to Carbon Capture (Compression) & Sequestration CC(C)&S system developers that we are in continuous contact with most of them. It appears that a pair of discrete-drive, single-stage compressors is preferred over a conventional integral gear compressor with a common bull gear drive. The HP or high pressure stage of that matched pair is shown below:

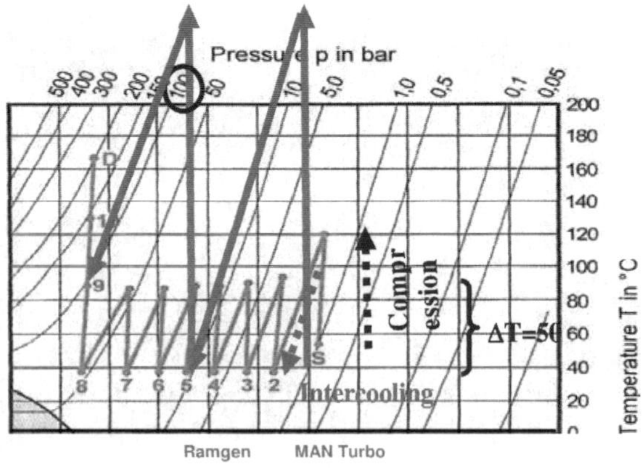

Ramgen Discrete-Drive HP Stage

At a minimum, this discrete-drive approach allows better matching of each stage to its specific process flow, including side-streams. In addition, the inherent variable speed capability of these optional drive systems provides operational flexibility that the developers see as very desirable. It should be noted that the inlet conditions to the first stage of compression are fixed by the process and relatively constant. The Inlet conditions on the subsequent stages are a function of the cooling medium, itself driven by the changing ambient conditions.

In simplistic terms, the first stage wants to run at constant speed, and subsequent stages before the next capture system flash level will want to run in a variable speed mode. The compressor, however, must be selected to meet the hot condition requirements and forcing multi-stage designs to operate at the same speed will drive one or the other off its optimum efficiency.

The Ramgen baseline drive utilizes a conventional low-speed motor and simple, single-step external speed-increasing gearbox, with other variations available to suit customer and contract requirements. High-speed permanent magnet motor drives or steam turbines are both feasible direct drive alternatives.

Ramgen intends to leverage high-speed motor development to the greatest extent possible by matching our rotor speeds and power requirements to existing high-speed motor designs and developments. Direct-drive steam turbine configurations would also offer the same degree of operational flexibility, and in addition to providing another opportunity for heat recovery. This product simplification with the use of existing driver systems substantially reduces development cost and risk.

This single-stage discrete drive approach also substantially reduces Ramgen's development cost and time. This schedule acceleration and design simplification will also allow Ramgen to submit proposals to participate in a number of the emerging opportunities to supply field demo units.

In addition, there are two, not-so-obvious, advantages to the discrete drive approach.

The first is that production times will not be dependent on the very limited and long lead time supply of precision, high-speed AGMA class 12/13 gearing, a chronic supply chain constraint.

The second is that discrete drive and integrally geared designs are rated differently. Geared units are nominally power rated, based upon predefined gear center distances and fixed design speeds.

Single stage units are rated on rotor diameters. The operating speed of a single-stage unit is adjusted within mechanical limits to meet the head or pressure required and the frame size is determined by the capacity required at that operating speed.

The important understanding is that as the stage is driven faster, it produces significantly more flow, and the rotor size can be reduced for a given capacity required. In practical terms if the HP stage is rated to the full 2200 psia nominal discharge pressure, the rotor will run faster and is likely to be two sizes smaller than it would be if it were rated at 1200 psia. The associated first cost savings is significantly more than the cost of the cryo-pump that would be required to boost the pressure from 1200 to 2200 psia.

Successful Ram 2 Test

The company has just completed a definitive aero test program where it demonstrated world-record level performance of a single-stage rotor, validating our advanced Computational Flow Dynamics (CFD) analysis and design tools on air, and proving a flow path directly traceable to the intended CO_2 compressor. The "as-built" rotor was predicted to produce a pressure ratio of 8.1:1, and the test resulted in a measured value of 7.8:1, and the CFD correlated almost exactly to the test day conditions.

The DOE has reviewed these test results and has authorized Phase II work to begin on a CO_2 specific design, suitable for pilot-scale field demonstration.

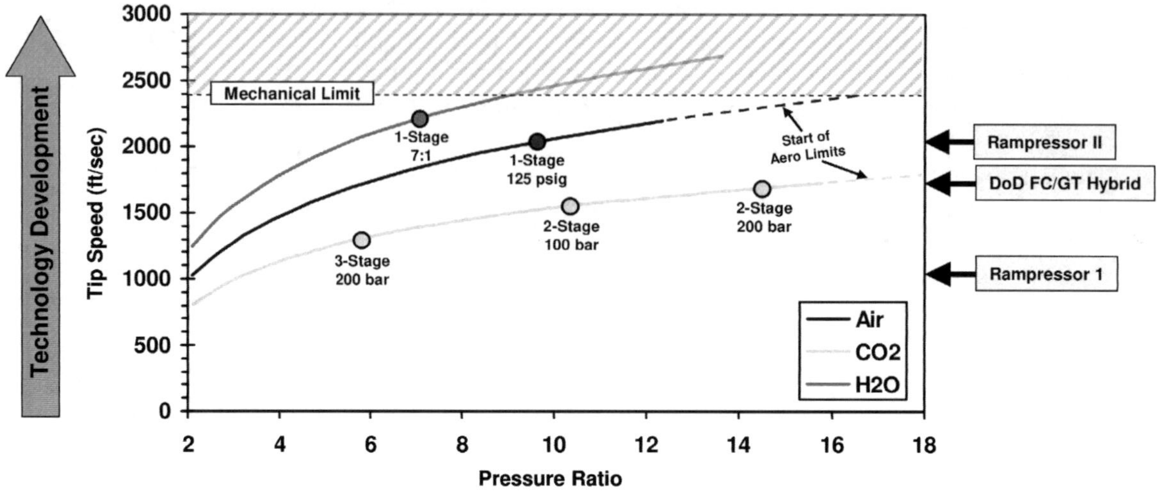

Of considerable importance is that the design tip speed of the 10:1 pressure ratio CO_2 compressor rotor is sufficiently low so as to allow for a shrouded rotor design concept, which eliminates the tip clearance and associated tip leakage effects and greatly simplifies the mechanical design.

Dresser-Rand Invests in Ramgen

On November 10, 2008, Dresser-Rand Group Inc. ("Dresser-Rand") (NYSE: DRC) announced that it had made an investment in Ramgen Power Systems, LLC, and further expects to support Ramgen's ongoing development work. Dresser-Rand's funding to help satisfy Ramgen's requirement with the Department of Energy's National Energy Technology Lab to obtain private matching funds for development.

Dresser-Rand is recognized by many scientists and engineers as the leading compression technology company in the world and they are the ideal partner for Ramgen in completing the development of its supersonic shock wave compression. Dresser-Rand has the credibility and capability required for the scale of roll out that will be necessary to make a difference with CO_2 emissions and climate change.

Dresser-Rand will bring its considerable resources to bear on "productizing" the Ramgen technology. Of particular interest is that Dresser-Rand does have a large scale test capability that could support testing of a commercial scale unit. D-R is also a world leader in the supply of steam turbine drives which are of considerable interest as prime movers for the CCS compressor application.

Ramgen Development Plan

Ramgen and D-R are currently blending development plans to leverage their individual skills and capabilities to accelerate its Phase II efforts.

The plan also includes specific efforts to characterize the CO_2 and its behavior in the supercritical regime in a series of static test rigs that are the equivalent of wind tunnel testing. These tests will also validate the impact of impurities contained within the gas, depending on application. Several of the key industry users have offered to assist us in identifying the range and variability of these components in their applications.

At the same time, Ramgen will be working with CC&S System developers to maximize heat integration with their respective processes. To date the CC&S developers have been designing their processes without much insight on compressor cost and performance. Most are now beginning to realize that a cost effective CO_2 compressor is key to their success and Ramgen is supporting their efforts to lower cost and improve efficiency.

The high-pressure ratio per stage capability of the Ramgen technology is the key enabling capability necessary to achieving this goal.

Commercialization Plan

Ramgen is currently planning to develop an HP-16 size unit. The HP-16 is the high pressure stage of a matched set that would be suitable for 250MW coal-fired power plant. That program is schedule to be completed in 2012, but test results that would allow us to offer commercial terms should be available in late 2011. We anticipate that by yearend 2011, we will be making firm commitments for projects with commercial operating dates scheduled for the 2014/2015 time frame.

That HP-16 would scale to an HP-24 to cover 500-600MW depending on assumptions. The matched LP stages, should they be required, would range from 24 to 40 inches.

We have investigated single train units for 800MW plants and consider them both feasible and within our capabilities to scale and deliver on the same commercial terms and schedule.

Some of the information contained in this document contains "forward-looking statements". In many cases you can identify forward-looking statements by terminology such ass "may," "will," "should," "expects," "plans," "anticipates," "estimates," "predicts," "potential," or "continue," or the negative of such terms and other comparable terminology. Forward-looking statements are only predictions and as such inherently include risks and uncertainties. Actual events or results may differ materially as a result of risks facing Ramgen Power Systems, LLC ("Ramgen") or actual results differing from the assumptions underlying such statements. These forward-looking statements are made only as of the date of this presentation, and Ramgen undertakes no obligation to update or revise the forward-looking statements, whether as a result of new information, future events or otherwise. Your decision to remain and receive the information about to be presented to you shall constitute your unconditional acceptance to the foregoing.

RAMGEN POWER SYSTEMS™

Ramgen Power Systems

2009 AIChE Spring National Meeting
April 26-30, 2009
Tampa Convention Center
Tampa, FL

0900-01293

Forward Looking Statement

RAMGEN POWER SYSTEMS™

Some of the information contained in this document contains "forward-looking statements". In many cases you can identify forward-looking statements by terminology such ass "may," "will," "should," "expects," "plans," "anticipates," "estimates," "predicts," "potential," or "continue," or the negative of such terms and other comparable terminology. Forward-looking statements are only predictions and as such inherently include risks and uncertainties. Actual events or results may differ materially as a result of risks facing Ramgen Power Systems, LLC ("Ramgen") or actual results differing from the assumptions underlying such statements. These forward-looking statements are made only as of the date of this presentation, and Ramgen undertakes no obligation to update or revise the forward-looking statements, whether as a result of new information, future events or otherwise. Your decision to remain and receive the information about to be presented to you shall constitute your unconditional acceptance to the foregoing.

0900-01293

Company

RAMGEN POWER SYSTEMS™

- Privately-held R&D company founded in 1992
- Focused on unique applications of proven supersonic aircraft technology
- Primary technology innovations
  - Supersonic stationary air & gas compressors
  - High velocity combustor
  - Supersonic expander
- Product embodiments
  - Two-stage 100:1 Pr CO2 Compressor
  - 30:1; 42% LHV ASCE Engine
  - Airborne APU
  - H_2 fuel combustor

US Army Corps of Engineers

0900-01293

RAMGEN POWER SYSTEMS™

Ramgen Compressor Technology

0900-01293

Shock Waves to Supersonic Inlets

RAMGEN POWER SYSTEMS™

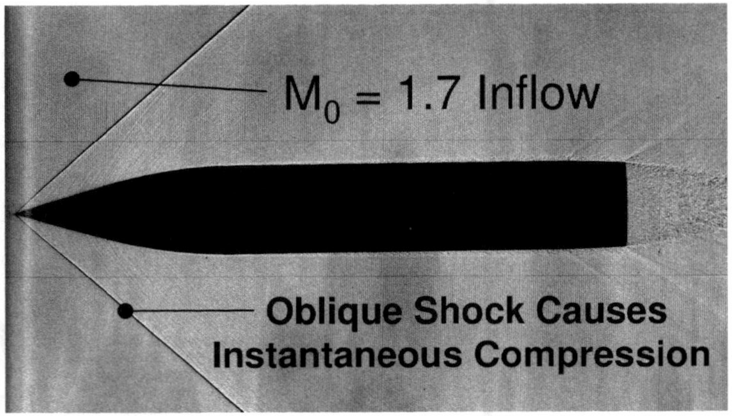

Schlieren Photo of Projectile with Shocks

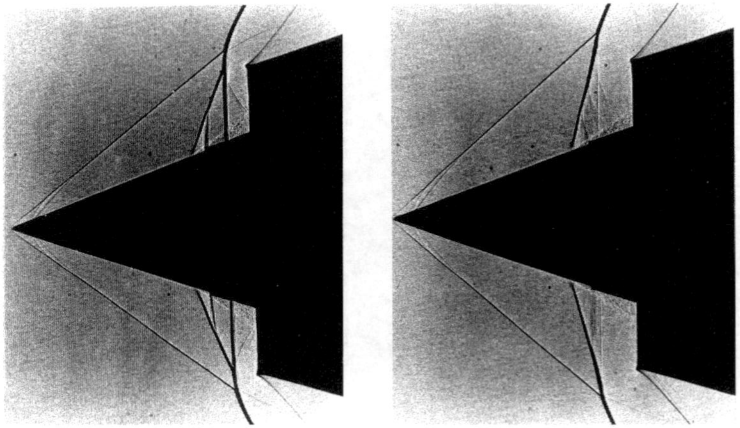

Schlieren Photo of Inlet Center-body and Cowl with Shocks

2-D Mixed Compression Inlet Model

- Initial External Shock System Followed by Internal Shock System
- Throat Bleed Slot For Inlet Starting
- Side Window For Schlieren Photography

0900-01293

F-15 2-D Planar Supersonic Inlet

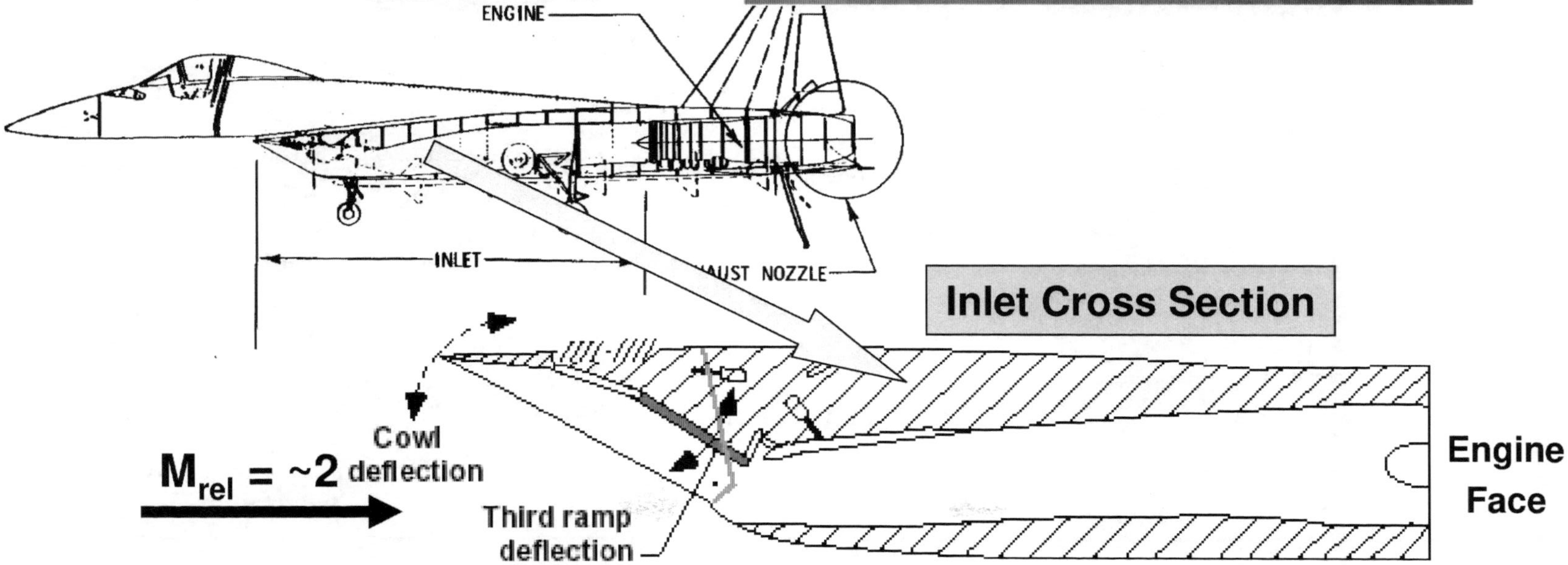

0900-01293

Rampressor Rotor Development

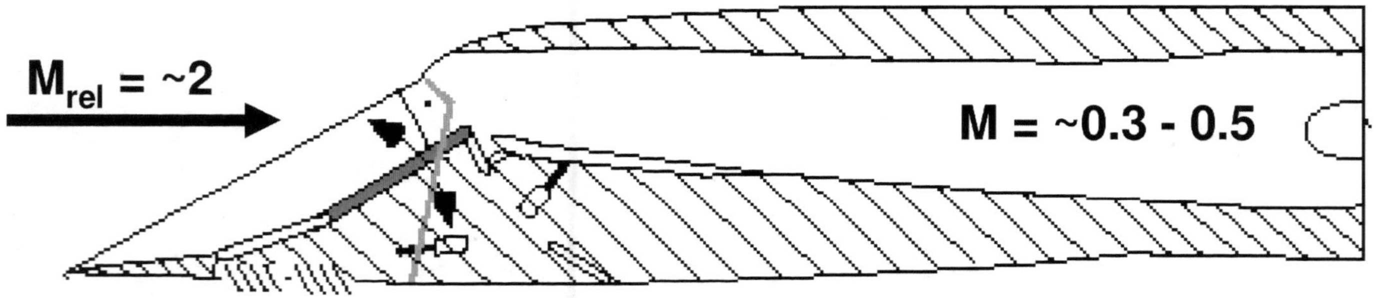

Supersonic F-15 Inlet

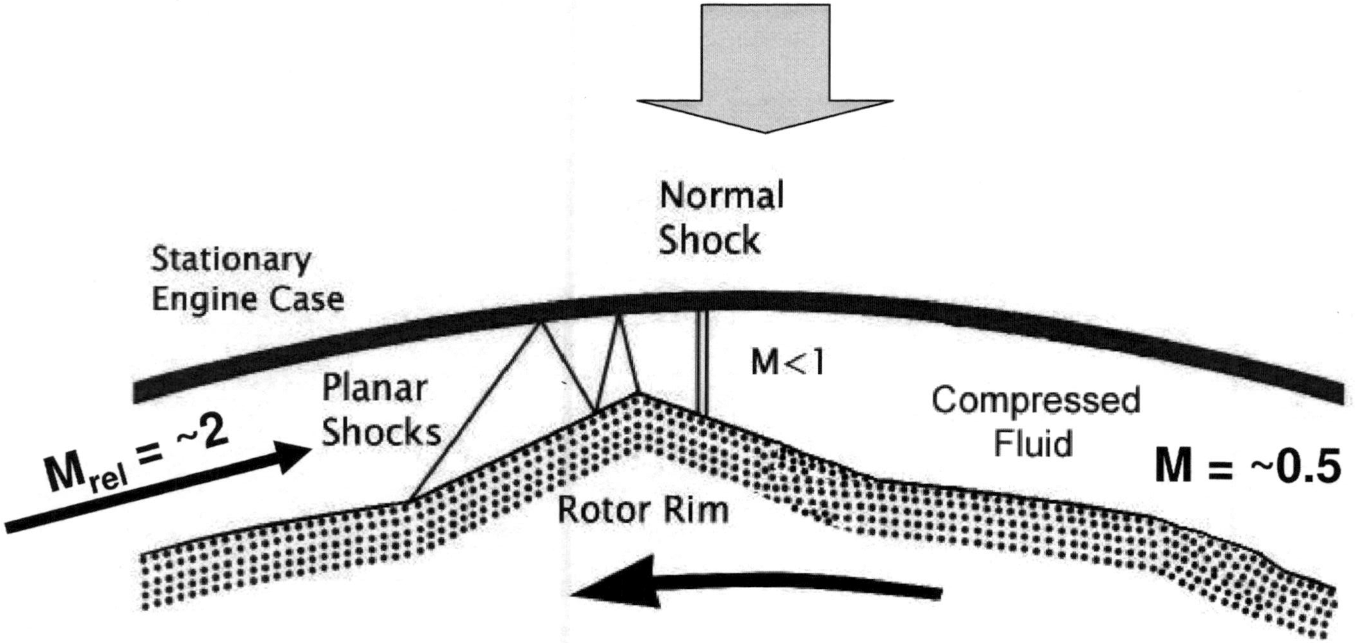

Rampressor Rotor

0900-01293

Typical Rotating Supersonic Flow Path

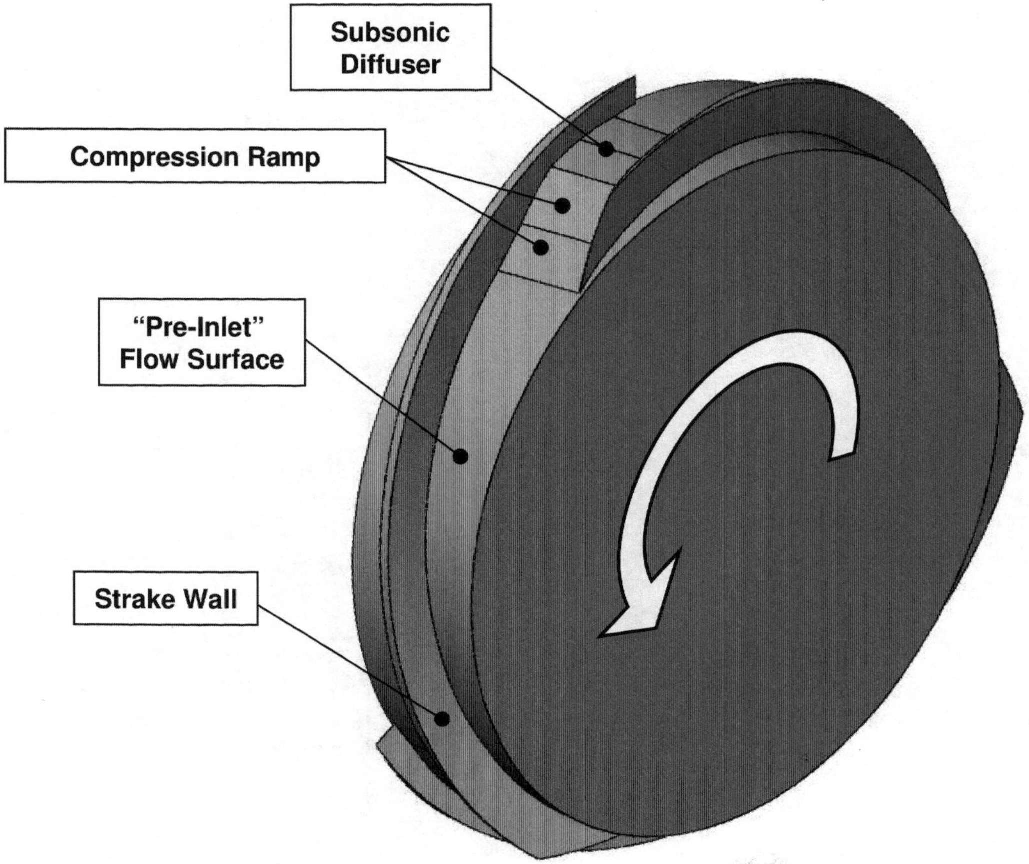

- **Rotor Flow Path:**
  - 3 Supersonic Compression Inlet Flow Paths On Disk Rim
  - High Efficiency, Compact Compression
  - Minimal Number of Leading Edges
  - Flow Path Geometry Similar For Different Pressure Ratios

- **Combination of Supersonic Flight Inlet & Conventional Axial Flow Compressor Aerodynamics:**
  - Rotor Rim Radius Change Produces Compression
  - 3 "Blades" (Strakes) Do Minimal Flow Work
  - Axial Inflow/Outflow

0900-01293

Compression Applications vs. Pr/Tip Speed

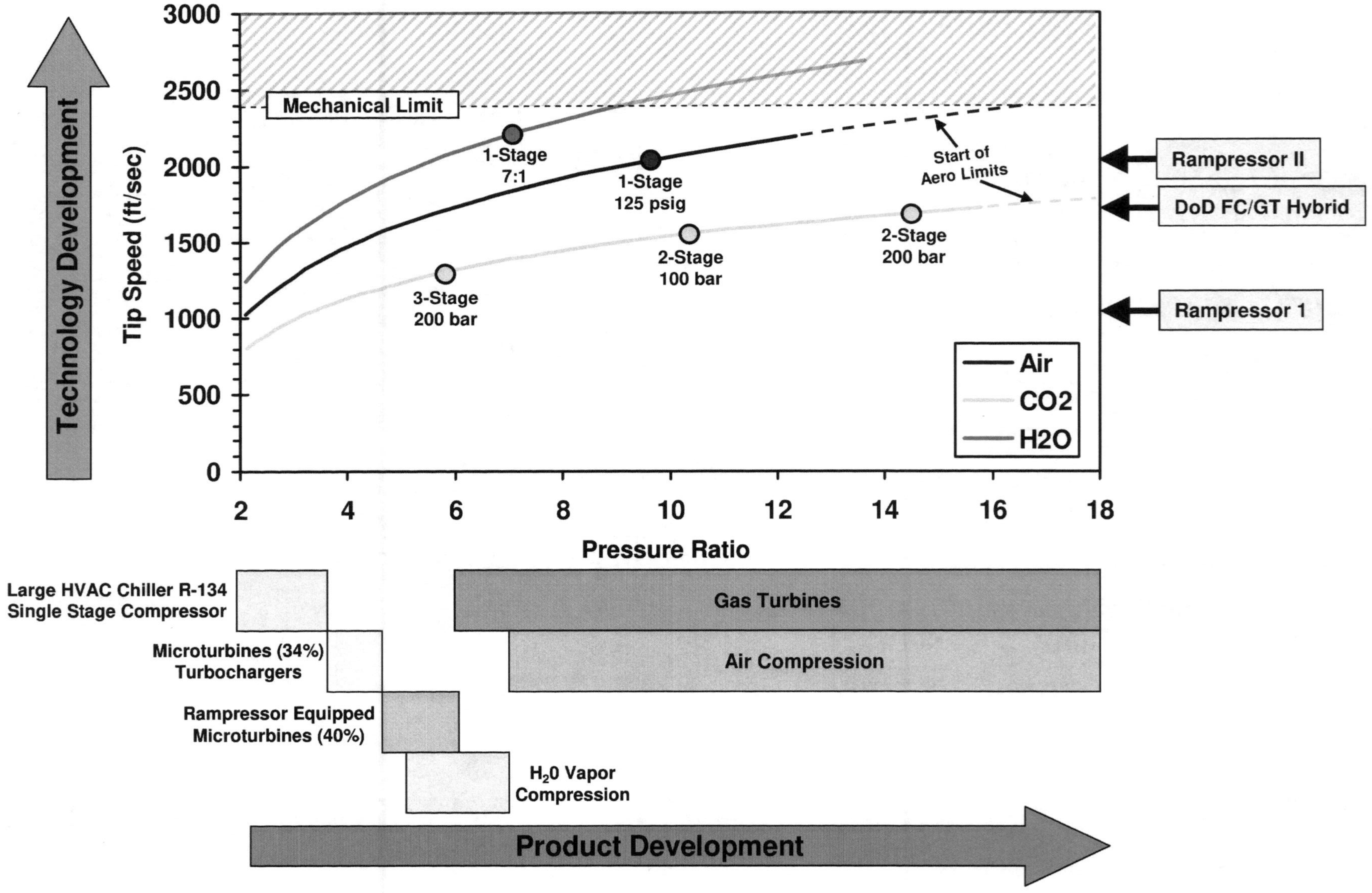

0900-01293

CO_2 Demonstration Program

- **Test plan & objectives**
  - Design & test CO2 compressor
    - 220 psia suction pressure
    - 2200 psia discharge pressure
    - CO2 working gas
  - Validate mechanical design
  - Validate drive system
  - Validate commercial level performance
  - Identify demo customer/site
  - Produce a ~3000 hp unit for field demo
- **Test article**
  - Three parallel flow paths
  - 10.2 inch 6Al-4Ti monolithic rotor
  - Shrouded option
  - Design speed 27,000 rpm
  - Tip speed 1400 fps
- **Analysis & Modeling Environment**
  - Numeca Fine/Turbo CFD
  - NASA Wind CFD
  - CADRE

- **CC(C)&S developers prefer single-stage discrete-drive design vs. common bull gear**
  - 150-250 psia suction pressures
  - Design point flow matching
  - Variable speed capability
  - Electric motor or steam turbine drive
- **Test facility**
  - Ramgen "Center of Excellence"
- **Demo Site**
  - TBD

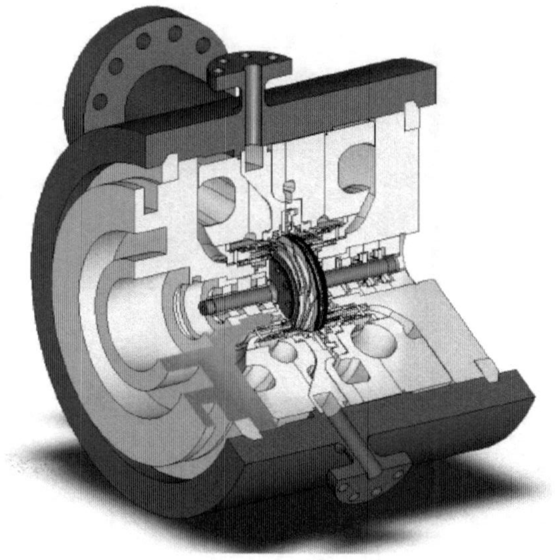

Ramgen Discrete Drive HP-Stage

0900-01293

Enter Dresser-Rand

RAMGEN POWER SYSTEMS™

Dresser-Rand Invests in Ramgen

- Dresser-Rand invests in Ramgen's "game-changing technology"
  - Support on-going CO_2 compressor development
  - Satisfy DOE matching funds requirement
  - Consistent with strategy to be technology leader in our industry
  - Extend served market into Electric Utility industry
  - Invest up to $49 million
    - Fund development & demonstration
    - Obtain an option to purchase assets

Dresser-Rand is consistently ranked among top three manufacturers in its served markets
  - Turbomachinery
  - Reciprocating compressors
  - Steam turbines
- #1 in North America
- Leading supplier of CO_2 compressors
- Global sales & service presence
- Strong products & brands
- Established customer base

0900-01293

Dresser-Rand Historical Overview

RAMGEN POWER SYSTEMS™

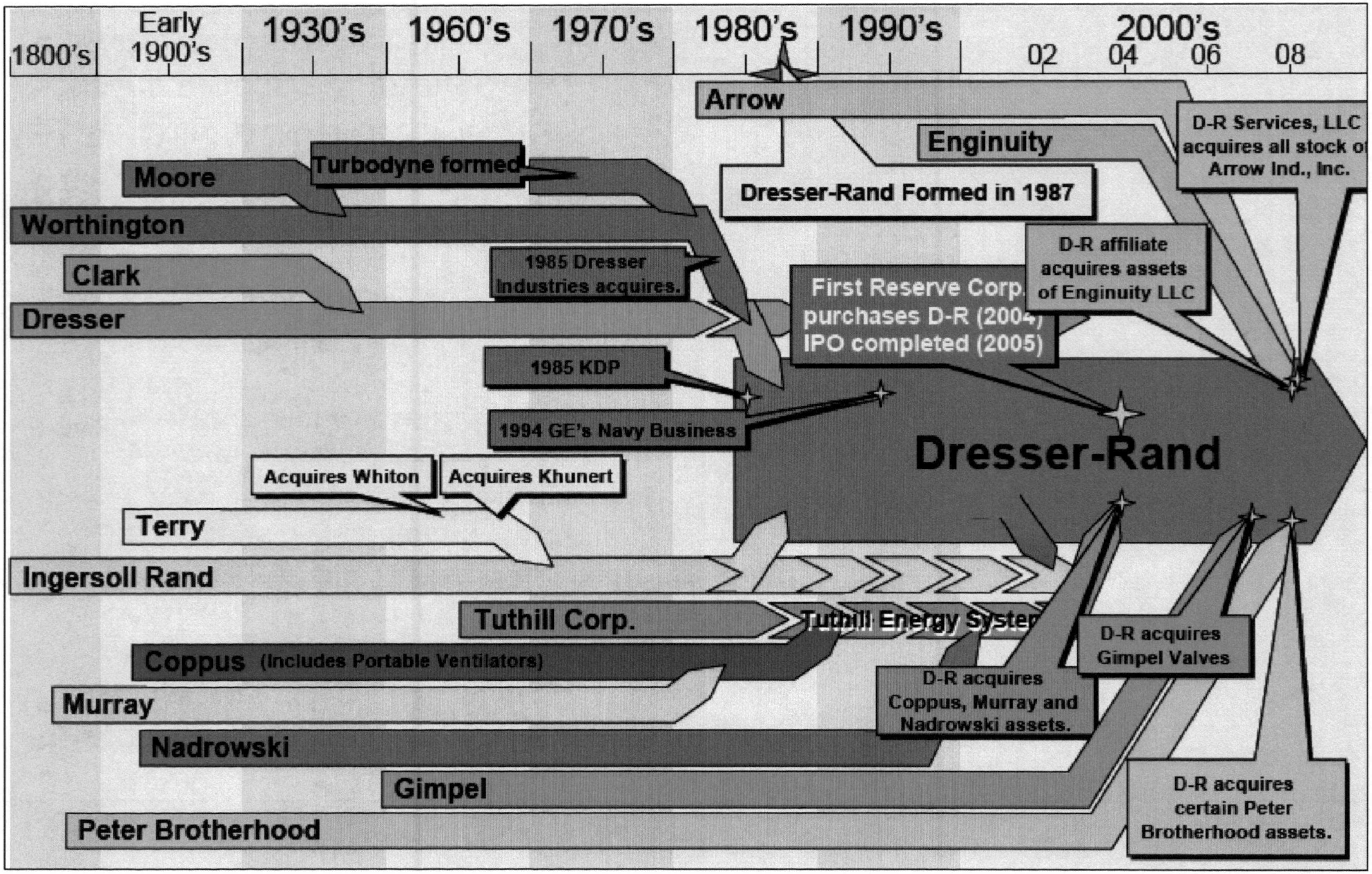

0900-01293

Dresser-Rand Heritage

RAMGEN POWER SYSTEMS™

DRESSER-RAND.

Serving the energy markets since 1840

0900-01293

Dresser-Rand's Global Presence

RAMGEN POWER SYSTEMS™

DRESSER-RAND.

0900-01293

RAMGEN POWER SYSTEMS™

Dresser-Rand Key Clients

STATOIL
Marathon Oil Corporation
bp
PEMEX
HESS
ConocoPhillips
Chevron
PETROBRAS
Chevron Phillips Chemical Company LLC
REPSOL YPF
ExxonMobil
ENEX
NDER MORGAN
PRAXAIR
PDVSA
VALERO ENERGY
PT Badak NGL
DOW
FLUOR
HOVENSA
LYONDELL
OXY
TEXAS GAS TRANSMISSION, LLC
FLINT HILLS RESOURCES
EQUISTAR
Transco
Panhandle Energy
WOODSIDE
TARGA
elpaso
dcp Midstream Pa

DRESSER-RAND.

Note: Partial list as of December 2007.

0900-01293

Products for All Served Markets

RAMGEN POWER SYSTEMS™

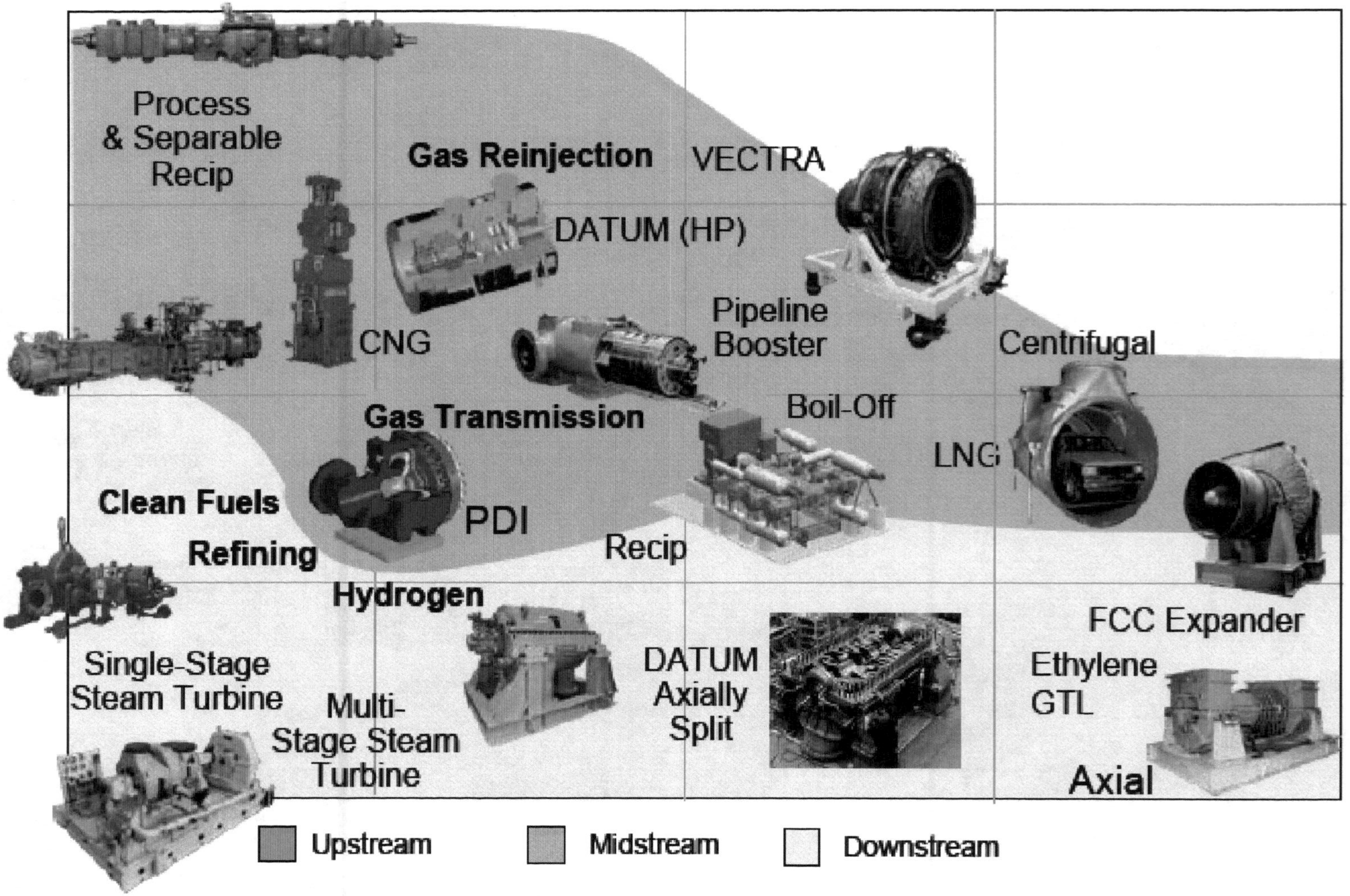

DRESSER-RAND.

0900-01293

World Class Test Facilities

RAMGEN™ POWER SYSTEMS

DRESSER-RAND.

0900-01293

CO_2 Demonstration Program

RAMGEN POWER SYSTEMS™

Develop HP-16 at ~10MW.... ...instead of 3000hp

- **Test plan & objectives**
  - Design & test CO2 compressor
    - 220 psia suction pressure
    - 2200 psia discharge pressure
    - CO2 working gas
  - Validate mechanical design
  - Validate drive syst
  - V
  - Id
  - P eld demo
- **Test**
  - Three parallel flow paths
  - 10.2 inch 6Al-4Ti
  - Shrouded option
  - Design speed 27.
  - Tip spee s
- **Analysis & Modeling Environment**
  - Numeca Fine/Turbo CFD
  - NASA Wind CFD
  - CADRE

- **CC(C)&S developers prefer single-stage discrete-drive design vs. com gear**
  - 150-250 psia s
  - p
  - e speed c
  - motor or stea
- **Test facilit**
  - a
  - o
  - B

Ramgen Discrete Drive HP Stage

0900-01293

HP-16 Accelerates Development

- Testing at a commercial scale
- Test article is more expensive, but technically less challenging
- Initial HP-16 testing scheduled 2:Q:2011
- Expect to be able to offer commercial performance guarantees and terms by 1:Q:2012
- HP-16 can support 200-250MW depending on CCS process and operating conditions
- HP-32 is the largest anticipated size at ~800MW
- LP anticipated stages range from LP-48 at ~800MW
- Projects of Interest
  - Existing coal-fired power plants
    - Greater than 400MW
    - Built after 1980
  - New Plants – COD 2014+
  - Coal-to Liquid plants

Compressor Designs

“Oops!” Hold the Press!!

RAMGEN POWER SYSTEMS™

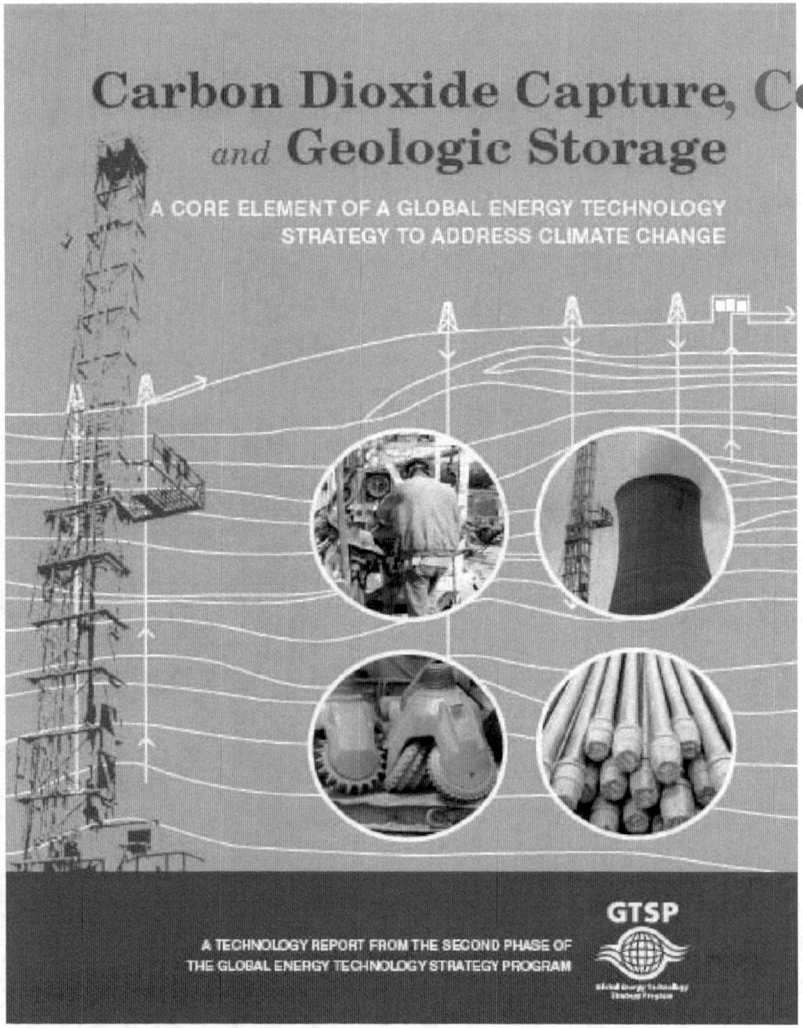

0900-01293

Compressors Are An After Thought!

RAMGEN POWER SYSTEMS™

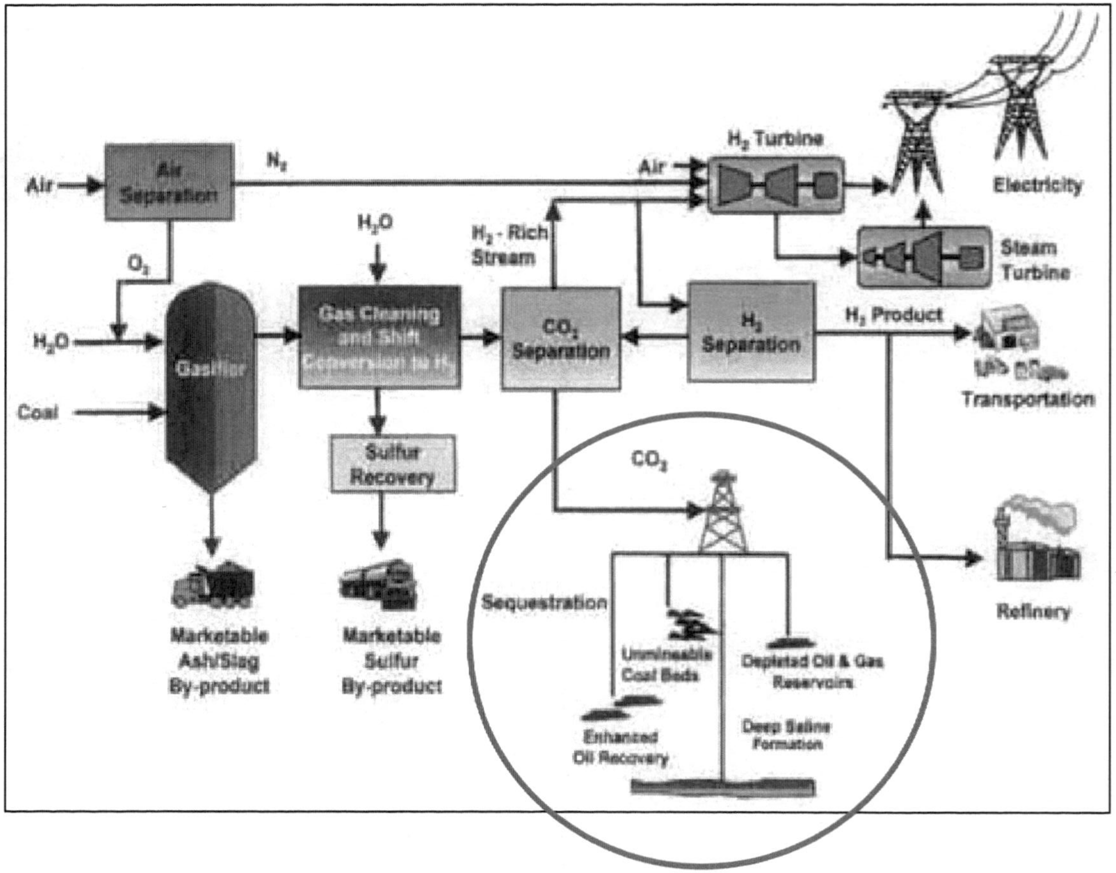

Sequestration

0900-01293

Fossil Fuel Power Plant – CC&S

RAMGEN POWER SYSTEMS™

- All fossil fuel power plants produce some level of CO2
- CO2 compressor power
  - Advanced pulverize coal – 8-12%
    - 600MW ⇨70MW ⇨93,000 hp
  - IGCC - 5%
    - 600MW ⇨30MW ⇨40,000 hp
  - CCGT – 8%
    - 400MW ⇨32MW ⇨43,000 hp
- 100 new power plants annually
  - $1.5 billion annual compressor market
- Retrofit opportunity
  - $0.7 billion annual compressor market

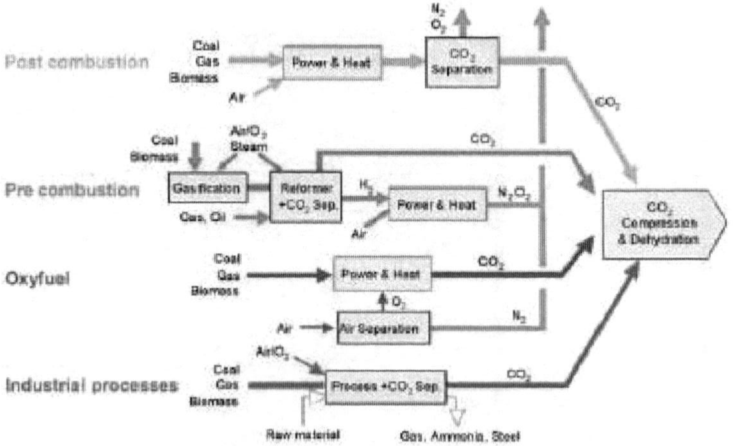

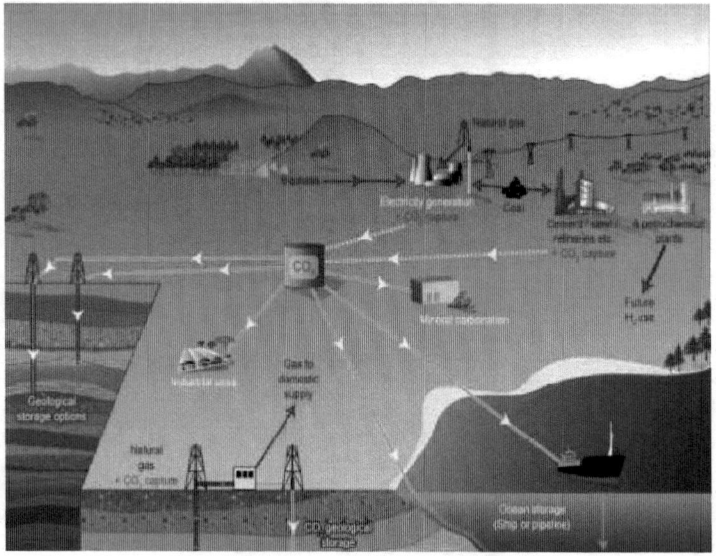

Over $2 Billion annual market opportunity

0900-01293

RAMGEN POWER SYSTEMS™

CCS Technologies

- Amine systems
  - Suction pressures – 15; 22; 25; 30 psia
  - Regeneration heat required
    - Conventional amines – 1550 Btu/lbm-CO_2
    - Advanced amines – 1200 Btu/lbm-CO_2
    - Really advanced amines – 800 Btu/lbm-CO_2
  - 8% parasitic power
  - Post combustion - New & Retrofit
- Ammonia-based systems
  - Suction pressures – ~ 200-300 psia
  - Regeneration heat required
    - Aqueous ammonia – 493 Btu/lbm-CO_2
    - Chilled ammonia – TBD
  - 4% parasitic power
  - Post combustion - New & Retrofit
- Selexol/Rectisol
  - Suction pressures 50, 150 & 300 psia with sidestreams
  - Regeneration heat required for the Claus Plant
  - 5% parasitic power
  - IGCC (new) only
- Oxy-fuel systems
  - Raw gas feed – 15 to 500 psia
  - Twin purified suction streams – ~150 & 300 psia
  - 12-13% parasitic power
  - New plants only
- Discharge pressures – 1200;1600; 2000; 2215; 2500; 2700; 2900 psia

0900-01293

CO_2 Compressor Configuration Choices

RAMGEN POWER SYSTEMS™

Ramgen Discrete Drive HP Stage

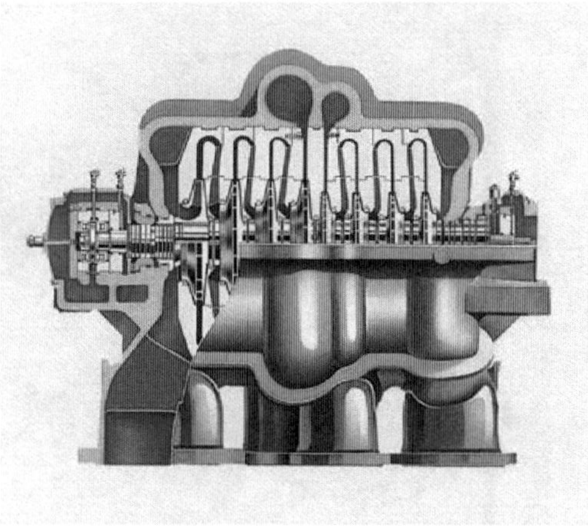

Hitachi Inline Process CO_2 Compressor

MAN Turbo Integrally Geared Compressor

MAN Turbo (Sulzer) Isotherm™ Compressor

0900-01293

Integrally-Geared Compressors

- **100:1 CO2 compressor is 8-stages/6 or 7-intercoolers**
- **Common drive thru an integral bullgear to stages**
  - 4 pinions, double horizontal split casing
  - Individual compression stages mounted on opposite ends of each pinion
  - At least one stage on each pinion is operating off its optimum specific speed.
  - Pr 1.7-1.8 per stage on CO2
  - Stage temperature rise approximately 100°F
  - Stage-1 impeller can be an open design at Pr of 2.0:1
  - Later stages suffer sharp aero efficiency falloff due to their small size
  - Even number of stages required to balance thrust loads
  - Frames are "torque limited" by the gearing and therefore "power rated"
- **External intercoolers between casings**
  - Low Pr per stage aero requires very low ΔP intercoolers
  - CO2 requires stainless steel intercoolers
  - Stainless steel has low a heat transfer properties
  - Concerns exist over 2-phase flow, aggravated by intercooling near critical point
  - Impurities add another level of uncertainty
- **Control – 30% turndown at constant pressure**
  - IGV's are common practice and can be applied on intermediate st:
  - System resistance ⇨constant speed; variable pressure
  - Full variable speed is not practical; variable speed trim possible
- **Cost**
  - $750/hp
  - Installation 50-75% depending on intercooler configuration
- **API oriented customers do not like integral gear drives.**

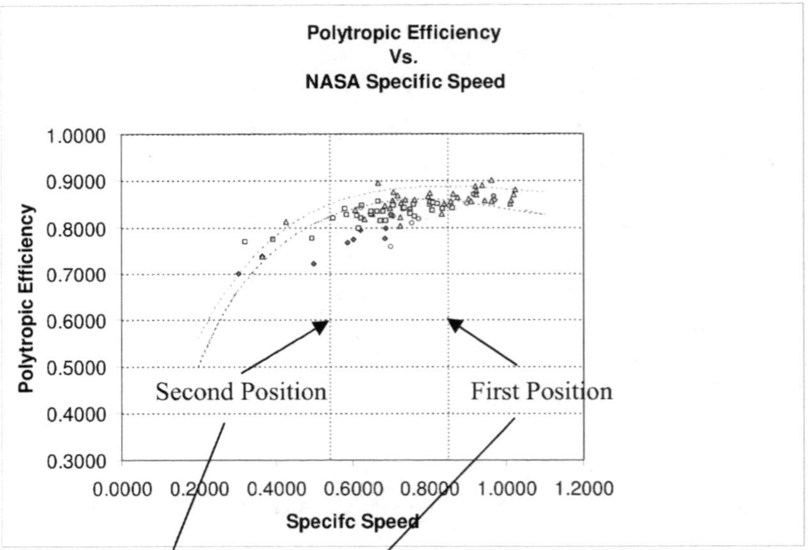

0900-01293

Ramgen CO_2 Compressor Product

RAMGEN POWER SYSTEMS™

- 100:1 CO2 compressor ⇨ 2-casings/2-stages
  - No aero Mach# limit
  - 10+:1 pressure ratio; 400°F temperature rise
  - 1400 fps tip speeds; Shrouded rotor design
- Single-stage, discrete-drive
  - Single stage per drive optimizes specific speed match
  - Simple single-step external gearbox or high speed direct drive
  - Lower mechanical losses
- Variable speed option
  - Match MW and temperature changes with speed changes
- Configuration adapts easily to match process requirements
  - Mismatched thru-flow
  - Side stream additions
- Active IGV Flow control on each stage
  - Match CO2 capture system constant pressure requirement
- Heat exchangers
  - Inter/aftercooler can be the CCS or power plant
  - "Compressor" heat exchanger cost can be eliminated
  - Eliminate or substantially reduce cooling tower requirement
  - Eliminate or substantially reduce cooling tower make-up water
  - 3x LMTD ⇨ heat exchangers with 1/3 the surface area
- 1/10th the physical size – facilitate space constrained retrofits
- 1/2 the installation cost

Ramgen Discrete Drive HP Stage

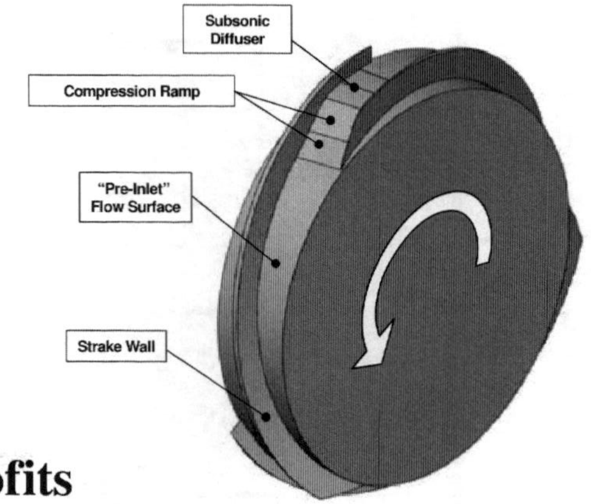

Ramgen Compressor Rotor

0900-01293

The "State-of-the-Art" is Expensive

- **10-stage 6000 hp**
  - **$8.0 million ⇨$1350/hp**
  - **Pr 200:1 ⇨1.70 per stage**

- **8-stage 20,000 hp**
  - **$15.0 million ⇨$750/hp**
  - **$23.0 million installed ⇨$1150/hp**
  - **Pr 143:1 ⇨1.86 per stage**

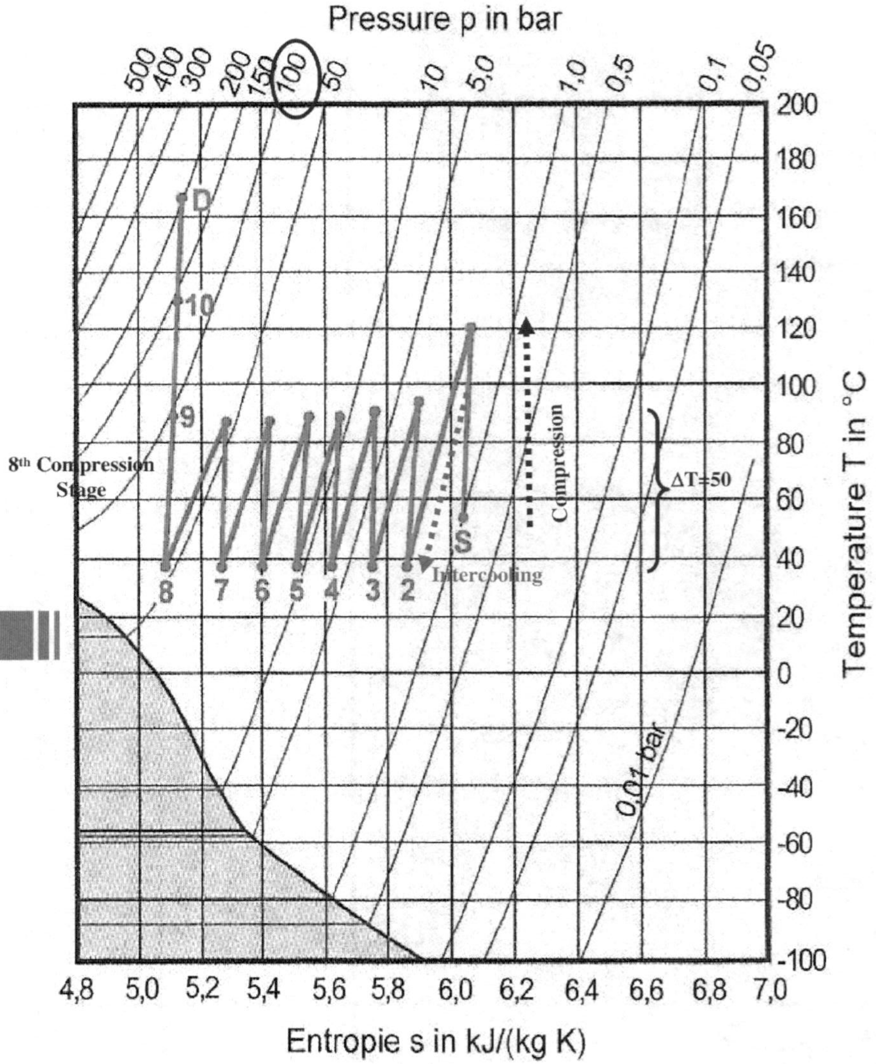

0900-01293

Siemens 24-XT-101 CO_2 Compressor

RAMGEN POWER SYSTEMS™

- **Siemens 24-XT-101**
  - Pr 62:1
  - 6-stages
  - Pr 1.99 per stage
- **10MW**
  - 13,400 hp
  - 220,264 lbm/hr
- **Weight**
  - 515,400 lbs

0900-01293

Inline Compressors

RAMGEN POWER SYSTEMS™

- 100:1 CO2 compressor is 4-casing/12-stages
- 3-4 stages per casing (body)
  - Stages are the same diameter and speed
  - Stage specific speed & efficiency is successively reduced thru the body
  - Pr 1.5-1.6 per stage on CO_2
  - Stage-1 can be an overhung, higher ratio, open design
  - Last stage specific speed/efficiency fall-off dictates shift to smaller stage & casing
  - Thru shaft reduces flow area amplifying Mach# constraint
  - Shrouded impeller design
  - Large diameter return channels exacerbate large diameter casing working pressure design/mfg challenges
- External gear drive
  - Operating speeds typically above synchronous motor speeds
  - 100:1 normally set up as an LP & HP train/2-casings each
  - 3 x 50% deployment = 12 casings/9 gearboxes/6 drivers
  - Mechanical loss add-on ~6%
- External intercoolers between casings
- Control – no turn down to match constant pressure CO2 system
  - System resistance ⇨constant speed; variable pressure
  - Constant pressure ⇨recirculation
  - Variable speed & IGV's not practical
- Cost
  - Equipment - $1000/hp
  - Installation +75%

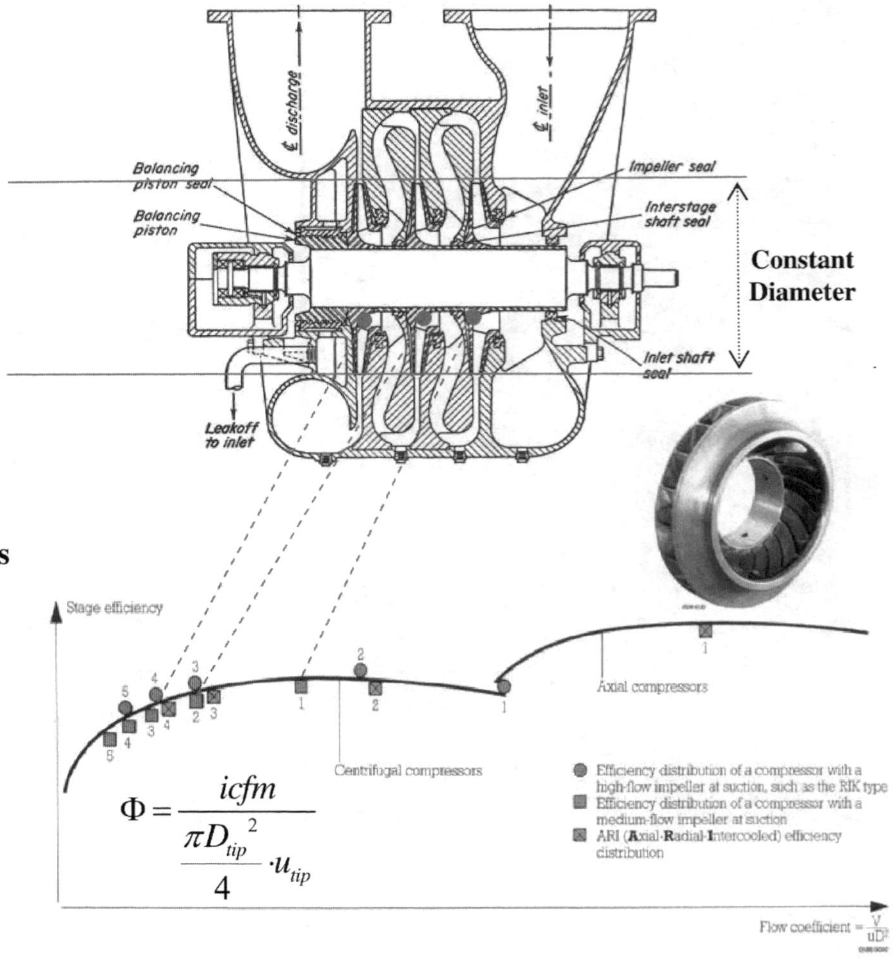

$$\Phi = \frac{icfm}{\frac{\pi D_{tip}^{2}}{4} \cdot u_{tip}}$$

0900-01293

Mitsubishi CO_2 Compressor

- **Recently announced**
- **High performance 11.7MW CO2 compressor**
  - 10 stages of compression
  - 2 intercoolers
  - 2 casings
- **LP casing**
  - 3 stages followed by and intercooler
  - 4stages followed by another intercooler
- **HP casing**
  - 3 stages to reach 203 bar Pr 100:1
- **The data in the associated technical publication indicates elevated inlet pressure of approximately 23 psia**

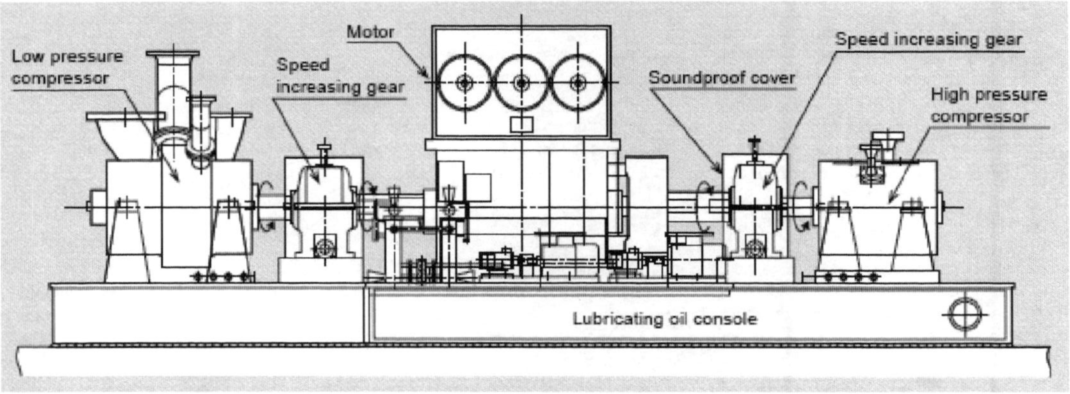

Fig. 1 CO_2 re-injection compressor train configuration
A both ends drive motor is installed in the center of the train, two speed increasing gears are provided at outer ends, with two compressors outside them.

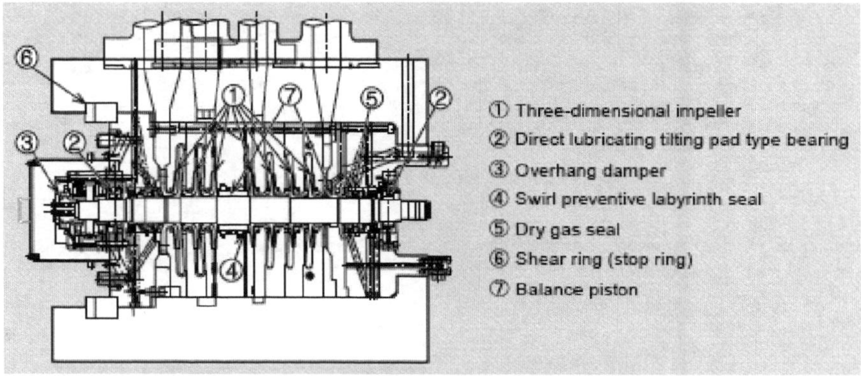

Fig. 2 A sectional view of CO_2 re-injection compressor
Required advanced technology, and health, safety and environment protection features are shown in a sectional view of the CO_2 re-injection compressor.

Mitsubishi CO2 Compressor

0900-01293

Compressed Air & Gas Handbook

can be handled with sufficient accuracy for most purposes when the unit is a typical single-stage air compressor. A little more discretion must be used on multistage compressors handling heavy gases, however, because fan-law deviation can become quite significant for speed changes as small as 10 per cent.

Choke Effect

The basic slope of the head flow curve has been discussed at some length, but the choke or stonewall effect that occurs at flows higher than design flow and which must be superimposed upon the basic slope (Fig. 11.19) has not yet been discussed.

Just as basic slope is controlled by impeller-tip vector geometry, the stonewall effect is normally controlled by impeller-inlet vector geometry. In Fig. 11.24, vector U_1 may be drawn to represent the tangential velocity of the leading edge of the blade similar to … of the inl… radially (… relative t… V. At des… shown.

...it is conventional practice to limit the Mach# to 0.85 or 0.90 at design flow.

Mach Number Considerations

The magnitude of V_{rel} compared to the speed of sound at the inlet pressure and temperature is called the relative inlet Mach number. It is the magnitude of this ratio that indicates stonewall effect in a conventional stage. While true stonewall effect should theoretically not be reached until the relative inlet Mach number is unity, it is conventional practice to limit the Mach number to 0.85 or 0.90 at design flow.

It is evident from Fig. 11.24 that, for a given rpm, the magnitude of V_{rel} will diminish with decreasing flow, since V is proportional to flow. If V_{rel} decreases, then relative inlet Mach number decreases, so the stonewall effect is normally not a factor at flows below design flow. It is also evident that at low flows the direction of V_{rel} is such that the gas impinges on the leading side of the blade, resulting in positive

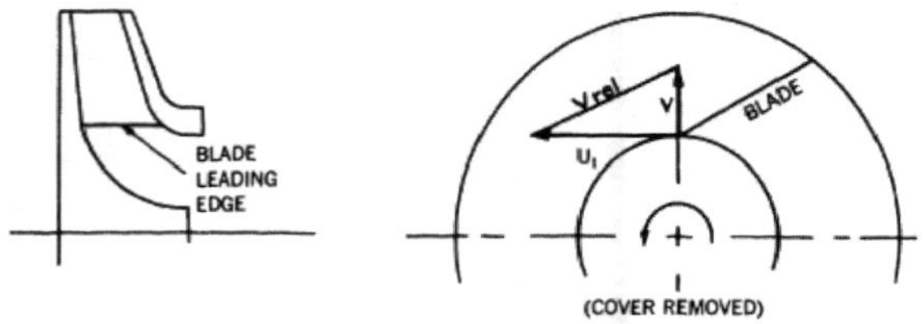

Figure 11.24 Impeller inlet geometry and velocity diagram.

$$C = \sqrt{\frac{kg\overline{R}T}{MW}}$$

incidence, a facto… …h values of positive incide…

Let us now … so also does V_{rel} and relati… …e trailing side of the blade, … …rved that high degrees of negative incidence tend to contribute to the stonewall problem as Mach number 1.00 is approached, presumably because of boundary layer separation and reduction of effective flow area in the blade pack.

Significance of Gas Weight

Since values of U_1 are typically in the 500-fps (152.4-m/second) range and values of V in the 250-fps (76.2 m/second) range, it is obvious that, since the speed of sound for air at 80 deg. F (26.7 deg. C) is 1140 fps (348 m/second), lighter gases suffer no true impeller stonewall problems as described, even at high overloads. Some head loss below the basic slope will be observed, however, in even the lightest gases, due in part to increased frictional losses throughout the entire stage and in part to the extreme negative incidence at high overloads.

The lightest common gas handled by conventional centrifugal compressors for which stonewall effect can be a definite factor is propylene with a sonic speed of 740 fps (225.7 m/second) at −40 deg. F (−40 deg. C). In order of increasing severity are propane at 718 fps (219 m/second) at −40 deg. F (−40 deg. C), butane at 630 fps (192.1 m/second) at −20 deg. F (−29 deg. C), chlorine, and the various Freons. The traditional method of handling such gases is to use an impeller of larger than normal flow area to reduce V, and run it at lower than normal rpm to reduce U_1, thus keeping the value of V_{rel} abnormally low. This procedure requires the use of more than the usual number of stages for a given head requirement and sometimes even requires the use of an abnormally large frame for the flow handled.

Inducer Impeller Increases Head Output

Much development work has been done in recent years toward the goal of running impellers at normal speeds on heavy gases in order to reduce hardware costs to those incurred in the compression of light gases. One approach has been to use inducer impellers (Fig. 11.25). The blades on this impeller extend down around the hub radius so that the gas first encounters the blade pack while flowing axially. Figure 11.25 shows the vector analysis at the inducer outer radius. Assuming that the inducer radius is the same as the leading edge radius of the conventional radial inlet impeller, the vector geometries of the two are identical.

The advantage of the inducer lies in the fact that, as we move radially inward along the blade leading edge, the value of U_1, and therefore of V_{rel} and Mach number, decreases. As we move along the leading edge of the conventional impeller, the vector geometry remains essentially constant. It can be seen, therefore, that while maximum Mach number for the two styles is the same, the average Mach number for the inducer

0900-01293

Ramgen CO_2 Compressor Product

- 100:1 CO2 compressor ⇨ 2-casings/2-stages
  - No aero Mach# limit
  - 10+:1 pressure ratio; 400°F temperature rise
  - 1400 fps tip speeds; Shrouded rotor design
- Single-stage, discrete-drive
  - Single stage per drive optimizes specific speed match
  - Simple single-step external gearbox or high speed direct drive
  - Lower mechanical losses
- Variable speed option
  - Match MW and temperature changes with speed changes
- Configuration adapts easily to match process requirements
  - Mismatched thru-flow
  - Side stream additions
- Active IGV Flow control on each stage
  - Match CO2 capture system constant pressure requirement
- Heat exchangers
  - Inter/aftercooler can be the CCS or power plant
  - "Compressor" heat exchanger cost can be eliminated
  - Eliminate or substantially reduce cooling tower requirement
  - Eliminate or substantially reduce cooling tower make-up water
  - 3x LMTD ⇨ heat exchangers with 1/3 the surface area
- 1/10th the physical size – facilitate space constrained retrofits
- 1/2 the installation cost

Ramgen Discrete Drive HP Stage

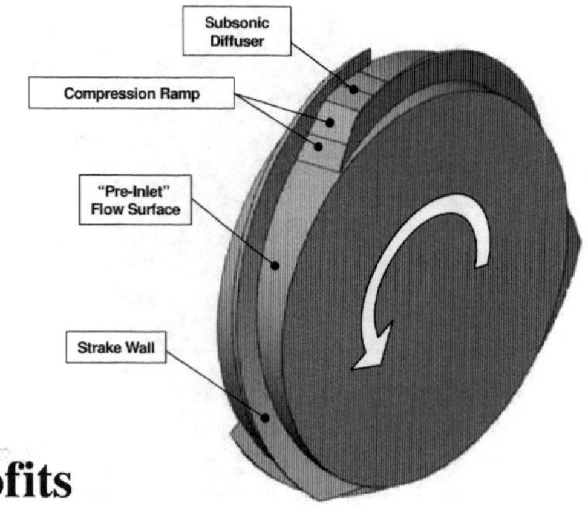

Ramgen Compressor Rotor

0900-01293

Reduce CC(C)&S COE Penalty

RAMGEN POWER SYSTEMS™

MAN Turbo CO_2 Compressor

- **10-stage 6000 hp**
  - **$8.0 million ⇨$1350/hp**
  - **Pr 200:1 ⇨1.70 per stage**
- **8-stage 20,000 hp**
  - **$15.0 million ⇨$750/hp**
  - **$23.0 million installed⇨$1150/hp**
  - **Pr 143:1 ⇨1.86 per stage**

Ramgen CO_2 Compressor

- **Pr 10+:1 per stage**
- **1/10th the physical size**
- **40% of the installed capital cost**
- **Same shaft input power requirements**
- **Recover of 80% of the input Btu at 500°F**
  - **Improve CCS efficiency**
  - **Reduce power plant de-rate**

0900-01293

HP Stage Line-up

RAMGEN POWER SYSTEMS™

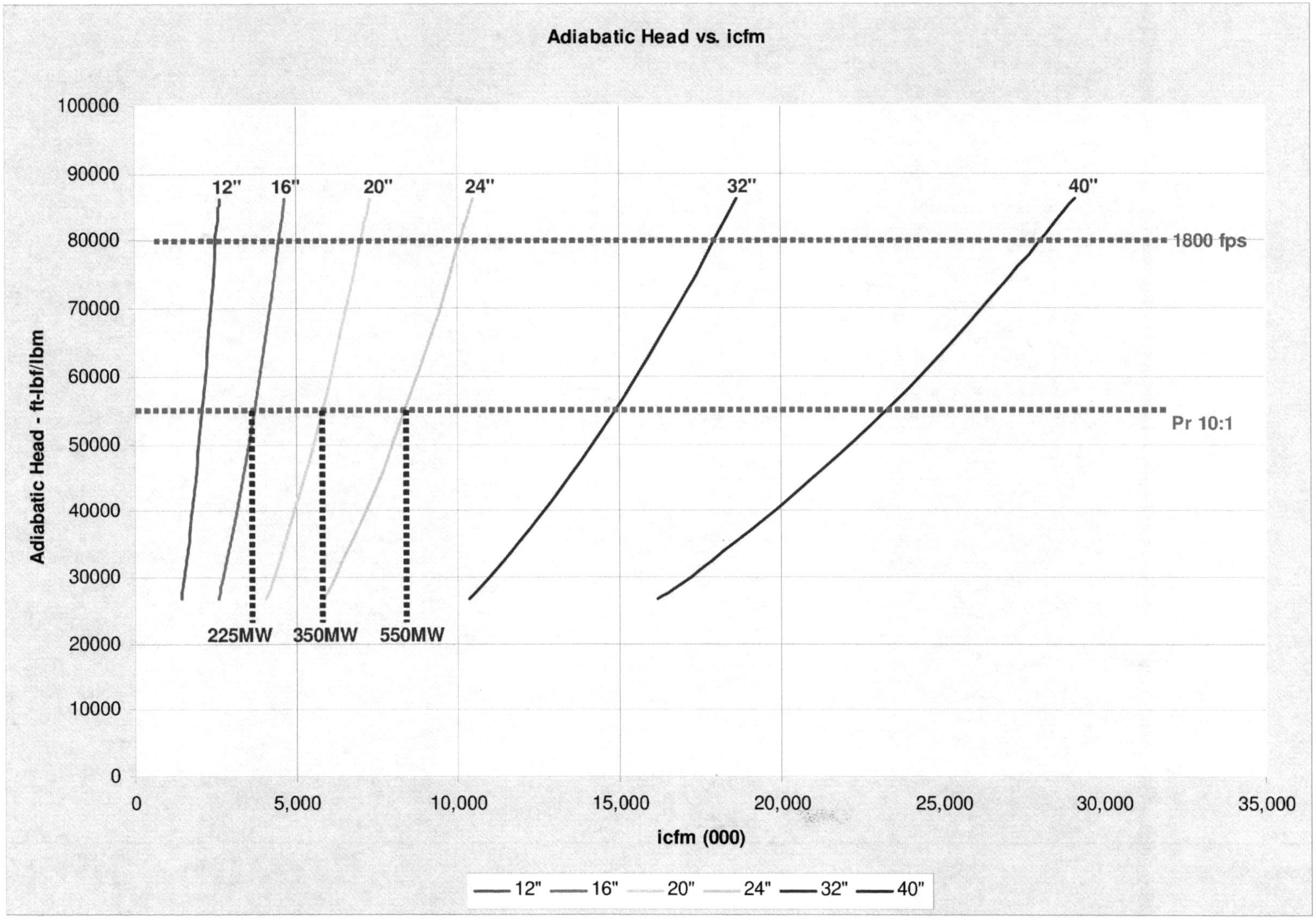

0900-01293

LP Stage Line-up

RAMGEN POWER SYSTEMS™

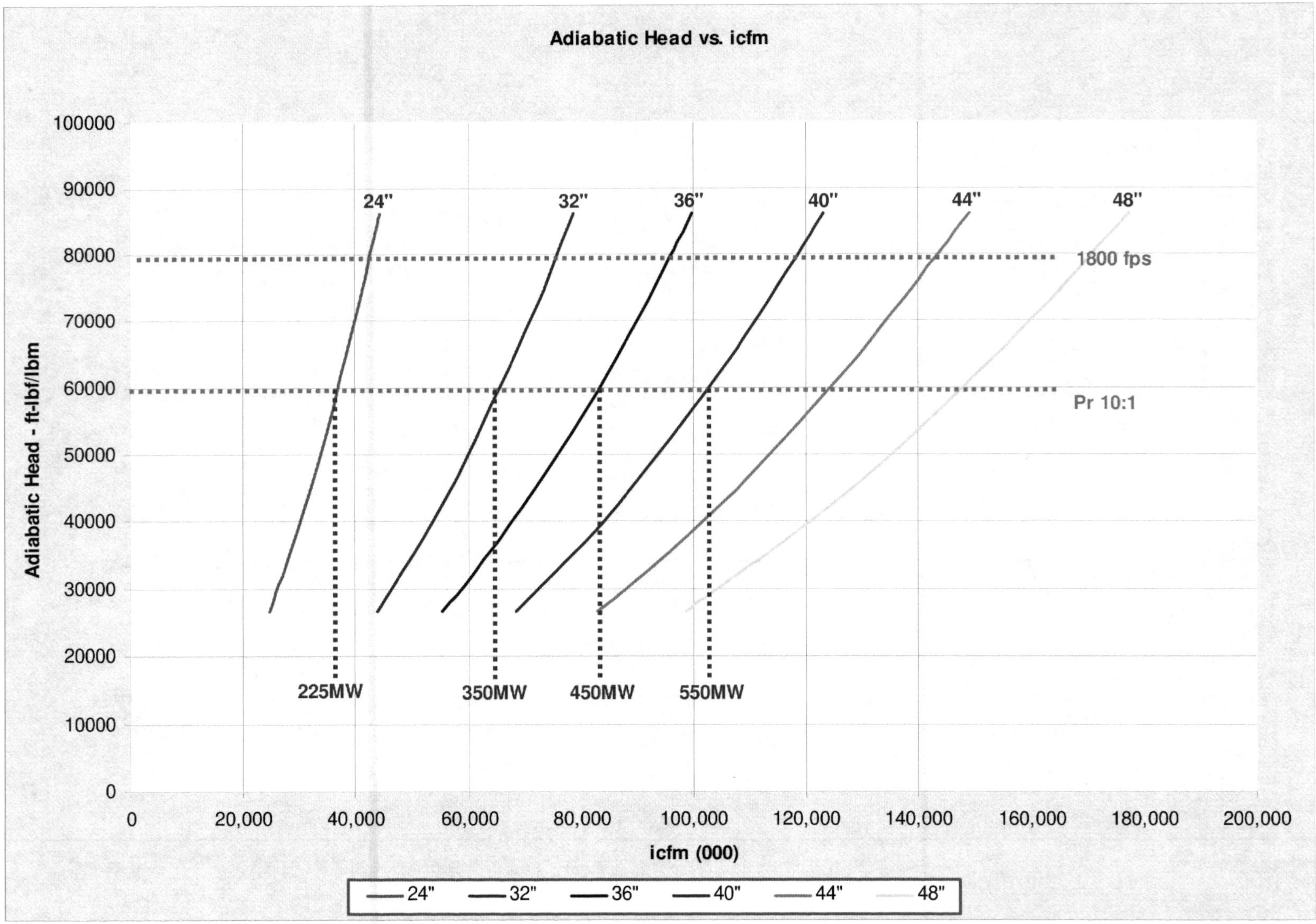

0900-01293

CC&S Systems Need to be Optimized w/ Compressor

RAMGEN POWER SYSTEMS™

- **Systems differ between PC and IGCC configurations and CC&S**
  - Differences not well understood or characterized for the compressor community
  - These are all different and in some cases outdated
  - Operating conditions and assumptions vary significantly
  - The Carbon Capture process developers operate in a proprietary environment and information is generally not available
- **Design "optimizations" are being done at the sub-system level....therefore not optimized**
- **Compressor assumptions not accurate**
- **Cost is not typically a constraint on efficiency objectives**
- **System Integration Opportunities**
  - **Heat recovery for solvent regeneration**
  - **Boiler feedwater**
  - **Pre-heater air**
  - **Coal drying**

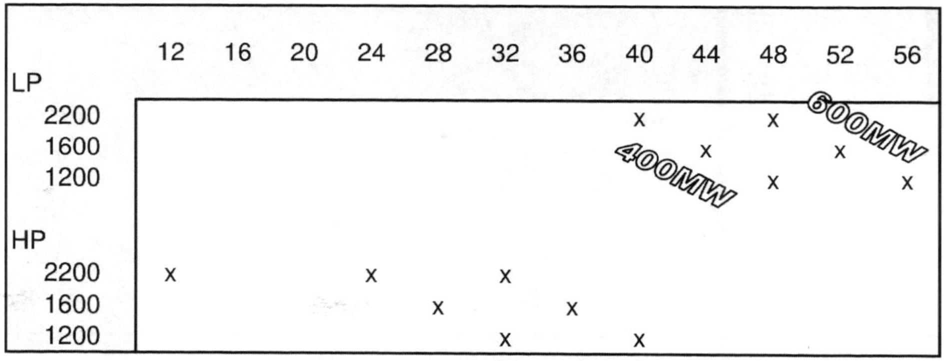

0900-01293

Compressor Power & Things That Affect It

- **The basic inputs**
  - **Gas composition, including moisture content**
  - **Mass flow**
  - **Inlet pressure**
  - **Inlet temperature**
  - **Discharge pressure**
- **Often forgotten**
  - **Cooling media & temperature**
    - **Air**
    - **Water-cooled**
    - **Process cooled**
  - **Interstage assumptions**
    - **Pressure drop**
    - **Design practice**
    - **Fluor estimate ΔP = P2^0.7/10; not to exceed 5 psi**
    - **Intercooler/heat exchanger approach temperature or Cold Temperature Difference – CTD**
    - **15°F CTD normal approach temperature**
  - **Mechanical losses**
    - **Compressor**
    - **Gearbox**
  - **Sparing philosophy (i.e., 2 x 50% + 1)**

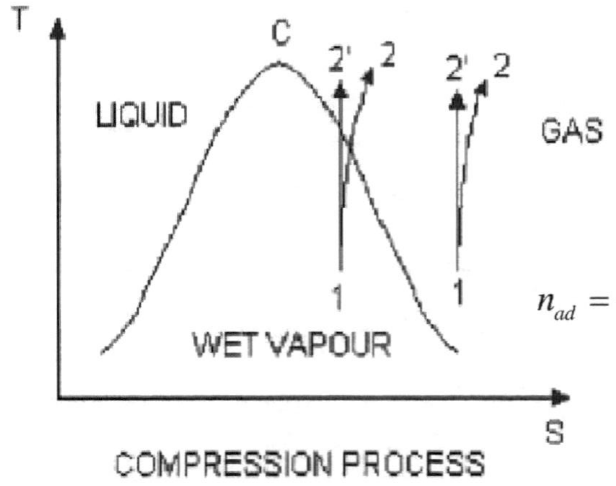

$$n_{ad} = \frac{h_{2'} - h_1}{h_2 - h_1}$$

- **CCS Application Specific Issues**
  - **Capture system flash levels & control requirements**
    - **Pressure**
    - **Mass flow additions**
  - **Water knockout**
    - **Process location (i.e., pressure)**
    - **Method – Glycol/Molecular sieve/PSA**
  - **CO2 compressor inlet pressure**
  - **Heat integration**
  - **Materials of construction**
    - **Heat exchangers**
    - **Piping**
  - **Discharge pressure**

Only the first stage is affected by the inlet conditions…..all the other stages are affected by interstage assumptions.

0900-01293

Baseline Case for Comparison

Data Provided

- Case 3 ASME TurboExpo Berlin - June 2008
- Case 12 in the Baseline Cost & Performance Study – May 2007
- Compressor 6-stage integrally geared design
- 84% isentropic efficiency all stages
- Inlet conditions 23.52 psia; 69°F inlet temperature; 92.4% RH
- Discharge conditions 2215 psia
- Cooling water 60°F
- Stage pressures
- 1,259,600 lbm/hr
- 2 units

Assumptions

- Intercooler approach temperature 9°F
- Interstage pressure drop DP = (P2^0.7)/10 ; but not greater than 5 psi
- Mechanical loss 1.5%
- Drying between stages 3 & 4
- Partial cooling between stages 5 & 6
- 46,900kW Published
- 46,898kW Calculated with these assumptions

	Stage					
	1	2	3	4	5	6
P1 - psia	23.52	52.00	113.01	248.00	545.00	1200.00
T1 - °F	69	69	69	69	69	100
P2 - psia	53.65	115.80	253.00	550.00	1205.00	2219.99
Pr	2.23	2.28	2.24	2.22	2.21	1.85

Baseline case does not use realistic assumptions

0900-01293

Compressor Power Sensitivity

RAMGEN POWER SYSTEMS™

23.52 psia

	69°F Interstage/60°F Water kW		100°F Interstage/85°F Water kW	
DOE Baseline Cost & Performance Study - Case 12 6-Stage Integral Gear @ Pr 2.2:1 per Stage	46,900	-	-	-
8-Stage Integral Gear @ Pr 1.8:1 per Stage Realistic Aero & 6% Mechanical Loss	59,300	26.4%	65,550	39.8%
Ramgen 2-Stage @ Pr 9.8:1 per Stage 3% Mechanical Loss	59,950	27.8%	64,600	37.7%

-16%

23.52 psig

	100°F Interstage/85°F Water kW	
	-	-
	-	-
Ramgen 2-Stage @ Pr 9.8:1 per Stage 3% Mechanical Loss	54,325	15.8%

Retrofit Capture Cost Assumptions

- **"Carbon Dioxide Capture from Existing Coal-Fired Power Plants"**
  - DOE/NETL 401/110907 – Revised November 2007
  - AEP/Alstom Conesville Unit #5
  - Base line & Case 1
- **Process Conditions**
  - P1 19 psia
  - T1 115 F
  - P2 2015 psia
  - Illinois #6 @ 1.80/mmBtu
  - 90% capture
  - 85% capacity factor
- **Financial Assumptions**
  - Make-up power 6.4 cents/kWh
  - Burden rate 2.28
- **Baseline Compressor Horsepower**

– CO2 compressor	31,262
– Propane refrigeration	23,321
– CO2 product pump	2,932
Total	57,515 hp
– Compressor only equivalent	56,800
– Analysis	56,800 hp

Description of Plant Retrofit for Incorporating Carbon Capture Technology

A simplified process flow diagram for the study unit, modified with the addition of the post-combustion amine-based capture system, is shown in below. This simplified diagram is applicable to each of the CO_2 capture cases included in this study. The operation and performance of the existing boiler, air heater, and electrostatic precipitator (ESP) systems are identical to the Base Case for all the capture cases investigated and are not affected by the addition of the post-combustion amine-(MEA)-based CO_2 recovery systems.

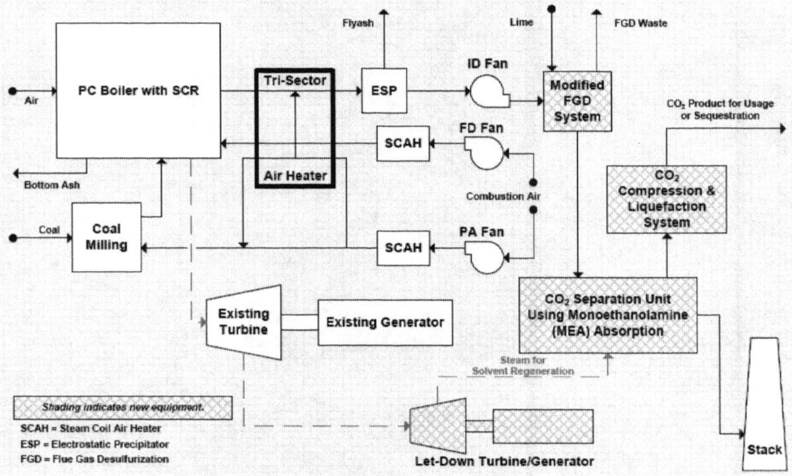

Simplified Process Flow Diagram for Power Plant Modified with the Addition of an Advanced Amine Based CO_2 Capture System

The flue gas desulfurization (FGD) system is modified identically for each of the cases with the addition of a secondary absorber to reduce the SO_2 content of the flue gas entering the new amine system to below 10 ppmv. Recovery of less than 90% CO_2 (Cases 2, 3, and 4 with 70%, 50%, and 30% recovery respectively) is accomplished by bypassing a fraction of the total flue gas stream around the new CO_2 absorber. Flue gas bypass was determined to be the least costly way to obtain lower CO_2 recovery levels.

0900-01293

Conventional CO_2 Compression

RAMGEN POWER SYSTEMS™

- CO2 compressor power
  - Advanced pulverize coal – 9.1%
  - 463MW ⇨42MW ⇨56,800 hp
- Capital Cost for 56,800 hp

– 2 x 50% operating units @ $1000/hp	57
– 1 x 50% spare	28
– Burdened Installation cost	109
– Total Cost	$194M
– $194M/303MW =	$640/kW

- Cost of Electricity (COE)

– Baseline w/o CCS	6.07
– Capture system	4.74
– Compressor	2.70
– Total cents/kWh	13.51
– Increase in COE for CCS	122%

- Cost per tonne

– Capture system	41
– Compressor	23
– Total	$64

Compression Costs are 36% of Total Cost/Mt of CO2

- Heat recovery – Btu/lbm-CO2

– Regeneration Heat	1548
– Heat recovery	0
Net Btu/lbm-CO2	1548

- Plant output

– Original rating	463
– De-rating @ 1548 Btu/lbm	160
– Net	303 MW
– Value @ 6.4 cents/kWh	$62M/year

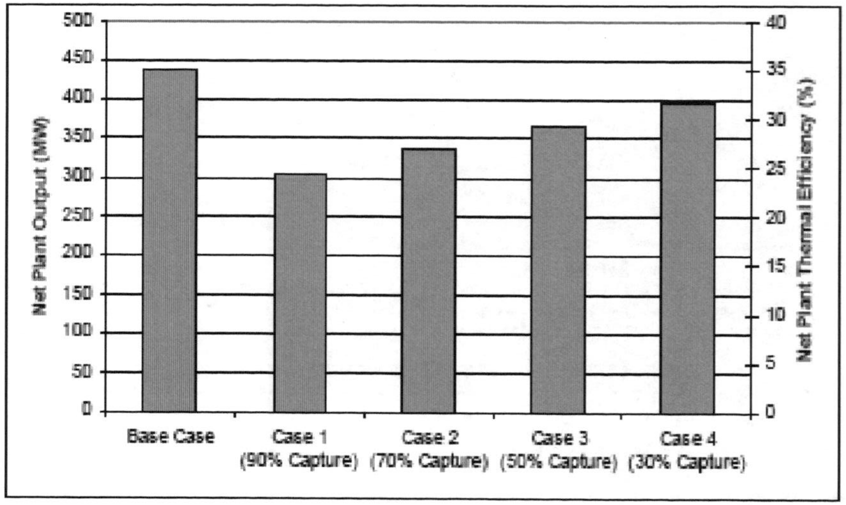

Figure ES-3: Plant Performance Impact of Retrofitting a Pulverized Coal-Fired Plant at Various Levels of Carbon Capture

0900-01293

Ramgen CO_2 Compression w/Advanced Amines

RAMGEN POWER SYSTEMS™

- CO2 compressor power
  - Advanced pulverize coal – 9.1%
  - 463MW ⇨42MW ⇨56,800 hp
- Capital Cost for 56,800 hp

– 2 x 50% operating units @ $400/hp	23
– 1 x 50% spare	12
– Installation cost	44
– Total Cost	$78M
– $78M/328MW =	$239/kW

- Cost of Electricity (COE)

– Baseline w/o CCS	6.07
– Capture system	4.05
– Compressor	1.43
– Total cents/kWh	11.55
– Increase in COE for CCS	90%

- Cost per tonne

– Capture system	38
– Compressor	13
– Total	$51

- Heat recovery – Btu/lbm-CO2

– Regeneration Heat	1200
– Heat recovery @ 230F	159
Net Btu/lbm-CO2	1141
– HR potential @ 100F	116

- Plant output

– Original rating	463
– De-rating @ 1548 Btu/lbm	135
– Net	328 MW
– Value @ 6.4 cents/kWh	$50M/year

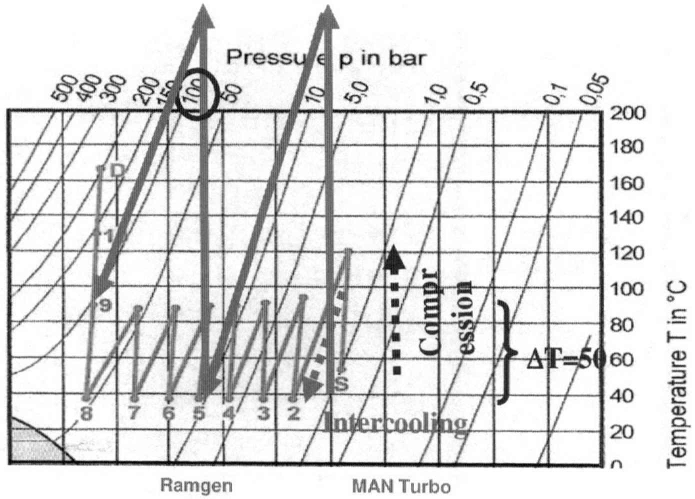

Rampressor Technology can save $116M or 7-10% in capital costs on a PC plant

0900-01293

Ramgen CO_2 Compression w/Advanced CCS

RAMGEN POWER SYSTEMS™

- CO2 compressor power
  - Advanced pulverize coal – 4.2%
  - 463MW ⇨20MW ⇨26,000 hp
- Capital Cost for 26,000 hp

– 2 x 50% operating units @ $400/hp	11
– 1 x 50% spare	5
– Installation cost	20
– Total Cost	$36M
– $36M/388MW =	$93/kW

- Cost of Electricity (COE)

– Baseline w/o CCS	6.07
– Capture system	2.02
– Compressor	0.47
– Total cents/kWh	8.56
– Increase in COE for CCS	41%

- Cost per tonne

– Capture system	22
– Compressor	5
– Total	$28

- Heat recovery – Btu/lbm-CO2

– Regeneration Heat	450
– Heat recovery @ 230F	93
Net Btu/lbm-CO2	357
– HR potential @ 100F	87

- Plant output

– Original rating	463
– De-rating @ 450 Btu/lbm	75
– Net	388 MW
– Value @ 6.4 cents/kWh	$22M/year

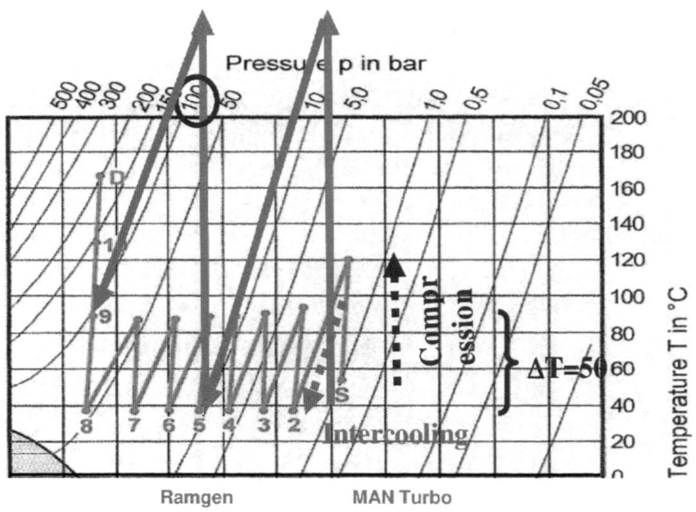

CC&S cost can be reduced by 56% from $64 to $28/tonne CO_2

0900-01293

The Problem is Energy Security and Climate Change

RAMGEN POWER SYSTEMS™

- Oil dependency threatens the U.S. national security & economic vitality
  - 65% of the USA oil requirements are imported
  - 75% of these come from governments categorized as unstable
- The U.S. has a 50-100 year supply of coal
- Capture & Storage (CC&S) is a requirement of Clean Coal
  - Capital cost penalty plus 30%
  - Operating cost penalty plus 25%
  - Plant output minus 5%
- CO_2 compressors - The largest single contributor to the CC&S capital & operating cost penalties

- Coal is the single largest contributor to Climate Change worldwide
- "The climate problem is a coal problem"

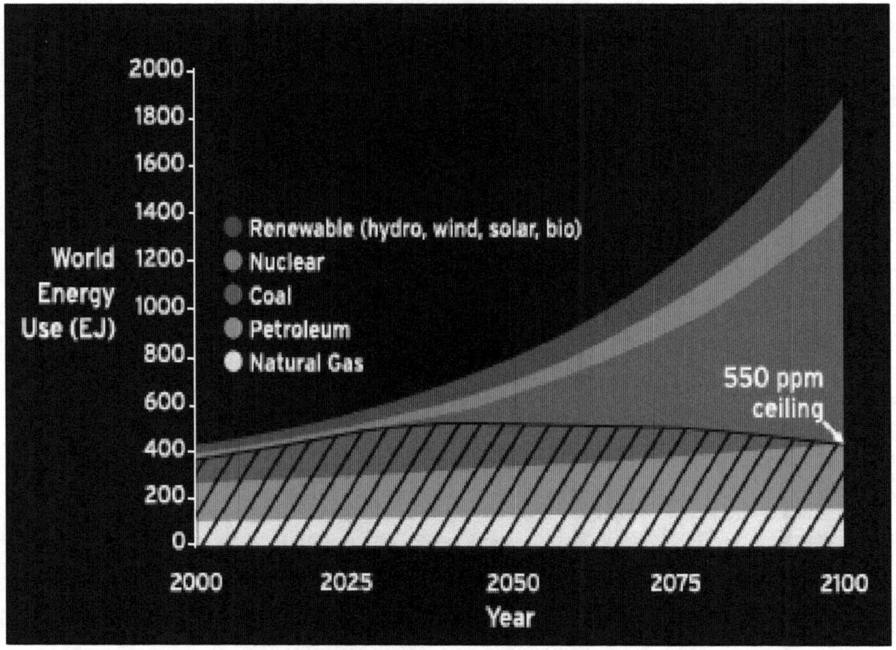

Cost effective CO2 Capture & Storage is essential to meet both requirements

0900-01293

"The Convenient Half-Truth"

RAMGEN POWER SYSTEMS™

	PC		SCPC		IGCC*		NGCC	
	w/out	with	w/out	with	w/out	with	w/out	with
Gross Power	583,315	679,923	580,260	663,445	770,350	744,960	570,200	520,090
Net Power	550,445	549,613	550,150	545,995	640,250	555,675	560,360	481,890
Coal Flowrate - lbm/hr	437,699	646,589	411,282	586,627	489,634	500,379	-	-
Natural Gas Flowrate - lbm/hr	-	-	-	-	-	-	165,182	165,182
Net Plant Heat Rate - Btu/kW-hr	9276	13724	8721	12534	8922	10505	6719	7813
Net Plant Efficiency - HHV%	36.8%	24.9%	39.1%	27.2%	38.2%	32.5%	50.8%	43.7%
Carbon Factor - lbm-CO2/mmBtu	203.3	203.3	203.3	203.3	196.7	196.7	118.5	118.5
Capacity Factor	85.0%	85.0%	85.0%	85.0%	80.0%	80.0%	85.0%	85.0%
Capture %	0.0%	90.0%	0.0%	90.0%	0.0%	90.0%	0.0%	90.0%
Capital Cost - $/kW	$1,549	$2,895	$1,575	$2,870	$1,813	$2,390	$554	$1,172
LCOE - $/kW-hr	$ 0.0640	$ 0.1188	$ 0.0633	$ 0.1148	$ 0.0780	$ 0.1029	$ 0.0684	$ 0.0974
CO2 lbm/MW-hr Net Output	1886	278	1773	254	1755	206	797	93
Capture % to Achieve 797 or 278 lbm/MW-hr	57.7%	71.4%	55.0%	68.7%	54.6%	61.4%	0.0%	70.0%
	797	797	797	797	797	797	797	278

Note: Baseline Report Cases 1 & 2

$$tons/year = (power_{net} \times 8760 \times capacity\ factor \times heat\ rate_{net} \times carbon\ factor)/10^6$$

$$CO_2\ lbm/MWh_{net} = heat\ rate_{net} \times carbon\ factor \times (1 - capture\%)/10^3$$

NETL Cost & Performance Baseline
NETL May 2007

0900-01293

RAMGEN POWER SYSTEMS™

Sierra Club "Hit List"

- Coal
  - $2.00/mmBtu
  - Converted at 40% HHV
  - Ultra Supercritical (USC) ~45% HHV
- Natural gas
  - $8-12/mmBtu depending on the season
  - Converted at 60% LHV
  - Converted at ~54% HHV
- LNG
  - Prices not generally available
  - Natural gas +10% for liquefaction
  - Natural gas +10% for transportation

	Projects	MW	Avg MW Size
Active	65	40540	624
Progressing	18	10228	568
Uncertain	14	10465	748
Upcoming	10	5290	529
Victory	74	47048	636
Total	181	113571	627

- Natural gas is 5x the cost....

  converted at 33% better efficiency
- Net effect is that natural gas is 4x the cost of coal
- LNG base power generation is ~5x the cost of coal

Sierra Club giving new meaning to the word "Victory"

0900-01293

Questions?

RAMGEN POWER SYSTEMS™

pete_baldwin@ramgen.com

425-726-7272 (c)

0900-01293

Ceramic and Coal: ITM Oxygen for Power Generation with Reduced CO2-Emissions

John M. Repasky, VanEric E. Stein, Phillip A. Armstrong
Air Products and Chemicals, Inc., 7201 Hamilton Blvd, Allentown, PA, 18195 USA

Correspondence should be addressed to:
J.M. Repasky, email: repaskjm@airproducts.com, tel: +1 (610) 481-5653
V.E. Stein, email: steinve@airproducts.com, tel: +1 (610) 481-4639
P.A. Armstrong, email: armstrpa@airproducts.com, tel: +1 (610) 481-8754

Prepared for Presentation at the AIChE Spring National Meeting
Session on Carbon Dioxide Capture and Separation
Tampa, FL April 26-30, 2009

Abstract

In partnership with the U.S. Department of Energy, an Air Products-led team is developing a new air separation technology - Ion Transport Membrane Oxygen - based on ceramic membranes that selectively transport oxygen ions when operated at high temperature. Under the influence of an oxygen partial-pressure driving force, the electrochemical ITM Oxygen process achieves a high-purity, high-flux separation of oxygen from air. This high temperature process lends itself to integration with advanced power generation processes that require oxygen as a feedstock, such as Oxy-Coal Combustion and IGCC, as well as for traditional industrial applications for oxygen and distributed power.

During Phases I and II of a three-phase program, the ITM Oxygen team established the feasibility of the ceramic membrane approach and designed and built commercial-scale membrane modules. A prototype facility designed to produce 5 tons-per-day of oxygen started up in 2005. Commercial-scale membrane modules continue to undergo rigorous testing in that unit. Larger-scale demonstration units are in the design phase, with commercialization projected in the early part of the next decade.

In previous studies, the very significant capital and operating cost reductions afforded by the technology have been reported for gasification, IGCC, and other energy-intensive applications. In this presentation, we will report additional flow sheet alternatives integrating ITM Oxygen with various unit operations, particularly toward reduced CO_2 emissions from power generation cycles. Results from pilot-scale testing of commercial-scale ITM Oxygen modules will be discussed, and a projected implementation schedule will be presented.

Introduction

Worldwide energy use is projected to grow significantly over coming decades as developing economies gather momentum. Coal will likely play a significant role in the production of that energy. To compete with other feed stocks and processes, and with the added complication of potential carbon constraints, power generation from coal will need to use the most advanced technologies available. Two of these are gasification and oxy-combustion which enable high efficiency power production and generate a CO_2-rich by-product stream suitable for capture. However, both gasification and oxy-combustion make use of large quantities of oxygen, which today is sourced from cryogenic air separation facilities. Cryogenic air separation is a reliable technology, but one which constitutes, for example, 15% of the cost of an integrated gasification combined-cycle (IGCC) power plant. Moreover, cryogenic technology is mature and is therefore unlikely to undergo future cost reduction coming from step-out technology improvements. A new air separation process is needed to support the role of coal in the future energy mix.

Background

In the late 1980's, Air Products identified a class of perovskite ceramic materials with high flux and separation selectivity to oxygen that can form the basis for cost-efficient air separation membranes. These materials separate oxygen from air at high temperature in an electrochemically driven process. The oxygen in air is ionized on the surface of the ceramic and diffuses through the membrane as oxygen ions, driven by an oxygen activity (partial pressure) gradient to form oxygen molecules on the other side. Impurities such as nitrogen are rejected by the membrane. In addition, these materials conduct electrons as well as ions, such that no external source of electrical power is required – the entire separation is driven by an oxygen partial pressure ratio across the membrane. Detailed descriptions of the materials and electrochemical processes can be found in references 1, 2, and 3. The air separation system that results from the use of such ceramic oxygen *ion transport membranes* (ITM's) produces a hot, pure oxygen stream and a hot, pressurized, oxygen-depleted stream from which significant amounts of energy can be extracted.

Because of the requirement to feed a high-pressure oxygen-containing gas to the membrane separator and the production of the hot high-pressure off-gas stream, ITM Oxygen systems advantageously integrate with turbo-machinery-based power cycles. Compressed air, obtained directly from an air compressor or extracted from the compression side of a gas turbine, is heated to 800-900° C, and supplied to the membrane. A portion of the oxygen is extracted by the membrane, and the resulting oxygen-depleted non-permeate off-gas (which can contain sufficient oxygen to support post-combustion, if desired) can be expanded in an expander or combustion turbine to recover useful work. Additional processing in a heat-recovery-steam-generator (HRSG) can result in an overall product mix of oxygen, power, and steam.

Thermodynamic and process economic analyses indicate that the ITM Oxygen process produces oxygen at significantly lower cost than conventional cryogenic processes. The high selectivity to oxygen separation afforded by the membrane and its rapid oxygen flux result in compact membrane devices with little pressure drop on the high-pressure air side, resulting in >30% reduction in the cost of oxygen. One implementation of an ITM Oxygen membrane in a power cycle is shown in Figure 1. A pressurized air stream is extracted from a gas turbine and fed to the ITM unit after passing through a direct combustor to raise the feed temperature to 800-900° C; oxygen-depleted off-gas from the membrane unit, at substantially the same pressure as the feed air, is reinserted into the combustion turbine for conversion of its thermal and hydraulic energy to shaft work. The compressed air not extracted from the gas turbine goes directly to the combustion turbine (not shown). More details on the integration of ITM Oxygen with gas turbines can be found in reference 4. By contrast, when air is fed to a cryogenic air separation unit (ASU) instead of the ITM Oxygen unit, heat in the compressed air stream is rejected and much of the pressure energy expended during the oxygen separation.

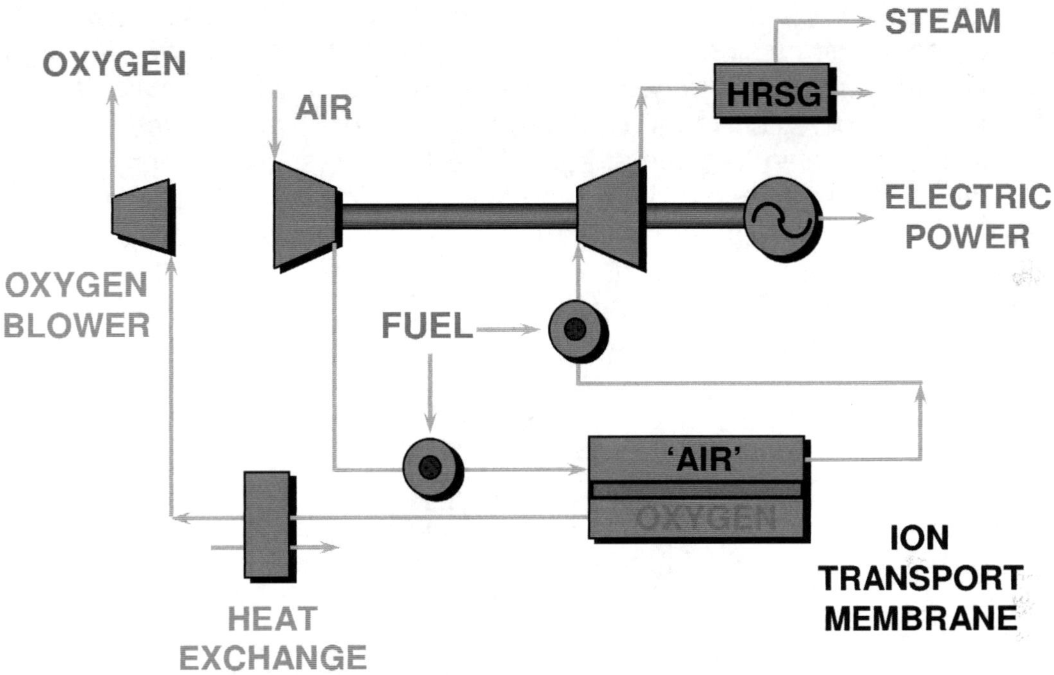

Figure 1. Simplified process flow diagram of an ITM Oxygen unit integrated into a power cycle.

Collaborative Development with the Department of Energy (DOE)

The potential for co-producing oxygen and power or producing oxygen with relatively little power input has interested the DOE and others in developing this technology further for advanced applications requiring low cost oxygen. Air Products and the DOE entered into a Cooperative Agreement in 1998 to develop ITM Oxygen membrane technology in a three-phase program. The overall goal of the program is to reduce the cost of oxygen by one-third over conventional cryogenic distillation-based technology.

The development effort has focused on identifying key technical objectives and addressing them comprehensively with a multi-disciplinary, multi-organizational team approach. The program addresses the following areas, among others: materials development, ceramic processing, integration of membrane systems with turbo-machinery, mechanical design, ceramic and system reliability, oxygen production performance, and process safety.

Phase 1 of the program, which focused on the technical feasibility of the approach, was completed in 2001. A material was chosen from a class of perovskite ceramics as the basis for further scale-up. The material has a combination of properties sufficient to meet commercial requirements for performance and operating life, including high oxygen flux, good material strength at high temperature, and resistance to chemical poisons such as sulfur. In addition, the material is amenable to standard ceramic processing techniques which facilitates the design, scale-up, and manufacture of multi-layer, planar wafer structures.

A schematic of the planar wafer design is shown in Figure 2. Four wafers are shown joined to a common oxygen withdrawal tube. Each wafer consists of a thin outer membrane layer through which the oxygen ions diffuse. The thin layer is supported by a porous layer which is itself supported by a slotted layer. Hot, high-pressure air flows between the wafers. Oxygen passes from the air outside each wafer, through the thin outer layer, through the pores of the porous layer and into the slots of the innermost layer. Oxygen is collected at the center of each wafer in a tubular region formed by the joined wafers, and passes out of the module through a ceramic tube sealed to a metal pipe. The high-pressure air on both sides of each wafer creates compressive stresses within the ceramic which stabilize the wafer. The planar design makes for a very compact separation device, while facilitating good gas phase mass transfer. All of the layers are made of the same ceramic material, and therefore expand and contract together during temperature changes.

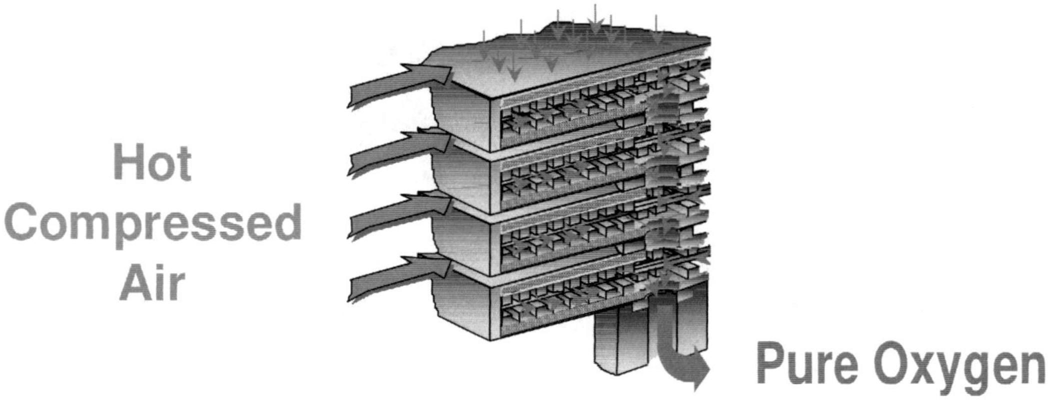

Figure 2. The heart of ITM Oxygen technology: planar membrane wafers. Four cut-away wafers are shown on a common oxygen withdrawal tube. Hot compressed air flows between each adjacent pair of wafers; oxygen is withdrawn at the bottom of the schematic.

The planar wafers were scaled to their full commercial dimensions and produced in volume on a pilot production line using standard ceramic tape-casting technology. The production activities established the feasibility of achieving low-cost production required to meet overall economic targets. Multi-wafer modules were constructed, and the modules were operated under commercially relevant high-temperature, high-pressure conditions in pilot-scale experiments. Oxygen production exceeded commercial flux and purity targets. Detailed process economic studies re-confirmed the overall, market-disruptive benefits of the technology (see reference 5).

The commercial concept of an ITM Oxygen separation vessel is shown in Figure 3. An array of multi-wafer modules (in green) is contained in a common flow duct and connected through a series of manifolds to an oxygen header below. Each commercial-scale module produces about one ton-per-day (TPD) of oxygen, and many modules are arrayed in parallel and series to meet the production requirements of a large tonnage oxygen plant.

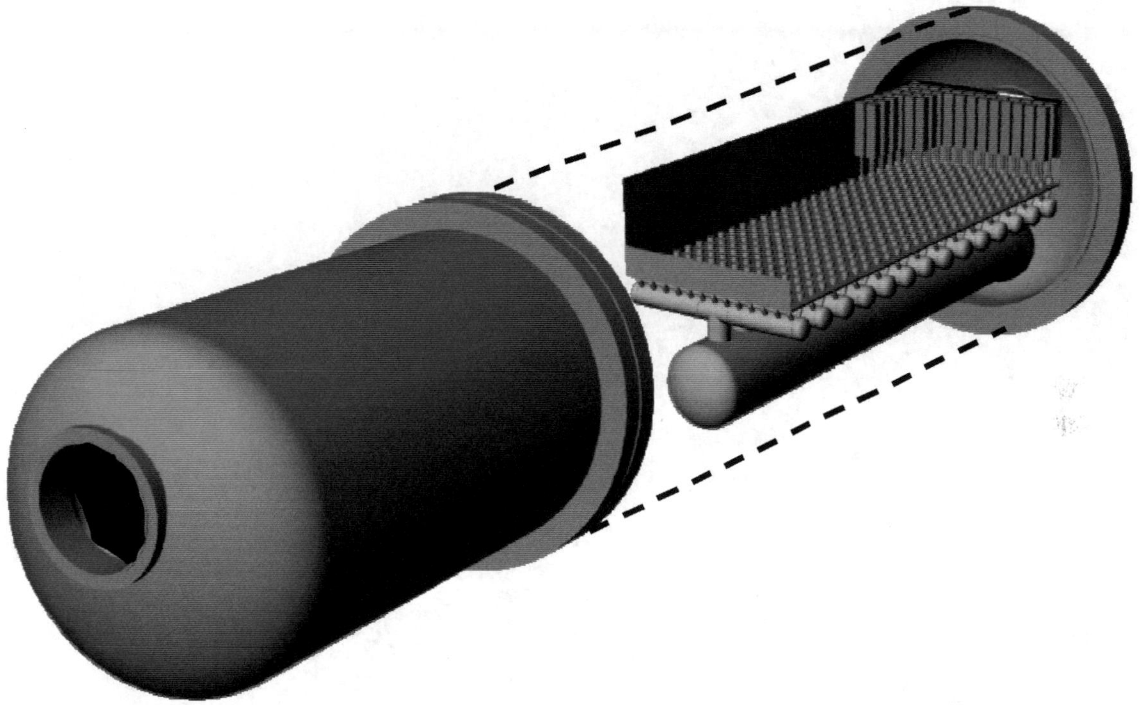

Figure 3. Schematic of commercial ITM Oxygen vessel showing ITM Oxygen modules (in green) arranged in a flow duct and connected to a common oxygen header. The vessel is shown only partially full of modules to expose the oxygen gas manifold tubes below.

In Phase 2, commercial-scale modules, capable of producing up to 0.5 TPD of oxygen were built by co-joining multiple wafers to form a unified ceramic device. The modules were also fitted with a terminating end cap and ceramic oxygen exhaust pipe. All ceramic joints are composed of the same ceramic material as the wafers, thus minimizing the potential for differential stresses caused by non-uniform expansion across the body of the device. Figure 4 shows four ITM Oxygen modules awaiting testing after fabrication.

The pilot plant or subscale engineering prototype (SEP) was designed, constructed and commissioned at Sparrows Point, Maryland during the final portion of Phase 2. This unit features a prototype ITM pressure vessel which holds six commercial scale modules in a 2-by-3 array. Nominal membrane operating conditions for the unit are 800-900° C, 200-250 psig feed pressure, and <1 atm permeate pressure. The SEP is equipped with sufficient flow capability to simulate feed gas velocities anticipated in commercial service. An aerial photograph of the SEP is shown in Figure 5.

Figure 4. Commercial-scale ceramic ITM Oxygen modules. Each module is capable of producing 0.5 TPD oxygen.

Figure 5. Aerial photograph of the ITM Oxygen SEP at Sparrows Point, Maryland.

Air Products carried out the first experiment in the SEP in January 2006. As of December 2008, the SEP has operated for total on-stream time of 515 days. Expected values of purity and flux have been measured during multiple runs, including oxygen purities of 99.9%. Various ceramic-to-metal seals have been tested, as have a range of startup and shutdown schedules designed to maximize ceramic reliability. These results have been published previously, and reference 6 contains more detailed discussion of the SEP configuration, test results, and future experimental plans.

Process Development

Process development has been conducted on the ITM Oxygen flow sheet and integrations, particularly toward reduced CO_2 emissions for power generation cycles. Process alternatives using ITM Oxygen technology have been developed with zero (or low) carbon dioxide emissions for oxy-firing, IGCC, and traditional air fired coal boiler applications. An extensive study on the use of ITM Oxygen for oxy-firing process boilers and heaters in a refinery to avoid CO_2 emissions was published previously in reference 7. Integrations are also attractive toward other industries, particularly those using high temperature processes or combustion (e.g. metals processing, pulp and paper, chemicals, petrochemicals, glass).

Process diagrams of ITM Oxygen processes are shown in Figures 1, 6, 7, 8, 9, and 10. Generically speaking, an oxygen-containing feed stream (e.g. air) is compressed, heated, and provided as feed to an ITM unit. Oxygen is removed from the oxygen-containing feed stream across an ion transport membrane. Oxygen product and non-permeate streams are removed from the ITM unit. Fundamental process optimizations were considered toward developing ITM Oxygen process alternatives.

- In one configuration, a portion of the heat required to heat the oxygen-containing feed stream to the ITM Oxygen unit is provided from the oxy-firing combustion of fuel with oxygen. The oxy-firing combustor may be separate or included in the heat exchanger equipment. Carbon dioxide produced from the oxy-combustion is recovered. Carbon dioxide, flue gases, or other fluids may be injected (e.g. recycled or from other processes) to the oxy-firing process, for example to manage the flame temperature or the heat release profile of the combustion process. Oxygen produced from the ITM unit may be used in the oxy-firing. Carbon dioxide may be recovered from the combustion gases individually or after combining with other oxy-fired combustion gases; for example, oxygen from the ITM Oxygen system may be used in an oxy-fired application, and heat and carbon dioxide may be recovered from multiple combustion gases in common equipment. Use of an expander for energy recovery from the non-permeate provides a cycle with no additional firing and associated emissions; and the expander may optionally be coupled to provide power for feed gas compression as indicated by the dashed lines in Figures 6, 7, and 8. A simplified example of an oxy-fired combustion process cycle to provide heat to an oxygen-containing feed stream (e.g. air) to an ITM Oxygen unit is shown in Figure 6. This ITM Oxygen configuration is particularly attractive toward stand-alone oxygen applications and oxy-fired end use oxygen applications, including oxy-fired pulverized coal (or other fuels) power generation.

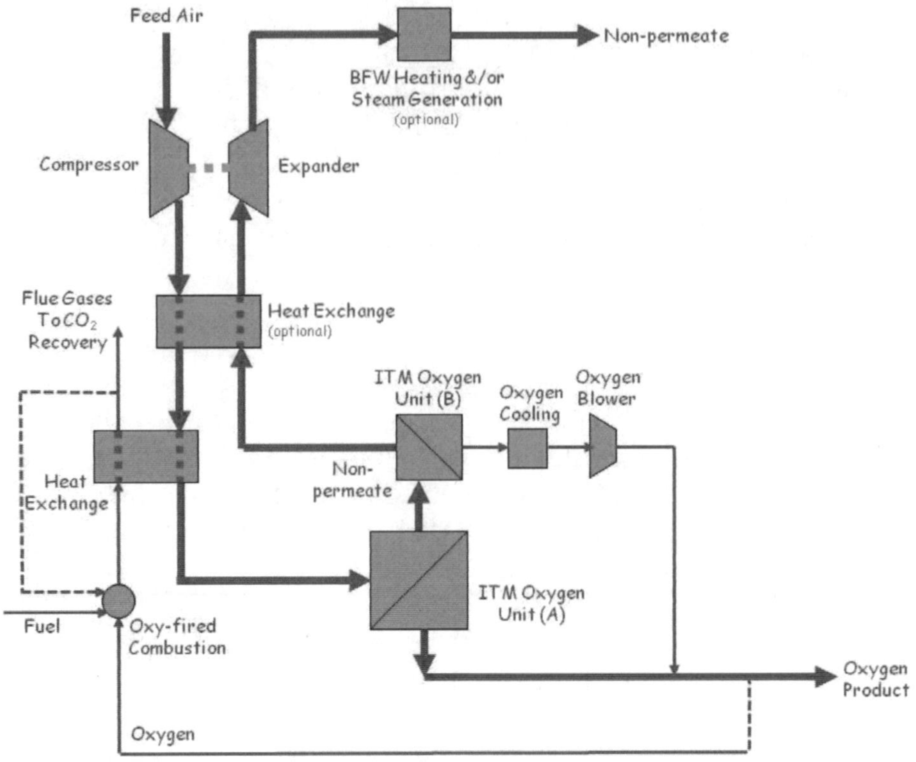

Figure 6. Simplified example of an oxy-fired combustion process cycle to provide heat to an oxygen-containing feed stream (e.g. air) to an ITM Oxygen unit. An (optional) two-stage ITM Oxygen cycle is shown. Oxygen produced from the ITM Oxygen units, along with flue gas recycle, are (optionally) used in the oxy-combustion. Energy in the non-permeate stream is (optionally) recovered to the feed air, to power generation through a letdown turbine, and to steam generation.

- In another configuration, a portion of the heat required to heat the oxygen-containing feed stream to the ITM Oxygen unit is provided from the combustion of essentially carbon-free fuel, for example hydrogen. A portion of the heat may be provided to the oxygen-containing feed stream from the combustion of essentially carbon-free fuel indirectly (e.g. via heat exchange) or directly (e.g. direct firing into the oxygen-containing feed stream and/or non-permeate). A simplified example configuration, shown in Figure 7, depicts several options for firing carbon-free fuel (e.g. hydrogen) into the feed air or non-permeate. This ITM Oxygen configuration is particularly attractive toward applications where essentially carbon-free fuel (e.g. hydrogen) is readily available, including refining, gas-to-liquids, coal-to-liquids, gasification, and IGCC power generation.

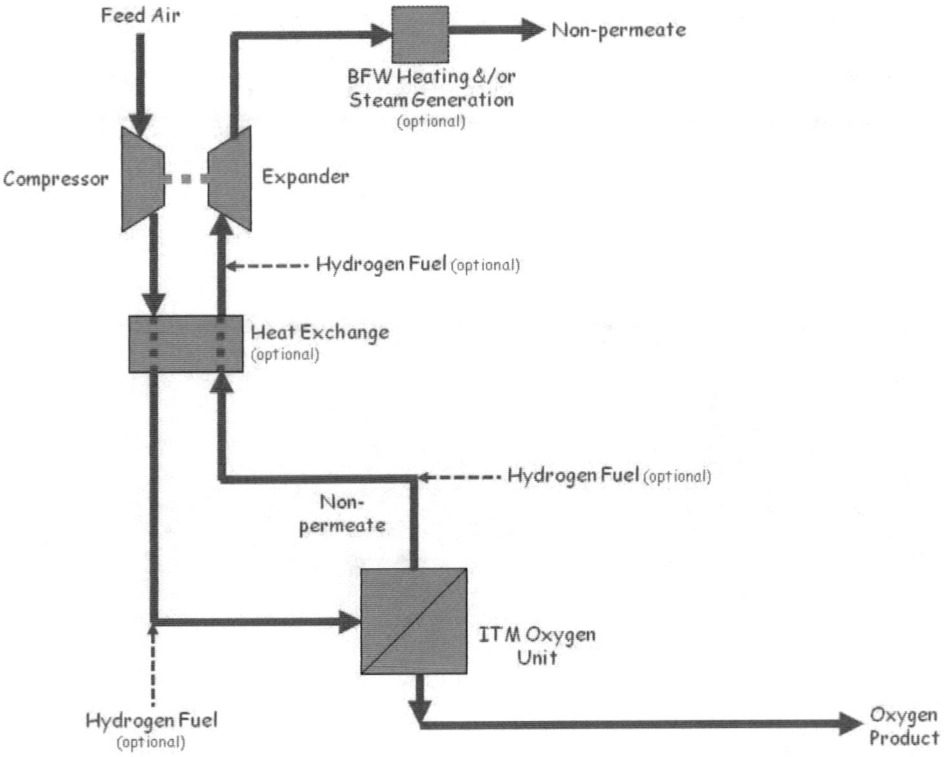

Figure 7. Simplified example of a process using the combustion of essentially carbon-free fuel (shown as hydrogen) to provide heat to an oxygen-containing feed stream (e.g. air) to an ITM Oxygen unit. Carbon-free fuel may be fired indirectly (not shown) or directly at essentially any point into the feed air or non-permeate.

- In another configuration, a portion of the heat required to heat the oxygen-containing feed stream to the ITM Oxygen unit is provided from the combustion of fuel, and carbon dioxide produced from the combustion is recovered (e.g. via an amine system). Heat may be provided from the combustion indirectly (e.g. via heat exchange) with carbon dioxide recovery from the combustion gases or directly (e.g. direct firing into the oxygen-containing feed stream and/or non-permeate stream) with carbon dioxide recovery from the oxygen-containing feed stream and/or the non-permeate stream. In a preferred arrangement, carbon dioxide is recovered from direct firing of fuel into the non-permeate, after heat exchange with the oxygen-containing feed stream and before pressure letdown (as the high pressure provides driving force for carbon dioxide recovery). Carbon dioxide may, of course, be recovered from the combustion gases individually or after combining with other combustion gases for common heat and carbon dioxide recovery. When combined with other combustion gases, carbon dioxide recovery at low pressure (e.g. from indirect low pressure combustion or from non-permeate after pressure letdown) may be preferred. For example, combustion gases associated with the ITM Oxygen process may be combined with combustion gases from a traditional air fired coal boiler and vented through common HRSG and/or carbon dioxide recovery equipment. In another combination, carbon dioxide may be recovered via multiple absorbers (e.g. MDEA absorber columns) integrated with a single stripper (e.g. MDEA stripper column). A simplified example illustrating several options of firing and carbon dioxide recovery to provide heat to an oxygen-containing feed stream (e.g. air) to an ITM Oxygen unit is shown in Figure 8. In this example, fuel may be fired indirectly or directly into the feed air or non-permeate. Carbon dioxide is recovered from the (optionally combined) combustion flue gases at low pressure (e.g. after non-permeate pressure letdown). This ITM Oxygen configuration is particularly attractive toward stand-alone oxygen applications and applications where carbon dioxide recovery is desired from multiple combustion sources, including traditional air fired coal boiler and retrofit applications.

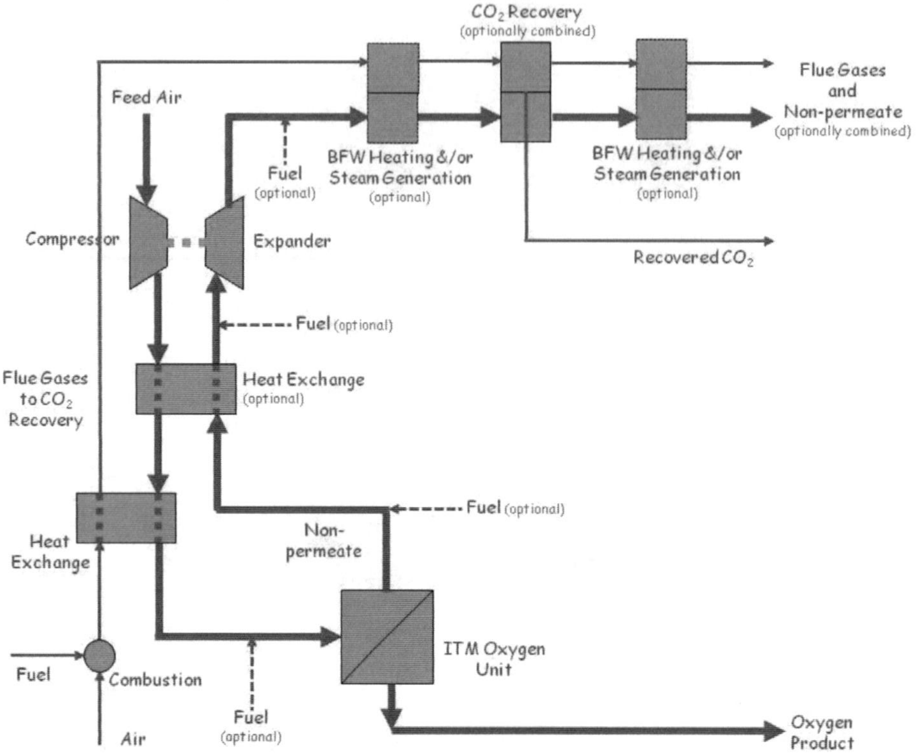

Figure 8. Simplified example illustrating several options of firing and carbon dioxide recovery to provide heat to an oxygen-containing feed stream (e.g. air) to an ITM Oxygen unit. Fuel may be fired indirectly or directly into the feed air or non-permeate. Carbon dioxide is recovered from the combustion flue gases.

- In another configuration, a portion of the heat required to heat the oxygen-containing feed stream to the ITM Oxygen unit is provided from electric heating. Preferably, the electricity may have been produced from a zero (or low) carbon dioxide emission process. This configuration is practiced in the SEP at Sparrows Point, Maryland, where the feed stream to the membranes is first heated by recuperative heat exchange with the non-permeate stream, then brought to final temperature by an electric induction heater. While perhaps not most thermodynamically attractive, this ITM Oxygen configuration may be suited toward low cost electricity scenarios and small scale installations.

- In another configuration, a portion of the heat required to heat the oxygen-containing feed stream to the ITM Oxygen unit is provided from heat exchange integration with another high temperature process. Preferably, the process providing heat to the ITM Oxygen process also produces zero (or low) carbon dioxide emissions. Combustion applications are of particular interest, including oxy-fired pulverized coal power generation (e.g. integration with furnace, HRSG, or CO_2-rich recycle streams), glass (e.g. integration with furnace, flue gas heat recovery), process furnaces (e.g. SMR), fired heaters, and HRSG's. Heat recovery from syngas is also of interest, including syngas produced from partial oxidation, auto-thermal reforming, catalytic partial oxidation, combi-reforming processes (e.g. SMR plus O_2-blown secondary reformer), ITM Syngas, and steam-methane reforming. Other high temperature fluids, including steam, are also of interest.

- The non-permeate stream from the ITM Oxygen unit contains appreciable heat energy, often even after appreciable heat recovery to the feed air and/or pressure letdown through an expander turbine to recover pressure energy. Heat may be recovered from the non-permeate to pre-heat BFW, produce steam, or superheat steam; and additional firing may even be introduced for such purposes. Heat recovery may, of course, be at any point in the non-permeate circuit, including before and after heat recovery to the feed air and before and after pressure letdown; or may provide the sole heat recovery from the non-permeate. For example, referring to Figures 6, 7, and 8, an attractive option can be to recover heat to preheat BFW and/or generate steam from the non-permeate after pressure letdown. Alternatively, the non-permeate circuit pressure and heat recovery may be optimized to provide expander inlet conditions (pressure and temperature) that directly yield an efficient cycle with low expander effluent temperature (e.g. 150° C, or lower). Recovery of heat to a steam system is particularly attractive toward stand-alone co-generation of oxygen and steam, as well as most power generation applications.

- Common ITM Oxygen cycles operate the ITM Oxygen oxygen product circuit at vacuum, then cool and recompress the oxygen using a blower. Low (e.g. vacuum) pressure on the oxygen product side of the ion transport membranes yields high flux and low required membrane area. While this remains an attractive process, it can also be attractive in some applications to operate the oxygen side of the ion transport membranes at higher pressure (e.g. 0.1 atmg, 1 atmg, up to 10 atmg, or more). For example, hot, low pressure oxygen may be exported directly to a combustion application (e.g. oxy-fired pulverized coal power generation, see references 8, 9). While the oxygen flux is reduced relative to the vacuum process (and membrane area increased), oxygen cooling and vacuum equipment (and associated power load) is no longer required and valuable heat energy may be exported with the oxygen product instead of being recovered in expensive equipment or lost to the atmosphere. Thus, the efficiencies of both the ITM Oxygen process and the oxy-fired combustion process are improved and HRSG equipment size and emissions are reduced. This is not possible from typical cryogenic ASU's which produce cold oxygen product. This staged configuration is particularly attractive toward low-pressure combustion applications, including oxy-fired pulverized coal power generation. Toward this application, hot oxygen may be injected directly into the burners, which may require modification to accept high temperature oxygen, or into a fuel gas, flue gas, or CO_2-rich (recycle) stream enroute to the burners. Close coupling of the ITM Oxygen unit with the end use oxygen application is preferred to minimize heat leak and high temperature (e.g. refractory-lined) piping.

- Furthermore, a staged ITM Oxygen cycle (see reference 10) including a first ITM Oxygen unit operating at sufficient oxygen product circuit pressure to directly export hot oxygen to an end use application (e.g. combustion) followed by a second ITM Oxygen unit operating at vacuum can be an optimal arrangement providing low overall membrane area, high oxygen recovery, reduced oxygen cooling and vacuum equipment sizing (and associated power load), and efficiency improvement in the end use application associated with use of hot oxygen. In this arrangement, the non-permeate from the first ITM Oxygen unit provides at least a portion of the oxygen-containing feed to the second ITM Oxygen unit. If the oxygen products from the two (or more) ITM Oxygen units are blended, the temperature of the blended oxygen product is reduced (relative to the first product), and the temperature is more readily accepted in downstream equipment. Left unblended, the multiple oxygen product streams may be exported to different applications. An example of an (optional) staged ITM Oxygen cycle is shown in Figure 6.

- Staged ITM processes may include any number of ITM Oxygen units, ITM Syngas systems (see references 11, 12), or other perovskite-based or membrane separation units arranged in parallel and/or series and operated at various pressure combinations across the membranes. For example, another attractive process alternative includes an ITM Oxygen unit followed in series by an ITM Syngas system. In this arrangement, the non-permeate from the ITM Oxygen unit provides at least a portion of the oxygen-containing feed to the ITM Syngas system. The ITM Syngas system, with it's reaction-driven driving force, maintains high flux even at very low feed gas oxygen content and partial pressure.

- The oxygen-depleted non-permeate from an ITM Oxygen unit (or ITM Syngas system) may be used in plant operations as an inert, blanketing, or purge gas (e.g. in place of nitrogen); or as diluent, NO_x control gas, or motive force fluid (e.g. in gas combustion turbine applications, in place of nitrogen).

- While gas turbine based ITM Oxygen cycles (see Figure 1 above) are typically optimized to produce excess power for export, cycles may be designed using expander(s) to be power deficit, power neutral, or power excess through various combinations of heat recovery and fuel firing. In power deficit cycles, needed shaft power may, for example, be provided by auxiliary electric motor or by direct steam drive, preferably using steam produced from the ITM Oxygen process and/or oxygen end use application. Although typically not as efficient as gas turbine cycles (due to expander inlet design temperature constraints), expander cycles may be attractive toward smaller scale applications (e.g. < 1000 TPD oxygen). An expander cycle is under consideration for the next phase ITM Oxygen Intermediate-Scale Test Unit (ISTU) at nominally 150 TPD oxygen capacity (discussed below).

- Fuel to an ITM Oxygen process may include gaseous fuel (e.g. natural gas, syngas, hydrogen, carbon monoxide, methane, purge or off gas from a process), liquid fuel (e.g. LPG, LNG, naphtha, methanol, hydrocarbons), or solid fuel (e.g. coal, petroleum coke). Gaseous fuel is generally preferred. Pumped liquid or solid fuels can be preferred for direct firing into elevated pressure streams (e.g. feed air, non-permeate).

- In certain instances, an essentially closed-loop steam cycle may be used to recover ITM Oxygen process energy. For example, a simple steam system may include: pump water to pressure, heat and produce steam (for example against hot non-permeate and/or oxygen product streams), use steam for direct drive oxygen-containing feed gas compression or power generation, cool/condense, return water to pump to re-circulate, repeat. Of course, fluids other than water may also be considered.

- In using ITM Oxygen in IGCC and other large energy applications utilizing gas turbines, it is desirable to avoid modifying the efficient, base load gas turbines to enable integration with the membrane unit. One method of accomplishing this is to introduce a second type of gas turbine for integration with the membrane unit, instead of integrating the ITM unit with the base load turbine. The primary role of the second type of gas turbine is to supply compressed air to the ITM unit and to recover a reasonable portion of the useful energy from the non-permeate stream. Generating power from the second type of gas turbine is of secondary importance. A simple configuration therefore involves the use of two turbines and an ITM Oxygen system, the purpose of the first (base load) turbine is to generate the bulk of the power, steam (through potential hot gas feed to the steam turbine generator), and/or shaft work required in an application; while the purpose of the second type of turbine is to integrate with the ITM Oxygen unit to produce oxygen and, secondarily, power, steam, and/or shaft work.

  The second type of turbine will in practical application be different from the base load power turbine. The second type of gas turbine should have a high degree of air extraction capability and accept the hot non-permeate stream with minimal modification to the turbine and with minimal non-permeate stream cooling/heating requirements. Because the power output of the second type of turbine is of secondary importance and will, in general, be significantly smaller than the base load machine, the efficiency of the second type of turbine may be reduced relative to the base load machine. The lower efficiency required of the second type of turbine reduces the requirement to utilize air at maximum efficiency within the turbine, while also reducing the requirement to fire the turbine at a high turbine inlet temperature. Ideally, the turbine inlet temperature of the second type of turbine would be the same or nearly the same as the operating temperature of the ITM Oxygen unit, thus minimizing the need for additional heat exchange and/or on-board turbine combustion.

  Figure 9 shows a simple schematic of an IGCC process integration using this method. The plant configuration is similar to the configuration of a typical IGCC plant with cryogenic oxygen supply, except that the ITM "oxygen plant" also produces power. A separate Heat Recovery Steam Generator (HRSG) associated with the second type of gas turbine is optional. The IGCC application is only an example; this method would, of course, apply to other applications in which power (with internal consumption of oxygen) or a combination of power and oxygen are desired products. Of course, the HRSG's may simply export steam, or include steam turbine power generation (HRSG-ST) for power export or steam/power co-generation. Fuel to the gasifier may include coal (shown), petroleum coke (shown), bio feed (e.g. corn, switch grass, animal waste; not shown), municipal waste (not shown), sludge (not shown), carbonaceous feed (not shown), or any other commonly known gasifier feed stocks.

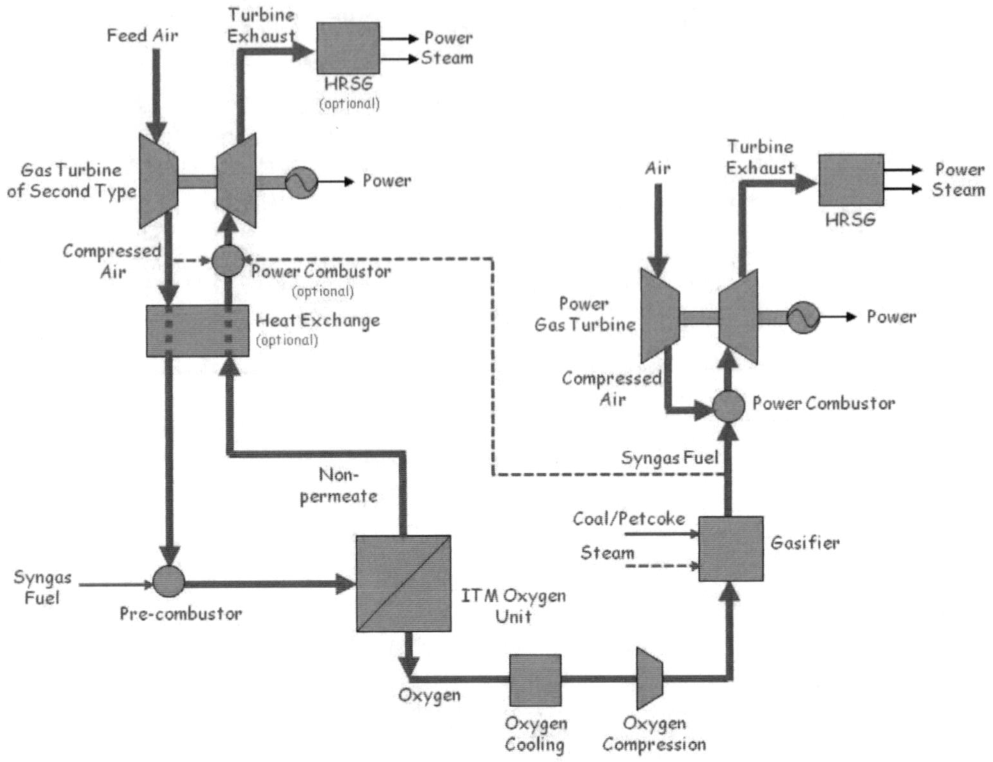

Figure 9. Example of a secondary gas turbine-based ITM Oxygen cycle, showing a dedicated gas turbine integrated with an ITM Oxygen membrane-based air separation unit feeding oxygen to a gasifier which provides syngas fuel to a power gas turbine in an IGCC configuration.

- Furthermore, the ITM Oxygen system and second type of gas turbine and the IGCC system (or other large energy application) may be integrated more extensively. For example, the exhaust from the second type of gas turbine can be combined with the exhaust from the primary power turbine and fed to a common HRSG system. This will save capital cost relative to the first configuration using two HRSG's. Other integrations are possible, including the following:
    1) use of the steam raised in cooling the hot oxygen generated by the ITM system in the main power turbine as auxiliary motive flow or in the main HRSG (or steam turbine) as supplementary steam flow;
    2) mechanically linking the primary gas turbine and the second type of gas turbine (for example, for start-up, or to use common equipment such as a generator);
    3) use of a (fractional) amount of extracted air from the primary power turbine as additional feed to the ITM unit;

4) use of a common chiller system for feed air to both the main power turbine and the second type of turbine;
5) others that will appear obvious to those skilled in the art.

Figure 10 shows an example in which a common HRSG is used for the two systems. Again, the HRSG may simply export steam, or include steam turbine power generation (HRSG-ST) for power export or steam/power co-generation. Fuel to the gasifier may include coal (shown), petroleum coke (shown), bio feed (e.g. corn, switch grass, animal waste; not shown), municipal waste (not shown), sludge (not shown), carbonaceous feed (not shown), or any other commonly known gasifier feed stocks.

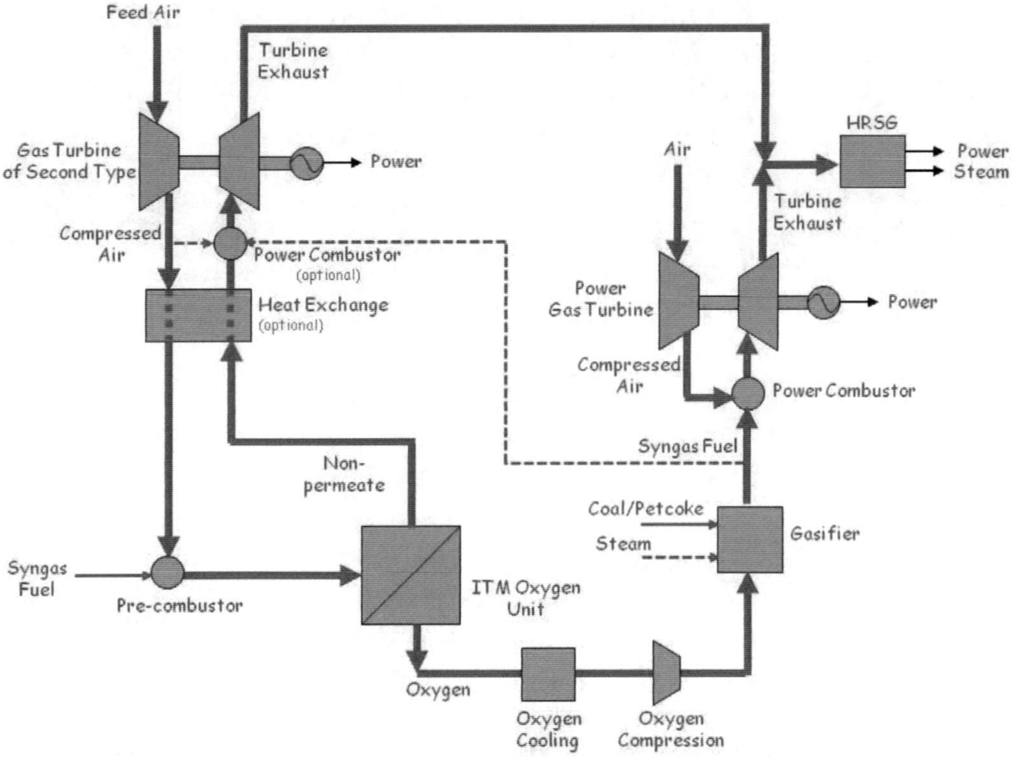

Figure 10. Example of an integration between the ITM Oxygen system and its dedicated gas turbine and the base load power turbine. The ITM Oxygen membrane-based air separation unit feeds oxygen to a gasifier which provides syngas fuel to a power gas turbine in an IGCC configuration. The two combined cycles are integrated through the use of single HRSG which accepts the turbine exhaust from both turbines.

Future work

The ITM Oxygen development program has transitioned from Phase 2, which culminated with successful testing of the SEP facility, into Phase 3, which will conclude with operation at a nominal 150 ton-per-day (TPD) oxygen scale in an Intermediate-Scale Test Unit (ISTU). The ISTU will provide design and scale-up data necessary for commercial introduction of similarly-sized units early in the next decade, as well as a nominal 2000-TPD test facility required for scale-up to clean energy plants by the middle of the next decade. In this development scenario, world-scale ITM Oxygen plants for IGCC and other large energy applications would be available by around 2015.

The ISTU is currently in the design and scope development stage, and several process configurations are still under consideration. Unlike the SEP, it will be a standalone commercial prototype integrated with dedicated turbo-machinery for air compression and non-permeate expansion for energy recovery. In one potential configuration this machinery will be an integrated gas turbine producing net power for export. In other potential configurations the machinery will be independent compressor and expander components. For configurations with an independent expander, various options are available for internal heat exchange to adjust the inlet temperature of the expander. The Air Products project development team continues to analyze these options with respect to cost and delivery schedule to determine the optimum arrangement for both the ISTU and potential subsequent development and commercial offerings.

A process flow diagram (PFD), heat and material balance simulation (HMB), and process and instrumentation diagram (P&ID) for the ISTU have been developed and modified several times based on vendor feedback. A preliminary hazards review has been conducted on the hot gas path portion of the plant with no major findings to impede ongoing design activities. The target schedule currently envisions construction in late 2009 for operation commencing in 2010.

Conclusions

Excellent results have been achieved in the pilot scale ITM Oxygen unit, confirming expected membrane performance in terms of flux and purity achieved. Multiple thermal cycles have been performed with unchanged product purity and the results of operability testing to date are very positive. These results support and confirm the excellent economic benefits originally projected for the ITM Oxygen technology. Design is in progress on the next step in scale-up of ITM Oxygen technology: a small commercial scale ITM test facility (100+ TPD) to provide a basis for scale-up to development facilities of the size required for initial deployment in coal gasification, oxy-combustion, and other clean energy applications. Process alternatives have been proposed to further improve efficiency and reduce CO_2 emissions, particularly toward large-scale power generation.

References

[1] D. Wright and R.J. Copeland. Report No. TDA-GRI-90/0303 prepared for the Gas Research Institute, September 1990.

[2] B.C.H. Steele. C. R. Acad. Sci. Paris, t.1, Serie II, 533 (1998).

[3] P.N. Dyer, R.E. Richards, S.L. Russek, and D.M. Taylor. "Ion Transport Membrane Technology for Oxygen Separation and Syngas Production," Solid State Ionics, 134 (2000) 21-33.

[4] H. Jaeger. "ITM Oxygen Offers 15% More Power and Over 10% Better Plant Efficiency," Gas Turbine World 2008 IGCC Reference Guide, pp. 34-38.
[5] V.E. Stein, E. Juwono, and E.P. Demetri. "The Impact of ITM Oxygen on Economics for Coal-based IGCC," Proceedings of the 27th International Technical Conference on Coal Utilization & Fuel Systems, Clearwater, Florida, 4-7 Mar 2002.
[6] L.A. Vratsanos, P.A. Armstrong, R.P. Underwood, V.E. Stein, and E.P. (Ted) Foster. "ITM Oxygen for the Real World: Pilot Scale Testing and Beyond", Proceedings of International Pittsburg Coal Conference 2008, Pittsburgh, USA, 29 Sept-2 Oct 2008.
[7] R. Allam, V. White, V. Stein, C. McDonald, N. Ivens, and M. Simmonds. "Revamping Heaters and Boilers to Oxyfiring – Producing Oxygen by ITM Technology," in Carbon Dioxide Capture for Storage in Deep Geologic Formations – Results from the CO2 Capture Project: Volume 1, Chapter 30, Elsevier Ltd., 2005.
[8] S. Nakayama and Y. Noguchi. "Pulverized Coal Combustion in O_2/CO_2 Mixtures on a Power Plant for CO_2 Recovery," *Energy Convers. Mgmt.,* 33(5-8) (1992) 379-386.
[9] K. Fogash and V. White. "Oxycoal Combustion: Opportunities and Challenges – Purification of Oxyfuel-Derived CO2," American Institute of Chemical Engineers, Proceedings of the 2007 Annual Meeting, Salt Lake City, USA, 4-9 Nov 2007.
[10] T.F. Yee, R.S. Srinivasan, and R.M. Thorogood. US patent #5447555.
[11] J.M. Repasky, C.M. Chen, M.F. Carolan, D.L. Bennett, C.M. Lowe, S.L. Heimel, and K.F. Gerdes. "ITM Syngas: Ceramic Membrane Technology for Lower Cost Conversion of Natural Gas," American Institute of Chemical Engineers, Proceedings of the 2007 Annual Meeting, Salt Lake City, USA, 4-9 Nov 2007.
[12] C.F. Miller, J.M. Abrardo, M.F. Carolan, J. Chen, E. Minford, W.E. Waldron, and J.M. Repasky. "ITM [Syngas] Ceramic Membrane Technology to Produce Synthesis Gas," Proceedings of 214th Electrochemical Society & PRiME 2008 Meeting, Honolulu, USA, 15 Oct 2008.

Acknowledgement

This technology development has been supported in part by the U.S. Department of Energy (DOE) under Contract No. DE-FC26-98FT40343. The Government reserves for itself and others acting on its behalf a royalty-free, nonexclusive, irrevocable, worldwide license for Governmental purposes to publish, distribute, translate, duplicate, exhibit and perform this copyrighted paper.

Disclaimer

Neither Air Products and Chemicals, Inc. nor any of its contractors or subcontractors nor the United States Department of Energy, nor any person acting on behalf of either:

1. Makes any warranty or representation, express or implied, with respect to the accuracy, completeness, or usefulness of the information contained in this report, or that the use of any information, apparatus, method, or process disclosed in this report may not infringe privately owned rights; or

2. Assumes any liabilities with respect to the use of, or for damages resulting from the use of any information, apparatus, method, or process disclosed in this report.

Reference herein to any specific commercial products, process, or service by trade name, trademark, manufacturer, or otherwise, does not necessarily constitute or imply its endorsement, recommendation, or favoring by the United States Department of Energy. The views and opinions of authors expressed herein do not necessarily state or reflect those of the United States Department of Energy.

106f New Technology Integration Approach for Energy Planning with Carbon Emission Considerations

Sajjad Ahmed[1], Juan Du[2], B. Erik Ydstie[3], Peter L. Douglas[4], and Ali Elkamel[4]. (1) PPG, One PPG Place , Pittsburgh, PA 15272, (2) Chemical Engineering, Carnegie Mellon University, 5000 Forbes Ave, Pittsburgh, PA 15213, (3) Department of Chemical Engineering, Carnegie Mellon University, 5000 Forbes Ave, Pittsburgh, PA 15213, (4) Chemical Engineering, University of Waterloo, 200 University Avenue West, Waterloo, ON N2L3G1, Canada

This paper considers the impact of new technology integration on the performance of a power grid consisting of a variety of power generation plants. As industry is confronted with the challenge of moving toward a clearer and more sustainable path of production, an approach to integrating new technologies into the existing process is needed to achieve industrial challenges. The new methodology is based on four major steps. If the improvement in the process is not sufficient to meet business needs, new technologies can be considered. Financial risk assessment and reliability risk analysis help alleviate risk of investment.

The reduction in carbon emissions is projected to be accomplished through a combination of fuel balancing, fuel switching, and switching to new technologies (carbon capture and sequestration). The fuel balancing technique is used to decrease carbon emissions by adjusting the operation of the fleet of existing electricity-generating stations. The technique of fuel-switching involves switching from carbon-intensive fuels to less carbon-intensive fuels, such as switching from coal to natural gas. Carbon capture and sequestration are applied to meet carbon emission reduction requirements. Power plants with existing technologies consist of fossil fuel stations, a nuclear station, hydroelectric stations, a wind station, pulverized coal stations, and natural gas combined cycle (NGCC) stations. Hypothesized power plants with new technologies include solar stations, wind stations, pulverized coal stations, and NGCC and integrated gasification combined cycle (IGCC) stations, with and without capture and sequestration.

The process is formulated as a mixed integer linear programming model and implemented in GAMS (General Algebraic Modeling System). The optimization model is applied to an existing Ontario Power Generation (OPG) fleet, the largest electric utility company in Ontario Canada. The goal is to meet a given CO2 reduction target while minimizing the cost of electricity. In addition, five planning scenarios are considered: a base load demand, a 1.0% growth rate in demand, a 5.0% growth rate in demand, a 10% growth rate in demand, and a 20% growth rate in demand. A sensitivity analysis study is done in order to investigate the effect of parameter uncertainties on coal price and natural gas price.

The optimization results demonstrate how to achieve the carbon emission/mitigation goal, while minimizing costs and also determining the configuration of the OPG fleet in terms of generation mix, capacity mix, and optimal configuration. Existing technologies, namely fuel balancing and fuel switching, can achieve reduction in carbon emissions by up to 40% in existing power plants for electricity demands with up to 20% growth. Carbon reductions of up to 60% can be achieved when new technologies are integrated for every scenario. Furthermore, cost of electricity decreases when new technologies are integrated in every scenario.

Sensitivity analysis results indicate that increases in coal and natural gas prices do not change the optimal generation mix and configuration.

Development of an integrated carbon capture, separation and geologic storage pilot: A technical research project of the Southeast Regional Carbon Sequestration Partnership

Gerald Hill, Southern States Energy Board, 6325 Amherst Court, Norcross, GA 30092, Richard Esposito, Southern Company, P.O. Box 2641, Birmingham, AL 35291-8195, Richard Rhudy, Electric Power Research Institute, 3412 Hillview Ave, Palo Alto, CA 94304, and Kimberly Sams, Southern States Energy Board, 6325 Amherst Court, Norcross, GA 30092.

The Southeast Regional Carbon Sequestration Partnership (SECARB) is one of seven regional carbon sequestration partnerships established in the United States of America, in cooperation with the United States Department of Energy (DOE). SECARB represents 11 southeastern states and is comprised of more than 100 research partners and participants. Managed by the Southern States Energy Board, the SECARB program is funded through DOE's National Energy Technology Laboratory and cost-sharing by field teams, host site companies/landowners and industry partners.

There are more than 900 large, stationary sources of CO_2 in the SECARB geographical region that contribute 31 percent of the Nation's CO_2 stationary source emissions. Approximately 40 percent of these sources are co-located with a suitable geologic storage formation. The southeastern United States is rich in storage resources, with a total of more than 2,367,215 million metric tons in oil and gas reservoirs, coal and shale basins and deep saline formations. Results from SECARB's field testing indicate that numerous thick, regionally extensive, high porosity saline formations with excellent thick shale confining zones exist within the region. These formations could potentially store CO_2 emissions for approximately 300 to nearly 1,200 years, should 100 percent of this CO_2 be captured and stored.

This paper will focus on the role of the regional partnerships in conducting integrated field tests and highlight results from the robust suite of field tests being implemented during Phase II and Phase III of the program across the Southeast. These field tests support the development of an integrated carbon capture, separation and geologic storage pilot, known as the "Anthropogenic Test," during which CO_2 will be captured at a stationary source and injected into a suitable deep saline formation, the lower Tuscaloosa Formation.

The lower Tuscaloosa Formation is one of many suitable carbon sequestration locations within the Gulf Coast Basin. SECARB has determined that the saline formations of the Gulf Coast are extensive and of regional significance as potential CO_2 storage sinks. The Gulf Coast Basin has the potential to become a major site, and perhaps the nation's predominant site, for carbon sequestration. The Gulf Coast of Alabama, Mississippi, Louisiana, and Texas is formed by the thickest sedimentary wedge in the onshore United States. This clastic sequence is comprised of late Mesozoic to Cenozoic age formations in six sub-basins. All formations are composed of fluvial, deltaic, shoreline, and marine sandstones that are of a suitable porosity and permeability for CO_2 sequestration. Potentially high CO_2 injection rates and high storage potential of the porous media are important criteria in selecting sequestration sites. Also, the formations are separated by regionally extensive thick impermeable marine shales that isolate fluids to ensure permanence of storage. Further compartmentalization and isolation of the storage volumes throughout the Gulf Coast is provided by structures, which include anticlinal four-way closures, roll-over anticlines, and fault bounded traps.

The SECARB team, led by the Electric Power Research Institute and Southern Company, conducted a Phase II Saline Reservoir Field Test at Mississippi Power Company's (a subsidiary of Southern Company) Plant Victor J. Daniel, a power generation facility capable of delivering over 1,000 megawatts of coal-fired electricity into the Jackson County power grid. The field test was conducted from October 2-28, 2008, with a total of 3,027 tons of natural CO_2 injected. The lower Tuscaloosa reservoir demonstrated excellent injectivity, as evidenced by an average wellhead pressure of 1,100 pound per square inch (psi) during the injection of 180 tons of CO_2 per day (about 3 MMcfd). Post-injection mechanical integrity testing (MIT) of the injection well was conducted on November 8, 2008. The MIT included the same set of tests that were run in July and were conducted to demonstrate that the wells remained mechanically sound after injection. The tests were successful, concluding that the well demonstrated mechanical integrity per Mississippi Department of Environmental Quality standards. Post-injection plume monitoring tools, vertical seismic profile, and a neutron log were deployed in December, 2008. The results will be assessed during early 2009. This data is of paramount importance to the design and implementation of the Anthropogenic Test.

The SECARB Phase III development project consists of two related field demonstrations; an "Early Test" that utilizes pipeline CO_2 and the "Anthropogenic Test" that will use coal-fired power plant CO_2. In addition to field test at Plant Daniel, the reservoir properties of the lower Tuscaloosa are being studied in the SECARB Phase II Stacked Storage Project led by the Bureau of Economic Geology (BEG) at The University of Texas at Austin. The Phase II injection field test has been underway since July 2008 at the Cranfield Oilfield, located near Natchez, Mississippi. On February 15, 2009, field monitoring equipment recorded a total of 500,000 metric tons of CO_2 injected. Pressure sensors were placed in a SECARB monitoring well within the injection zone at 10,226 feet and above the confining layer at 9,859 feet. During the period of July 2008 until February 2009 pressure in the injection zone increased by approximately 1,200 psi while pressure in the above-zone monitoring area remained nearly constant. The data support the hypothesis that CO_2 injected below confining layers can be stored safely and monitored for potential leakage.

The Phase III Early Test will expand the Detailed Area of Study of the current Cranfield test and take advantage of ongoing CO_2-enhanced oil recovery (EOR) efforts by the field operator, Denbury Resources, International, during 2009 and 2010. This presents SECARB with the opportunity to monitor the large-volume injection of approximately 1.5 million tonnes of CO_2, injected over a 1.5-year period, in the down dip water leg of the oil reservoir in order to test commercial and experimental monitoring, verification, and accounting (MVA) protocols. Both field tests at Cranfield will provide important data in preparation for the Anthropogenic Test.

The Anthropogenic Test's research and development objectives will be to: 1) further test the lessons learned at the Plant Daniel and Cranfield field tests, with respect to CO_2 storage, flow and immobilization mechanisms; 2) evaluate injection and storage conditions at a third geographic point in the same, or a similar, regionally extensive saline reservoir to help establish ultimate CO_2 storage capacity; 3) evaluate how best to integrate and minimize the impacts of CO_2 captured from a power plant on the operation of the power plant, transportation infrastructure and long-term storage formations; 4) understand how the reservoir architecture (the interplay between the reservoir flow units, seals and baffles) can be used to optimize storage and to minimize the areal extent of the plume; and 5) evaluate the impact of captured CO_2 from power plants on the geochemistry of the saline water in the lower Tuscaloosa Formation.

The Anthropogenic Test site design, for this development stage experiment, will be linked to the daily rate of delivered CO_2 produced by a capture-separation system applied to a slip stream of stack gases from the operating coal-fired power plant. The capture-separation technology was selected in late 2008, and its design and fabrication will occur in 2009 and 2010 respectively, with start of injection slated for 2011. SECARB plans to inject CO_2 for four years, followed by two or more years of post-injection monitoring. Capture volumes are anticipated to range from 100,000 to 150,000 tonnes per year. Based upon injection rates that have been recorded in the Phase II small-volume field tests, it is anticipated that one injection well will be sufficient to handle these daily volumes.

The CO_2 from the capture facility is anticipated to be at injection pressure and will be transported through well site metering equipment to confirm the rate, pressure and temperature of the CO_2. A deep, subsurface monitoring well will be constructed similarly to the injection well. However, the monitoring well will include additional in-well monitoring equipment, such as pressure, temperature and possible geochemistry probes.

With regard to subsurface activities, SECARB is planning an extensive monitoring, verification and accounting program for its field activities. The Phase III Anthropogenic site will be well instrumented with multiple sensor arrays. Standard off-the-shelf technologies will be tested in carbon sequestration applications. In addition, novel new tools and techniques will be tested and evaluated. SECARB has proposed an extensive program to monitor performance during the 10-year Phase III project. In addition, the industry partners who are hosting the injection tests have an interest in supporting MVA activities for extended periods.

A key element in the design of an injection plan for a coal fired power plant is the reservoir architecture, or interplay between the reservoir flow units, seals and baffles. Reservoir models developed during Phase II work at Plant Daniel demonstrate the influence of reservoir architecture on the aerial extend of CO_2 plumes. Saline formations with simple architecture allow buoyant CO_2 to rise towards an upper confining layer and "pancake" out. Formations with complex architecture, such as baffles, extend the pathways of CO_2 and increase pore space contact in the vicinity of the injection well. As a result, the injection plume is column-shaped and makes more efficient use of pore space within a smaller footprint. CO_2 plume modeling data from the Plant Daniel injection is being processed.

SECARB's design of the injection plan will be complemented by the design of the CO_2 capture, separation, compression, and transportation systems. A key focus of the Anthropogenic Test will be to evaluate how best to integrate and minimize the impacts of CO_2 capture on the operation of the power plant and transportation infrastructure. Design of the CO_2 related systems will take into account the challenges that exist due to the duty cycle of power plants. Capture and separation units will be subjected to numerous starts and stops as power plants are dispatched. Changes in flue gas volume and chemistry must be factored into systems design, taking into account the optimum parameters for both pipeline and injection specifications. Capture and separation units must be integrated with NO_X, SO_X, and particulate removal systems. Contingency plans must be developed that will not result in forced outages of power plants due to single contingency failures in the capture train or injection field. Southern Company and EPRI have developed preliminary designs and test plans that address capture train contingency issues. SECARB is working with DOE-NETL on an assessment of regional CO_2 pipeline capacity needs and with its industry partners on strategies for overcoming single contingency failures. Experience gained in the Phase III Anthropogenic Test will assist in addressing many real world concerns associated with the commercialization of CO_2 capture, transportation, and storage systems.

106h Two-Stage Stochastic Programming and Financial Risk Management for Energy Planning with Uncertainty

Sajjad Ahmed[1], Juan Du[2], B. Erik Ydstie[2], Peter L. Douglas[3], and Ali Elkamel[3]. (1) PPG, One PPG Place , Pittsburgh, PA 15272, (2) Department of Chemical Engineering, Carnegie Mellon University, 5000 Forbes Ave, Pittsburgh, PA 15213, (3) Chemical Engineering, University of Waterloo, 200 University Avenue West, Waterloo, ON N2L3G1, Canada

This paper proposes a new methodology to include financial risk management in the framework of two-stage stochastic programming for energy planning under uncertainties in demands and fuel price. A deterministic mixed integer linear programming formulation is extended to a two-stage stochastic programming model in order to take into account random parameters that have discrete and finite probabilistic distributions. Thus the expected value of the total cost of power generation is minimized, while the carbon emission reduction constraint is satisfied. Furthermore, the so-called "conditional value at risk", a risk measure in financial risk management, is incorporated within the framework of two-stage mixed integer programming.

The proposed new methodology is implemented in an existing Ontario Power Generation (OPG) fleet with existing technologies and new technologies individually, while reducing carbon emissions under uncertain factors. The new methodology is analyzed under the integrated mode, which minimizes the cost of power generation through a power grid consisting of power generating plants. Taking into account environmental and economic issues, new technologies are integrated into the existing technologies such that much stricter carbon reduction requirements can be achieved and cost of electricity decreased dramatically.

Existing technologies are fuel switching and fuel balancing. Fuel balancing is used to decrease carbon emissions by adjusting the operation of the fleet of existing electricity-generating stations. Fuel switching involves switching from carbon-intensive fuels to less carbon-intensive fuels, such as switching from coal to natural gas. New technologies are carbon capture and sequestration. Power plants with existing technologies consist of fossil fuel stations, nuclear stations, hydroelectric stations, a wind station, pulverized coal stations, and natural gas combined cycle (NGCC) stations. Hypothesized power plants with new technologies include solar, wind , pulverized coal, NGCC and integrated gasification combined cycle (IGCC) stations, with and without capture and sequestration.

The optimization model is implemented in GAMS (General Algebraic Modeling System) and solved using CPLEX, a commercial solver available in GAMS. The computational results demonstrate the effectiveness of the proposed new methodology.

Abstract for Spring AIChE Meeting 2009

Title: "Activated Carbon Injection for Mercury Control: Overview"

Author: Sharon Sjostrom (sharons@adaes.com)
Vice President, Technology
ADA Environmental Solutions
8100 SouthPark Way, Unit B
Littleton, CO 80120
(303) 734-1727

The power industry in the U.S. is faced with meeting state imposed regulations, as well as expected federal legislation, to reduce the emissions of mercury compounds from coal-fired plants. Regulations are directed at both new units and the existing fleet of more than 1,000 boilers. Full-scale evaluations of the commercial feasibility of activated carbon injection (ACI) for mercury control in coal-fired plants have been underway in North America since 2001 through DOE, EPRI and industry-funded projects. Commercial injection systems began to be sold into the power generation industry in 2005 in response to state regulations and new source requirements under the Clean Air Mercury Rule (CAMR) and ACI is now considered the most robust technology for mercury control at many coal-fired units. As of late 2007, mercury control systems had been ordered for 80 units. On February 8, 2008, the U.S. Court of Appeals for the District of Columbia vacated the Clean Air Mercury Rule (CAMR), removing any federal regulations that require monitoring or control of mercury from electric generating units. However, many state and voluntary mercury control programs are requiring the power industry to respond quickly to meet implementation schedules. In addition, it is expected that new, possibly stricter federal regulations will be implemented in the future.

Successful widespread implementation of ACI for mercury control throughout this industry requires an in-depth understanding of the many parameters that can affect performance. In addition, continued development efforts are required. Data obtained from full-scale mercury control demonstrations indicate that several parameters influence the effectiveness of PAC for mercury control including:

- PAC distribution
- PAC/gas residence time
- Gas temperature
- Flue gas constituents.

PAC distribution is determined by the injection grid design, which can be affected by the ability to install ports in select locations, the injection location and mixing in the duct at the injection location, the particle size of the sorbent injected, and the amount of conveying air used to enhance distribution. Residence time is a function of the configuration of the plant and ability to inject well upstream of the particulate collection device as well as the type of particulate collection device (electrostatic precipitator (ESP) vs. fabric filter (FF)). The flue gas temperature is primarily determined by plant design and operating factors. Flue gas constituents typically present that can increase the effectiveness of PAC are halogens and hydrogen halides (primarily chorine and bromine), which are present in the flue gas from either the coal or from coal or flue gas additives introduced to increase the baseline concentration. In low-halogen flue gas, halogen-treated PAC can be very effective at capturing mercury. Flue gas constituents that decrease the effectiveness of PAC include SO_3 and NO_2. Figure 1 summarizes the results of the Phase I and Phase II test programs sponsored by the Department of Energy with significant cost sharing from EPRI and industry participants.

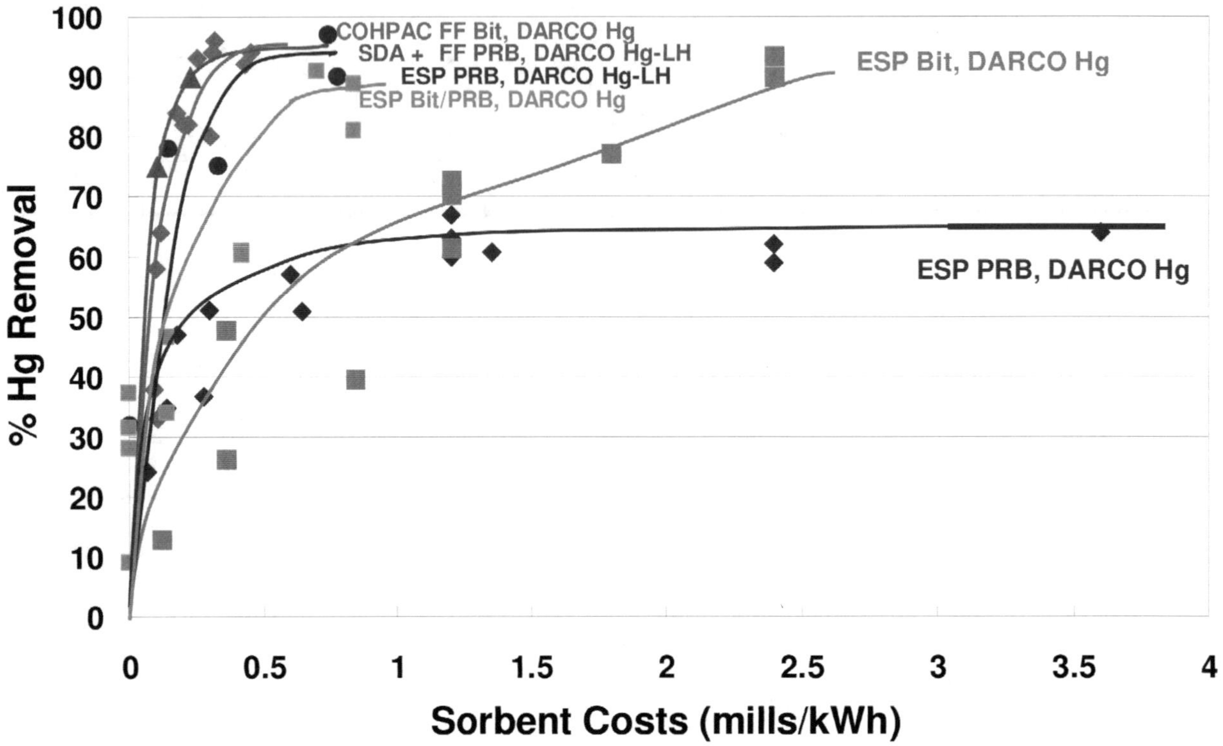

Figure 1. Compilation of Results from DOE Phase I and Phase II Mercury Control Programs.

Although much has been learned about the effectiveness of mercury control using PAC, the long-term feasibility of mercury control is influenced by additional factors including the following: 1) impacts of technologies to control other pollutants, 2) the ability of carbon and equipment manufacturers to continue improving the effectiveness of ACI, 3) the ability of carbon manufacturer to assure the quality of delivered carbon, 4) the impact of PAC on fly ash sales for use in concrete, and 5) the availability of sufficient quantities of PAC to meet the needs of the industry.

Increased levels of SO_3 and NO_2 are often associated with technologies designed to reduce the emissions of other flue gas constituents. For example, SO_3 is used to "condition" the flue gas to improve particulate capture in ESPs. Across the U.S., approximately 25 GW of power is produced from units firing PRB and low-sulfur bituminous coal that inject SO_3 to improve ESP performance. Typically, the SO_3 concentration downstream of injection is below 15 ppm. SO_3 levels may also be elevated if a selective catalytic reduction (SCR) is in-service to increase NOx removal and if the catalyst converts sufficient SO_2 to SO_3. The native mercury removal at these sites is typically lower than at sites without elevated SO_3 levels. Some plants are using or considering dry sorbent injection (DSI) using sodium sesquicarbonate, or trona, to lower SO_2 emissions. NO_2 is produced as a secondary reaction that occurs when trona reacts with SO_2 in flue gas. Preliminary data suggests that NO_2 impacts PAC effectiveness when present in similar concentrations to SO_3.

Techniques to improve the effectiveness of PAC for mercury control are being developed. For example, chemicals to replace SO_3 for flue gas conditioning that do not detrimentally impact PAC performance are under evaluation. For sites firing high sulfur coal or when alternate SCR catalysts are not available, the most effective option evaluated to-date is lowering the SO_3 concentration using techniques such as trona injection. As a result of decreased SO_3 levels, the mercury removal can be significantly increased. Note that the introduction of trona can also result in other balance-of-plant issues. Reducing the PAC particle size has also improved PAC effectiveness for mercury control. During recent tests on units firing western subbituminous coal from the Powder River Basin (PRB), milling PAC[1] resulted in a reduction in PAC requirements of over 50%[2]. Further tests are necessary to determine if the PAC amounts can be further reduced, and the resulting effect on mercury removal.

Fly ash from the combustion of coal provides a valuable raw material for use in the production of concrete. In 2006, over 72 million tons of fly ash were produced in the U.S., 46% of which was used in concrete, concrete products, and grout with another 6% used to make cement.[3] The remainder was placed in landfills or used in other applications. Air entrained in the concrete matrix improves the durability of the concrete over freeze/thaw cycles. Carbon in fly ash is typically not desirable. PAC in fly ash from a plant firing western coal, which typically contains very little unburned carbon, can change the fly ash from a saleable product to a potential landfill liability because activated carbon can absorb the organic air-entraining additives. Fly ash land filling costs are significant and can become one of the largest operating costs for plants after labor and fuel.[4] Options to preserve ash sales, while using ACI for mercury control, include separating the PAC-laden ash from the bulk of the fly ash by using EPRI-patented techniques such as TOXECON™ [5] or TOXECON II™ [6]; reducing the amount of powdered activated carbon (PAC) required through techniques such as on-site milling, or use of a specialized ash compatible activated carbon. These specialized PAC sorbents are fairly new to the market and are being evaluated for the mercury control effectiveness and their impact on concrete properties. Another option being evaluated is the use air entraining agents that are not impacted by PAC.

Ensuring the quality and consistency of delivered PAC is critical to maintaining mercury removal targets. During demonstration programs, PAC deliveries of consistent quality are typically experienced. However, when key characteristics change, such as the density of the bulk material, the bromine level, the particle size, or the hardness of the sorbent there may be significant impacts to the mercury removal, quantity of sorbent required, calibration of the feed equipment, or conveying system material properties may be affected. PAC quality control was studied in 2007 under an EPRI funded project. Parameters such as particle size, surface area, pore size distribution, halogen content, and iodine number were compared with field removal data. Analytical and other measurement techniques to assess PAC as-delivered on a truckload basis were also investigated.

The growth of ACI as a mercury removal option is also dependent on regulations of mercury emissions. ADA estimates that upcoming federal and state regulations will result in tripling of the annual U.S. demand for activated carbon to nearly 1.5 billion pounds from approximately 450 million pounds, requiring rapid expansion of production capacity. Today activated carbon is currently used primarily for cleaning up contaminants in drinking water. In addition, U.S. activated carbon production plants are already operating at near-capacity. ADA is currently fabricating the largest activated carbon facility in the U.S. Other manufacturers are also discussing expansion of their existing production capability.

An overview of the effectiveness of activated carbon for mercury control will be presented including the range of expected control levels, costs, and balance-of-plant issues as well as recent developments to reduce overall control costs for many common air pollution control configurations and overcome complications caused by some new control configurations. An update on carbon supply and progress on ADA's activated carbon manufacturing facility will also be provided.

References

1. Durham, Michael and Cameron Martin. "Apparatus and process for preparing sorbents for mercury control at the point of use," ADA Environmental Solutions, LLC. U.S. Patent 7361209, April 22, 2008 (filed April 2, 2004).
2. Sjostrom, Sharon et al. (2008). "TOXECON II™ and Other Options for Preserving Ash Sales," Presented at the Power Plant Air Pollutant Control MEGA Symposium, Baltimore MD, August 25-28. Paper #08-A-172.
3. ACAA 2006 CCP Production and Use Survey, August 24, 2007, http://www.acaa-usa.org/associations/8003/files/2006_CCP_Survey_(Final-8-24-07).pdf (Accessed July 2008)
4. Wen, Haifang. High Carbon Fly Ash Finds Uses in Highway Construction, *ASH at Work* Issue 2, 2008.
5. Chang, Ramsay. "Method for removing pollutants from a combustor flue gas and system for same" Electric Power Research Institute. U.S. Patent 5,505,766; November 9, 1996 (filed July 12, 1994).
6. Chang, Ramsay. "Method and apparatus for removing particulate and vapor phase contaminants from a gas stream" Electric Power Research Institute. U.S. Patent 7,141,091; November 28, 2006 (Filed December 17, 2003).

113b MetalliFix: A Novel Scrubber Additive for the Capture and Sequestration of Oxidized and Elemental Mercury

Mike Cooper, CoaLogix, 11701 Mt. Holly Road , Charlotte, NC 28214

As regulatory requirements for control of mercury emissions from coal fired power plants become better understood by the utility industry it is becomming clear that addressing the issue of scrubber re-emission will be critical component of an overall compliance strategy. MetalliFix is a novel, three component chemical mix which is fed as slurry to a wet FGD in order to capture and sequester Hg+2 and Hg0 as insoluble mercury sulfide. When converted to mercury sulfide, Hg+2 is no longer able to reduced to Hg0 in the scrubber slurry, and, therefore, re-emission is significantly reduced or eliminated. This paper presents data from recent testing conducted by EERC on the impact of MetalliFix on scrubber mercury emissions from a Texas lignite coal generated flue gas. Preliminary indications from these tests show that MetalliFix essentially eliminated what had been a significant level of re-emission, and may have brought about a direct reduction in elemental mercury emissions as well.

113c Slipstream Tests of Palladium Sorbents for High Temperature Capture of Mercury, Arsenic, Selenium and Phosphorous from Fuel Gas

P. Vimalchand[1], Frank C. Morton[1], Subhash Datta[1], Robert C. Lambrecht[1], Evan J. Granite[2], Henry W. Pennline[2], Dennis Stanko[2], Hugh Hamilton[3], Liz Rowsell[3], Stephen Poulston[3], Andrew Smith[3], and Wilson Chu[4]. (1) Southern Company, P.O. Box 1069, Wilsonville, AL 35186, (2) U.S. Department of Energy, National Energy Technology Laboratory, PO Box 10940, M/S 58-106, Pittsburgh, PA 15236-0940, (3) Johnson Matthey Technology Centre, Blounts Court, Sonning Common, Reading , RG4 9NH, United Kingdom, (4) Johnson Matthey, 1397 King Road, West Chester, PA 19380

In gasification for power generation, the removal of mercury and other trace elements such as arsenic, selenium and phosphorus by sorbents at elevated temperatures preserves the high thermal efficiency of the integrated gasification combined cycle system. Unfortunately, most sorbents will display poor capacity for elemental mercury at elevated temperatures.

Palladium is an attractive sorbent candidate for the removal of mercury and the trace elements from fuel gases at elevated temperatures. The Power Systems Development Facility (PSDF) at Wilsonville, Alabama is a large-scale test facility established to develop and improve advanced coal-based technologies for power generation and chemical production. The palladium sorbents were recently tested for an extended period of time in a slipstream of fuel gas at the PSDF. These results will be described, and future testing will be discussed.

113d Palladium Sorbents for High Temperature Capture of Mercury, Arsenic, Selenium and Phosphorus from Fuel Gas

Evan J. Granite[1], Henry W. Pennline[1], Dennis Stanko[1], John P. Baltrus[1], Hugh Hamilton[2], Liz Roswell[2], Stephen Poulston[2], Andrew Smith[2], and Wilson Chu[3]. (1) U.S. Department of Energy, National Energy Technology Laboratory, PO Box 10940, M/S 58-106, Pittsburgh, PA 15236-0940, (2) Johnson Matthey Technology Centre, Blounts Court, Sonning Common, Reading , RG4 9NH, United Kingdom, (3) Johnson Matthey, 1397 King Road, West Chester, PA 19380

In gasification for power generation, the removal of mercury by sorbents at elevated temperatures preserves the high thermal efficiency of the integrated gasification combined cycle system. Unfortunately, most sorbents will display poor capacity for elemental mercury at elevated temperatures. Previous experience with sorbents in flue gas has allowed for judicious selection of potential high temperature candidate sorbents. The capacities of many sorbents for elemental mercury from nitrogen, as well as from different simulated fuel gases at temperatures from 400 -700°F, were determined. The simulated fuel gas compositions contain varying concentrations of mercury, arsine, hydrogen selenide, phosphine, carbon monoxide, hydrogen, carbon dioxide, moisture, and hydrogen sulfide.

Palladium is an attractive sorbent candidate for the removal of mercury from fuel gases at elevated temperatures. In addition, recent results suggest that palladium also has excellent potential for arsenic, phosphorus and selenium capture from fuel gases, making it capable of multi-pollutant capture. A license agreement has been signed by the United States Department of Energy and Johnson Matthey for further development of the sorbents. Future sorbent development for trace metal capture from fuel gases will be discussed.

Surface Characterization of Palladium-Alumina Sorbents for High-Temperature Capture of Mercury and Arsenic from Fuel Gas

John P. Baltrus[1], Evan J. Granite[1], Henry W. Pennline[1], Dennis Stanko[1], Hugh Hamilton[2], Liz Roswell[2], Stephen Poulston[2], Andrew Smith[2], and Wilson Chu[3].
(1) U.S. Department of Energy, National Energy Technology Laboratory, Pittsburgh, PA, USA, (2) Johnson Matthey Technology Centre, Reading, UK, (3) Johnson Matthey, West Chester, PA, USA

Introduction

Coal gasification offers a pathway for converting coal into electricity, hydrogen, and other valuable energy products. Coal gasification electric power plants are now operating commercially in the United States and in other nations, and it is predicted that coal gasification will be at the heart of future generations of clean coal technology plants.

In order to improve the benefits of coal gasification, sorbents composed of noble metals supported on alumina have been examined for their efficiency in removing mercury and other toxic metals from fuel gas.[1] Palladium is an attractive sorbent candidate for the removal of mercury from fuel gases at elevated temperatures. In addition, recent results suggest that palladium also has excellent potential for arsenic, phosphorus and selenium capture from fuel gases, making it capable of multi-pollutant capture. Ideally, the sorbent should also be minimally affected by other impurities such as hydrogen sulfide.

Initial studies to find the best sorbent were somewhat empirical, screening a wide variety of noble metals and supports. These initial sets of studies were focused on measuring sorbent capacities and the affects of adsorption temperature. The studies have recently been expanded to the research described herein, which focuses on understanding the composition of the reacted sorbents and the mechanisms that are affecting their performance. Such studies will hopefully identify the key factors influencing palladium's capacity for the toxic elements in fuel gas. These studies of the sorbent's composition initially focus on the adsorption the single toxic elements, mercury or arsenic, and later include studies of simultaneous exposure to both elements and the effects of hydrogen sulfide. X-ray photoelectron spectroscopy (XPS) was the primary characterization method employed; sorbent powders and model surfaces of palladium on alumina crystals were examined

Experimental

Sorbents were prepared by impregnation of a high surface area (140 m^2/g) alumina powder. Details of the preparation method can be found elsewhere.[2] Model surfaces of Pd/alumina were prepared by vacuum deposition of an approximate 0.5 µm-thick Pd film on alpha alumina oxide pieces that are made for vacuum deposition (Alfa Aesar).

XPS analysis was conducted on a series of sorbent powders that were exposed to synthetic fuel gas of composition 5% CO_2, 35% H_2, 45% CO and balance N_2 (SFG-A) in a separate reaction chamber attached to the spectrometer. This permitted transfer of the sorbent between the reaction chamber and the spectrometer without exposure to air. The powders were reacted for 1 h at the specified temperature under a gas flow of 30 cc/min. The internal volume of the reactor was less than 5 cc. The samples were purged with nitrogen before and after the reaction. Vacuum-deposited Pd/alumina was treated with SFG-A to which 240 ppb Hg and/or 3.8 ppm AsH_3 were added. Mercury adsorption capacities of the sorbent powders were measured in a simulated fuel gas of somewhat different composition: 2% CO_2, 10% H_2, 14% CO, 1% H_2O, 0.2% H_2S, 180 ppb Hg and balance N_2 (SFG-B). The reactions with fuel gas containing Hg, AsH_3 and/or H_2S were conducted for 6 h in a packed-bed reactor described elsewhere to avoid contamination of the spectrometer chamber and attached reactor.[1]

XPS measurements were carried out using a PHI 5600ci instrument. The XPS instrument employed monochromatic Al Kα X-rays and the pass energy of the analyzer was 58.7 eV unless noted otherwise. Elemental surface concentrations were calculated from peak areas based on manufacturer-provided sensitivity factors.

Results and Discussion

The first set of experiments focused on determining what factors influence the adsorption of mercury by Pd/Al_2O_3 sorbents. The adsorption of Hg from SFG-B, which contains H_2O and H_2S typically found in fuel gas, was measured for a set of such sorbents containing up to 20% Pd. It was found that Hg adsorption increased linearly up to a loading of 5% Pd (0.0252 Pd/Al) and then leveled out at higher loadings.

Although a previous study of Hg adsorption by a set Pd/Al_2O_3 sorbents did not find much correlation between Hg adsorption and dispersion of Pd, that study focused on sorbents with Pd loadings of 8.5% or less.[2] That study also used hydrogen chemisorption to measure the dispersion of Pd even though it requires a pre-reduction of the sorbent at elevated temperature, which can reduce the dispersion of Pd if it is not in strong interaction with the support.

The study described herein relies on measuring the relative dispersion of Pd using XPS and a model developed by Kerkhof and Moulijn (K-M model) that relates XPS metal-to-support intensity ratios to the bulk metal-to-support atomic ratio.[3] A thorough discussion of the model and some selected applications are described elsewhere.[3-5] The model predicts that there will be a linear relationship between the bulk metal-to-support atomic ratios and the XPS metal-to-support intensity ratios. Deviations from the model are indicative of changes in dispersion.

The XPS Pd 3d/Al 2p intensity ratios for the set of sorbents used in this study are plotted in Figure 1. It can be seen that the intensity ratios lie close to the values predicted by the K-M Model at low loadings, but become increasingly divergent at higher loadings. The observed positive divergence is indicative of repartitioning and growth of the Pd particles outside of the pores of the alumina.[6] It is also noted that exposure of the sorbents for 1 hr to increasingly higher temperatures in SFG-A causes a decrease in the XPS Pd/Al intensity ratio, more so for higher loadings of Pd. This indicates that the Pd, which is relatively poorly dispersed and does not have much interaction with the support at higher Pd loadings, undergoes sintering in the presence of synthetic fuel gas. The sintering is most severe at the highest Pd loading. Conversely there is relatively less sintering at the lowest Pd loadings where Pd is initially better dispersed.

The dispersion of Pd as measured by XPS correlates well with the sorbents' capacity for Hg. The loadings of Pd where the steep linear increase in Hg adsorption with the bulk Pd/Al ratios was observed are those where the XPS results indicate that Pd is relatively well dispersed and sintering is minimal. The higher loadings of Pd, where the adsorption of Hg levels off, are those where Pd is poorly dispersed and undergoes significant sintering on exposure to fuel gas. Therefore, a well-dispersed Pd phase is able to absorb Hg more easily than Pd trapped inside of larger particles.

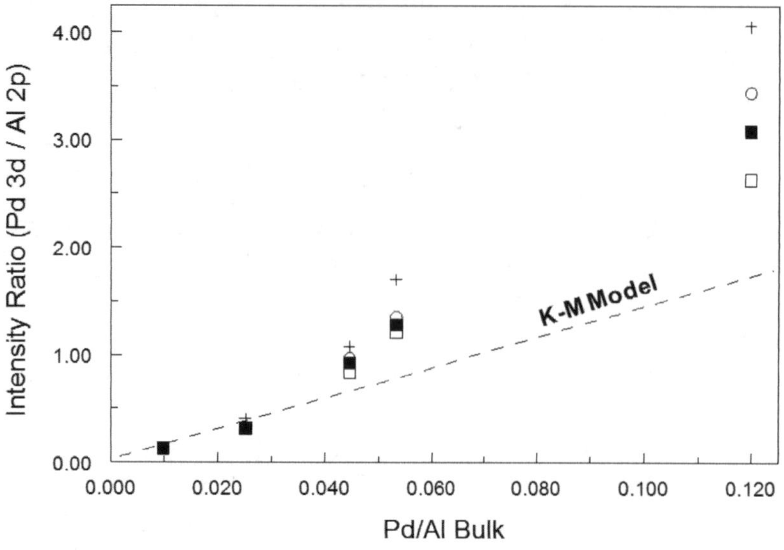

Figure 1. XPS Data for Pd/Alumina Sorbents Treated with SFG-A. (+) Untreated, (○) 204°C, (■) 288°C, (□) 371°C. Dashed Line – Values predicted by Kerkhof-Moulijn Model for Untreated Pd/Alumina.

The results described above indicate that temperature has an effect on Hg adsorption because it alters the dispersion of Pd. Experiments with model Pd films on alumina crystals demonstrated that a more direct effect of temperature on Pd adsorption exists. XPS was used to measure the Hg/Pd atomic ratios of a Pd/alumina film as a function of sputtering depth. To avoid the influence of surface contamination, the average atomic ratios over the depth of 34-102Å are reported in Table 1 as a function of treatment temperature in SFG-A. The greatest capacity for Hg was found at the lowest temperature and the Hg solubility in the film decreased as the temperature increased.

Table 1. XPS-Determined Hg/Pd and As/Pd Atomic Ratios for Pd Films on Alumina Crystals Exposed to SFG-A + 240 ppb Hg or AsH_3.

Temperature (ºC)	Hg/Pd	As/Pd – XRD As-Pd Phase
204	0.0413	0.240 - Pd_8As_3
288	0.0292	0.258 (phase 1), 0.398 (phase 2)
371	0.0225	0.403 - $AsPd_2$
204 (SFG + H_2S)	0.1060	0.163 - Pd_8As_3

Since H_2S is a well-know poison for Pd and is a constituent of fuel gas, it is of particular concern when using Pd-based sorbents for the adsorption of Hg. Its affect on palladium's ability to absorb Hg was measured at 204°C, the temperature at which maximum adsorption of Hg was observed. As shown in Table 1, the Hg adsorption capacity of the sorbent increased significantly in the presence of H_2S. Further information about the role of sulfur in the adsorption of Hg was obtained by curve fitting the XPS S 2p spectrum. Two major sulfur-containing species were detected: a mercuric sulfide with a S $2p_{3/2}$ binding energy of 163.1 eV and a palladium sulfide with a S $2p_{3/2}$ binding energy of 163.1 eV. The sample was sputter profiled to determine its composition as a function of depth over the top 120Å. Initially,

Pd sulfide was the major sulfur-containing species on the surface. X-ray diffraction was able to detect Pd_4S in the sample. As the sample was sputtered, mercuric sulfide became the dominant species while the Pd $3d_{5/2}$ binding energy shifted to reflect the presence of metallic Pd in addition to Pd sulfide. The results suggest that while sulfur bonds with Pd on the surface, it does not interfere with the adsorption of Hg. Sulfur also bonds with Hg below the surface, which may enhance the stability and sorption capacity for Hg.

Studies were also carried out in order to understand how the same sorbents absorb arsenic, another toxic element and catalyst poison. Arsenic is typically present in fuel gas at higher concentrations than mercury, even though it is less volatile than Hg, due to its to greater abundance in the feed coal. Alumina crystals coated with Pd films were reacted with SFG-A containing AsH_3. XPS was used to measure the As/Pd atomic ratios over the same 34-102Å depth that Hg was measured in the experiments described earlier. The results are reported in Table 1 along with the As phases detected by XRD. The interaction of Pd with As was significantly different in some respects than the interaction observed between Pd and Hg. An increase in As capacity was observed at the higher temperature, as opposed to the decrease in capacity observed with Hg. This was due to the formation of different As-Pd phases over the temperature range studied; Pd_8As_3 at low temperature and $AsPd_2$ at high temperature. At an intermediate temperature of 288°C, two areas with distinctly different As/Pd atomic ratios were detected; the exact phases could not be detected due to limitations of the XRD analysis. It is reasonable to assume that there is a transition between the low and high-temperature phases at that temperature. In contrast to the enhancement in Hg adsorption that was observed in the presence of H_2S, it was found that H_2S inhibited As absorption.

An experiment was also conducted in which a Pd-coated crystal was exposed to SFG-A containing both AsH_3 and Hg at 204°C. The resulting XPS As/Pd atomic ratio was 0.338, which was higher than the ratio measured in the absence of Hg. It was also found that the adsorption of Hg was significantly reduced compared to what was measured at that temperature when the Pd film was reacted with SFG containing Hg alone. Analysis of the reacted film by XRD identified the presence of Pd_8As_3. Although the adsorption of Hg was minimal, it may have altered the stoichiometry of As-Pd interaction enough to promote the adsorption of As and form a more As-rich phase. A strong preference for As adsorption at the expense of Hg is consistent with the results of treatments of Pd/Al_2O_3 powders. For example, the atomic ratio of As/Hg adsorbed by a 10% Pd/Al_2O_3 sorbent treated with SFG-B at 204°C was found to be 7.5. Also the Hg capacity of that sorbent was reduced by approximately 36% compared to the capacity that was measured when As was not present. Similarly, measurements of the capacity of a 20% Pd/Al_2O_3 sorbent for the elements when present individually in nitrogen showed that the sorbent's capacity for As was 35 times greater than for Hg on an atom basis.

Conclusions

This study shows both a temperature effect and dispersion effect on the adsorption of toxic metals. Maximum adsorption of Hg occurs at 204°C while maximum adsorption of As occurs at 371°C. The presence of H_2S in the fuel gas enhances Hg adsorption, but inhibits the adsorption of As. When Hg and As are both present in a fuel gas, Hg adsorption is depressed while As adsorption is enhanced. This does not necessarily pose a problem for use of Pd/Al_2O_3 sorbents for the simultaneous adsorption of these elements because both can be accommodated if the adsorption bed is sufficiently deep to remove As in the first stage and Hg in the second stage.

References

1. Granite, E.J.; Myers, C.R.; King, W.P.; Stanko, D.C.; Pennline, H.W. *Ind. Eng. Chem. Res.* **2006**, *45*, 4844-4848.
2. Paulston, S.; Granite, E.J.; Pennline, H.W.; Myers, C.R.; Stanko, D.P.; Hamilton, H.; Rowsell, L.; Smith, A.W.J.; Ilkenhans, T.; Chu, W. *Fuel* **2007**, *86*, 2201-2203.
3. Kerkhof, F.P.J.M; Moulijn, J.A. *J. Phys. Chem.* **1979**, *83*, 1612-1619.
4. Stranick, M.A.; Houalla, M.; Hercules, D.M. *J. Catal.* **1987**, *103*, 151-159.
5. Hoffman, D.P.; Proctor, A.; Houalla, M.; Hercules, D.M. *J. Phys. Chem.* **1991**, *95*, 5552-5557.
6. Delannay, F.; Houalla, M.; Pirotle, D.; Delmon, B. *Surf. Interface Anal.* **1979**, *1*, 172-174.

AUTHOR INDEX

Abbasian, Javad 147
Adebiyi, Ibrahim 27
Ahmed, Sajjad 232, 236
Alesi, W. Richard 145, 148
Al-Hamamre, Zayed 71
Alikina, Galina M. 64
Armstrong, Phillip A. 215
Atwood, David A. 150
Baldwin, Pete 160
Baltrus, John P. 243, 244
Barg, S. 98
Baune, M. 98
Beloshapkin, Sergei 64
Berry, David A. 28, 38
Blasucci, Vittoria M. 158
Bokerman, Gary 5
Bolhar-Nordenkampf, Johannes 83
Botas, J. A. 120
Carpenter, Stephen 91
Ceder, G. 153
Chen, De-Li 149
Chen, Liquan 55
Chen, Rongrong 63
Choi, Sunho 136
Chu, Deryn 63
Chu, Wilson 242, 243, 244
Cooper, Mike 241
Couling, D. 153
Czernichowski, Piotr 2, 4
Darmon, Adeline 7
Datta, Subhash 242
Dilek, Cerag 158
Douglas, Peter L. 232, 236
Drese, Jeffrey H. 136
Du, Juan 232, 236
Dvorak, Gregory T. 97
Ebner, Armin D. 133
Eckert, Charles 158
Elangovan, S. 2, 4
Elkamel, Ali 232, 236
Ellzey, Janet L. 54
Esposito, Richard 233
Evdou, Antigoni P. 105
Fauth, Daniel J. 136
Fedorova, Yulia 64
Fierro, J. L. G. 120
Fischer, C. 153
Fridman, Alexander 1, 3
Frost, Lyman 2, 4
Fulcheri, Laurent 7
Gallagher, M. J. 3
Gao, Xiaomei 6
Garceau, Nathaniel 53
Gómez, G. 120
Gonzalez-Aguilar, Josè 7
Granite, Evan J. 242, 243, 244
Gray, McMahan L. 136, 145
Green, William H. 153
Gribble, David A. 29
Guil-López, R. 120
Hamilton, Hugh 242, 243, 244
Hart, Ryan 158
Hartvigsen, Joseph J. 2, 4
Hassanzadeh-Khayyat, Armin 147
Haynes, Daniel J. 28, 38
Henton, Lee M. 29
Hill, Gerald 233
Hofbauer, Hermann 83
Hollist, Michele 2, 4
Hsu, Andrew T. 63
Huang, Cunping 27
Huang, Lihong 63

AUTHOR INDEX

Hüppmeier, J. 98
Hurley, John P. 97
Huttenhower, Hillary A. 158
Jain, Anubhav 153
Jentzsch, Nicholas R. 29
Johnson, J. Karl 149
Jones, Christopher W. 136
Junaedi, Christian 16
Karan, Kunal 39
Kitchin, John R. 145, 148
Kolbitsch, Philipp 83
Kundu, Arunabha 39
Lambrecht, Robert C. 242
Lee, Dong Hyun 159
Lee, Sung Jun 156
Levin, Leonard 151
Li, Deyi 131
Li, Jigang 55
Li, Mingwei 6
Li, Yongdan 113
Lin, Kunyi Andrew 157
Linkous, Clovis 27
Liotta, Charles L. 158
Liu, Jin-Chen 149
Marx, Klemens 83
Mastanduno, Rich 16
McNally, Dustin P. 97
Mehrotra, Amal 133
Mestre, Veronica Llopis 158
Mezentseva, Natalia V. 64
Moheisen, Ragab M. 17
Morton, Frank C. 242
Mundschau, M. V 29
Muradov, Nazim 5, 27, 53
Nalbandian, Lori 105
Nelson, Sid 152
Nguyen, Ty 155
Ni, Changjun 131
Nickens, Anthony 2, 4
Otterstrom, Gary 132
Pan, Liwei 131
Park, Ah-Hyung Alissa 157, 159
Parmar, Rajesh D. 39
Patel, Nikhil 97
Pelipenko, Vladimir. 64
Pennline, Henry W. 242, 243, 244
Peppley, Brant. A. 39
Petitpas, Guillaume 7
Pizarro, P. 120
Plassmeyer, Paul N. 29
Polevich, A. 3
Pollet, Pamela 158
Poulston, Stephen 242, 243, 244
Proell, Tobias 83
Qiao, Ailing 113
Qiu, Xinping 55
Rabinovich, A. 3
Ramasamy, Karthikeyan K. 27
Rankin, Rees B. 149
Rasmussen, Christian D. 17
Repasky, John M. 215
Rhudy, Richard 233
Ritter, James A. 133
Ross, Julian 64
Roswell, Liz 242, 243, 244
Roychoudhury, Subir 16
Rui, Zebao 113
Sadykov, Vladislav A. 64
Salavani, Reza 17
Sams, Kimberly 233
Schoegl, Ingmar 54
Serrano, D. P. 120

AUTHOR INDEX

Seyed-Reihani, Seyed-A. 153

Shaaban, Aly 17

Shamsi, Abolghasem 91

Shekhawat, Dushyant 28, 38

Simone, Laura 159

Siriwardane, Ranjani V. 91

Sjostrom, Sharon M. 137, 237

Smith, Andrew 242, 243, 244

Smith, Colin H. 54

Smith, Franklyn 5, 27, 53

Smith, Mark 28, 38

Smorygo, Oleg 64

Spivey, James J. 28, 38

Stanko, Dennis 242, 243, 244

Stein, Vaneric E. 215

Steinberg, Meyer 48

Stevens, James 27

Stevens, Robert W. 91

Sun, Jie 55

Thöming, J. 98

Thurgood, Christopher P. 39

Tian, Ye 113

Tian, Yiling 6

Tikhov, Sergei 64

T-Raissi, Ali 27

Ulyanitskii, Vladimir Yu. 64

Usoltsev, Vladimir 64

Vimalchand, P. 242

Vostrikov, Zakhar 64

Vyhmeister, Eduardo 158

Walsh, Dennis 16

Wang, Shudong 131

Wang, Yuguang 55

Wendt, Jost O. L. 156

Xie, Hong-Bin 149

Xie, Jian 63

Ydstie, B. Erik 232, 236

Yuan, Zhongshan 131

Zaspalis, Vassilis T. 105

Zhang, Bo 149

Zhang, Cheng 131

Zhang, Ke 113

American Institute of Chemical Engineers
3 Park Avenue
New York, New York 10016-5991